Prefectural / City Map of Japan

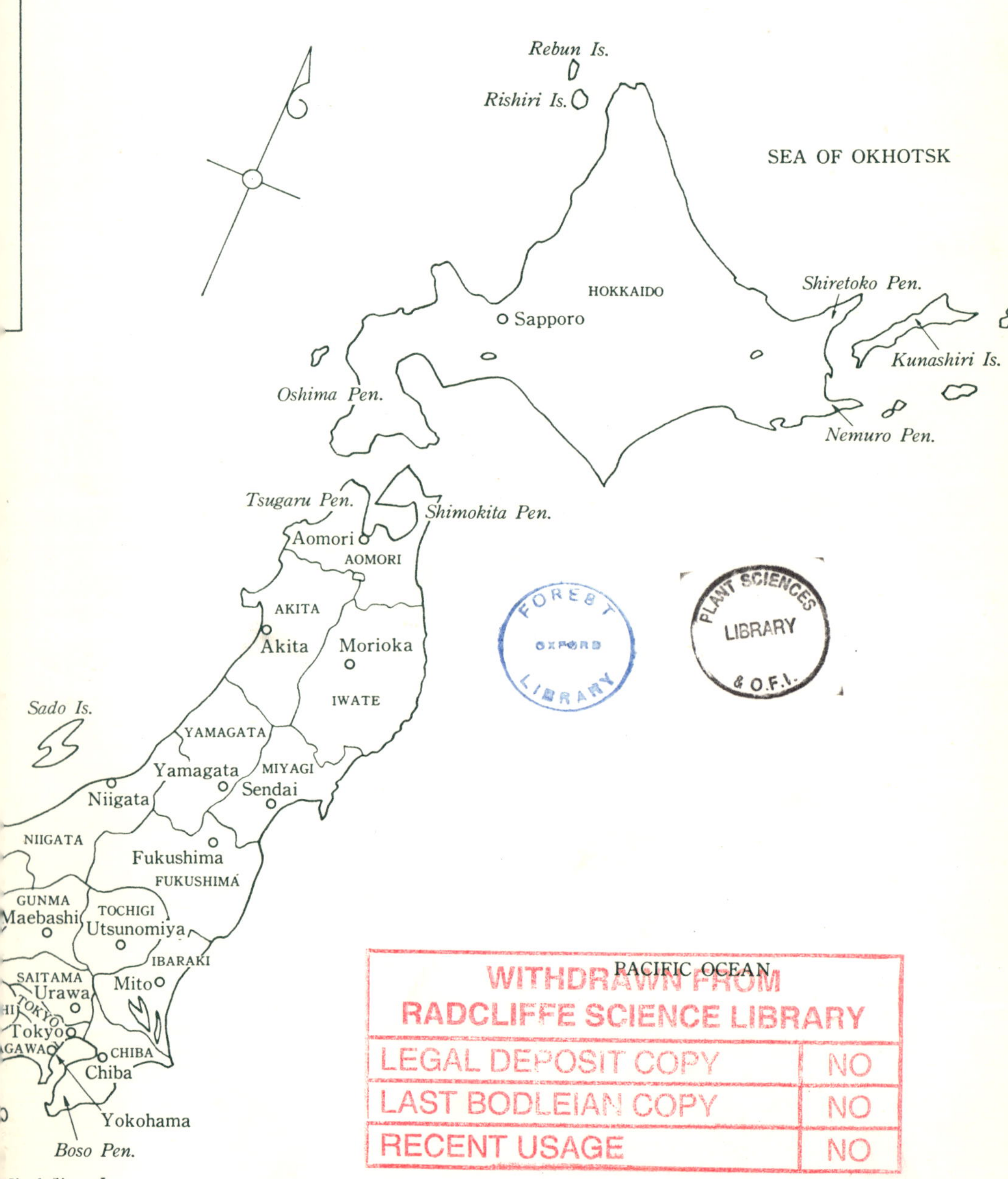

THE FLORA AND VEGETATION OF JAPAN

Ed. M. Numata, Dept. of Biology
Faculty of Science, Chiba University, Japan

KODANSHA LIMITED

Tokyo

ELSEVIER SCIENTIFIC PUBLISHING COMANY

Amsterdam · London · New York

KODANSHA SCIENTIFIC BOOKS

Copublished by
KODANSHA LIMITED
2-21 Otowa 2-chome, Bunkyo-ku, Tokyo 112
and
ELSEVIER SCIENTIFIC PUBLISHING COMPANY
335 Jan van Galenstraat, P.O. Box 211, Amsterdam
AMERICAN ELSEVIER PUBLISHING COMPANY, INC.
52 Vanderbilt Avenue, New York, N.Y. 10017

ISBN: 0-444-99880-2
KODANSHA EDP NO.: 3045-298216-2253(0)

PRINTED IN JAPAN

Contributors

Kazuo ISHIZUKA, Faculty of Liberal Arts, Yamagata University,
1-4-12 Koshirakawa-cho, Yamagata-shi 990, Japan

Fumio MAEKAWA, Emeritus Professor of the University of Tokyo,
1-13-12 Shimizu, Suginami-ku, Tokyo 167, Japan

Makoto NUMATA, Dept. of Biology, Faculty of Science, Chiba
University, 1-33 Yayoi-cho, Chiba-shi 280, Japan

Tsunahide SHIDEI, Dept. of Agriculture, Kyoto University,
Sakyo-ku, Kyoto-shi 606, Japan

Kuniji YOSHIOKA, Biological Institute, Tohoku University,
Aramaki Aza Aoba, Sendai-shi 980, Japan

Contents

Contributors, v

Preface, ix

1. Geographical Background to Japan's Flora and Vegetation **1**
1. General geography of Japan and its relationship to the flora (F. MAEKAWA) 2
2. Climate and the distribution of vegetation zones (T. SHIDEI) 20
3. Soil types and their plant productivity (T. SHIDEI) 27

2. Origin and Characteristics of Japan's Flora (F. MAEKAWA) **33**
1. The geohistory and past flora of Japan as a key to the present flora 34
2. The present flora: its general features and regional divisions 58

3. Forest Vegetation Zones (T. SHIDEI) **87**
1. General survey 88
2. Principal natural forest zones 90
3. Special/local forest types 104
4. Man-made forests 109

4. Grassland Vegetation (M. NUMATA) **125**
1. Grassland types and classification 126
2. Grassland types and their relation to macroclimate 128

3. The course of grassland successions in Japan 129
4. Detailed descriptions of grassland types in Japan 134

5. Maritime Vegetation (K. Ishizuka) **151**
1. Macroclimate and the distribution of maritime vegetation 152
2. Vegetation of salt marshes 153
3. Vegetation of sandy beaches and coastal sand dunes 158
4. Vegetation of coastal cliffs and scree 168

6. Mountain Vegetation (K. Ishizuka) **173**
1. Altitudinal vegetation zones 174
2. The montane zone 178
3. The subalpine zone 188
4. The alpine zone 196

7. Aquatic and Wetland Vegetation (K. Yoshioka) **211**
1. Aquatic vegetation of lakes and ponds 213
2. Aquatic vegetation of rivers 217
3. Swamp forest 217
4. Marshes 220
5. Paddy fields and irrigation ponds 222
6. Fens and bogs 223

8. Volcanic Vegetation (K. Yoshioka) **237**
1. Mt. Usu and Showa-shinzan 238
2. Mt. Komagatake 241
3. The Yakebashiri lava flow on Mt. Iwate 244
4. The Sainokawara lava flow on Mt. Zao 245
5. The Urabandai mud flow on Mt. Bandai 247
6. Lava flows on Mt. Fuji 249
7. Mt. Asama 250
8. Izu-Oshima Is. 253
9. Miyakejima Is. 255
10. Mt. Aso 256
11. Sakurajima Is. 257
12. Solfataras and vapour fumaroles 259
13. Conclusion 264

9. Conservation of Flora and Vegetation in Japan
(M. Numata) **269**

Index of Species **278**

Subject Index **293**

Vegetation Map of Japan

Map of Geographical Localities in Japan

Prefectural/City Map of Japan

Preface

Almost five years have passed since I was first consulted by Mr. Y. Haga of Kodansha (Publishers) Ltd. about the possibility of preparing a book in English on Japan's flora and vegetation. I agreed with his proposal immediately, for two reasons: (1) Japan lacks any strong tradition in natural history and so there is no suitable reference work available on the subject, and (2) I have on several occasions received invitations to give lectures outlining Japan's flora and vegetation during my periodic visits overseas.

The first seeds of natural history in Japan were sown in ancient times but they did not grow up and flourish. In fact, until the Meiji Restoration (1868), Japanese natural history was firmly under the influence of herbals prepared on the Chinese mainland. At that time, excellent studies were carried out by a number of botanists, mostly on medicinal plants. In the Meiji Era, botany and zoology were established as modern scientific disciplines based on the principles developed in the West. However, the accumulation of data on the field biota, i.e. based on orthodox natural history, was noticeably lacking, and most botanical data in the immediate post-Meiji period simply followed the newly imported Western systems. The commencement of organized ecology was even more delayed. In fact, it is only in the last 50 years that professional ecologists have carried out active, broad-based studies in Japan, and during this time our knowledge of the nature of the flora and vegetation has vastly increased. An outline of the fruits of these labours can now be given.

Japan constitutes a long north-south chain of islands having four principal horizontal climax forest zones; (1) subarctic evergreen coniferous forest, (2) cool-temperate broad-leaved deciduous forest, (3) warm-temperate broad-leaved evergreen forest, and (4) subtropical broad-leaved evergreen forest, together with corresponding altitudinal forest zones such as subalpine evergreen coniferous forest, montane broad-leaved deciduous forest, and lowland broad-leaved evergreen forest in central Japan. In alpine regions, there is also a zone of scrub, grassland and rocky desert which corresponds to the arctic zone. Thus, only two of the major biogeographic regions of the world, the polar and the tropical, are unrepresented in present-day Japan. Nevertheless, for a complete study, it is very important that we elucidate the past history and origins of Japan's flora and vegetation based on palaeobotany

and palaeoecology.

For the purpose of adequately surveying the flora and vegetation of Japan, the present volume has been divided into nine separate chapters. The guiding themes in selecting the contents were (1) to emphasize the physiographic and historical background of the present flora, and (2) to detail its characteristics, displaying both the horizontal range of vegetation (from the farthest north to the farthest south) and its vertical range (from the coast to the mountain-tops), together with local variations according to differences in hydrologic and edaphic conditions. Thus, the descriptions consider the present status *vs.* geological history, latitude *vs.* altitude, mesic *vs.* hydric environments, volcanic *vs.* non-volcanic environments, etc.

As editor for the present volume, I wish to express my appreciation and thanks to the respective authors for their contributions, and would also like to take this opportunity to credit Mr. A. J. Smith of Kodansha for his invaluable efforts over the past year in the preparation of the English manuscripts which comprise this book. Also, since this project represents the first, perhaps rather ambitious, attempt to survey the flora and vegetation of Japan from the viewpoint of natural history, we await with interest the reader's criticisms and comments.

January 1974 M. Numata

1

Geographical Background to Japan's Flora and Vegetation

Fumio MAEKAWA*[1] (Section 1)
Tsunahide SHIDEI*[2] (Sections 2, 3)

1. General Geography of Japan and Its Relationship to the Flora
2. Climate and the Distribution of Vegetation Zones
3. Soil Types and Their Plant Productivity

*[1] *Emeritus Professor of the University of Tokyo, 1-13-12 Shimizu, Suginami-ku, Tokyo 167, Japan*
*[2] *Department of Agriculture, Kyoto University, Sakyo-ku, Kyoto-shi 606, Japan*

1. General Geography of Japan and Its Relationship to the Flora

On opening a map of the world, we can readily distinguish several island-arcs extending along the eastern seaboard of the Eurasian continent. These stretch from the tip of the Kamchatka Peninsula (U.S.S.R.) to the Philippines far to the south, and appear superficially as three long arcs whose convex sides face the Pacific Ocean. Their total combined length is *ca.* 5000 km, and they are often referred to collectively as a "festoon" island chain.

Following World War II, Japan's territory was restricted to the central part of this island chain, and is at present made up of the four principal islands, Hokkaido (formerly Yezo), Honshu (formerly Hondo), Kyushu and the smaller Shikoku, together with their numerous accompanying islets, plus two southerly situated archipelagos, the Ryukyus and Bonin Islands (see Table 1). However, from a regional floristic point of view, we must also consider the neighbouring Kuril Islands to the northeast, Sakhalin to the north, the Korean Peninsula to the west, Formosa to the southwest, and the Marianas to the south. Indeed, a simple examination of the regional arrangement of the islands at the Asian continental border allows us to picture several possible transmigrational routes of plants during recent geological time (see Fig. 1). Aleutian elements probably came from Alaska and/or Kamchatka to both Hokkaido and the alpine regions of Japan, via the Kuril Islands. Siberian elements travelled comparatively easily to Hokkaido, via Sakhalin. Manchurian and Korean elements entered western Japan, in general using the Tsushima and Iki Islands as stepping stones. Several species invaded the Ryukyus from Formosa and southern China, while Micronesian species entered the Sulphur Islands (part of the Bonin Islands) from the Marianas. Nevertheless, the palaeogeography of the Japanese islands should of course be taken into account when considering such transmigrational routes in detail (see Chapter 2).

TABLE 1. Dimensions of Japan's principal islands and archipelagos

	Area (1000 km²)	Length measured along curve (km)
Hokkaido	77.9	—
Honshu	228	*ca.* 1400
Shikoku	17.8	—
Kyushu	35.7	—
Izu-Shichito~Bonin Islands	—	*ca.* 1200
Ryukyus	—	*ca.* 1100

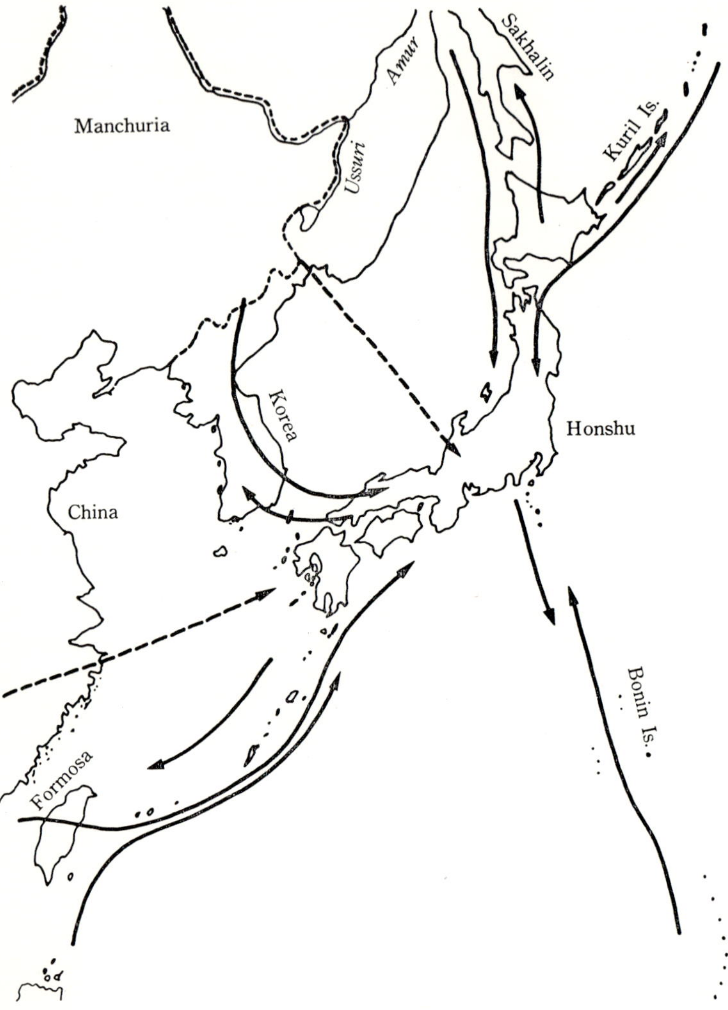

FIG. 1. Map showing transmigrational routes of plants between Japan and its environs.

In general terms, the islands of Japan can be said to occupy the eastern border of the Asian continental shelf, and their position is thus critical from the viewpoint of plate tectonics. Simply speaking, the region represents the point at which the Pacific plate creeps slowly and steadily under the Asian continental plate after the long transmission from the east. The resulting

upheaval at the junction of the plates has given rise to several mountain ranges and numerous volcanoes, parts of which are sufficiently elevated to appear as high peaks. As the fore-deep on the southeast front, there are two elongate trenches, the Japan Trench∼Izu–Ogasawara Trench (deepest point, the Ramapo Deep, east of Torishima; depth, 10,680 m) and the Ryukyu Trench (deepest point, 7413 m). According to the latest theory on the nature of these trenches, a double-arc system is proposed instead of the earlier three-arc "festoon" idea mentioned above. Both of the two open-V-shaped arcs distinguished (the so-called West Japan and East Japan arcs) face southeastwards and are bent backwards to abut against each other in the "Fossa Magna" region of central Honshu (see Fig. 2). The West Japan arc stretches from central Honshu through Kyushu to the Ryukyus in the south, while the East Japan arc extends from Hokkaido through northern Honshu to the Izu Islands and Bonin Islands in the south. These ideas thus explain quite satisfactorily the general structure and geographic characteristics of the region, including the frequent earthquakes, numerous volcanoes, steep, highly-folded mountains, narrow plains, complicated coastline, etc.

Looking next towards the inner side of these island arcs, we can distinguish three inland seas, the Sea of Okhotsk, the Sea of Japan and the East China Sea. Of these, the Sea of Japan has the greatest extent of sea-bottom at or below the 2000 m isobath, the limits of which lie roughly parallel to the surrounding coastline. The region is thus considered as part of the ocean floor, the deepest recorded depth being 3712 m. The Sea of Japan also has two famous banks near its centre, the Yamato and Kita-yamato Banks, which are characterized by the presence of rounded pebbles or gravel on their eroded upper surfaces. This fact suggests that the two banks are in fact remnants of former terraces, and that therefore they probably had some direct influence in the establishment of Japan's present flora. Nevertheless, the details remain unknown.

The Sea of Okhotsk is situated to the northwest of the Kuril Islands, and in the vicinity of these islands its 3000 m isobath closely parallels the trend of the island arc. The East China Sea resembles the Sea of Okhotsk in its general form, also having a deep bottom towards the southeast (i.e. northwest of the Ryukyus), but it is in general much shallower than the Sea of Okhotsk. The depth of the western half of the East China Sea does not exceed 200 m, where the Yangtze Kiang and Old Hwang Ho rivers provide large quantities of mud and coarser debris to build up the sea-bottom. It thus seems very probable that plant migrations from China to Japan (and vice versa) have occurred easily and frequently in the past.

The southern shores of Japan are washed by a strong, warm current, the so-called Japan Current (or *Kuro-shio* in local terminology) (see Fig. 3). Its

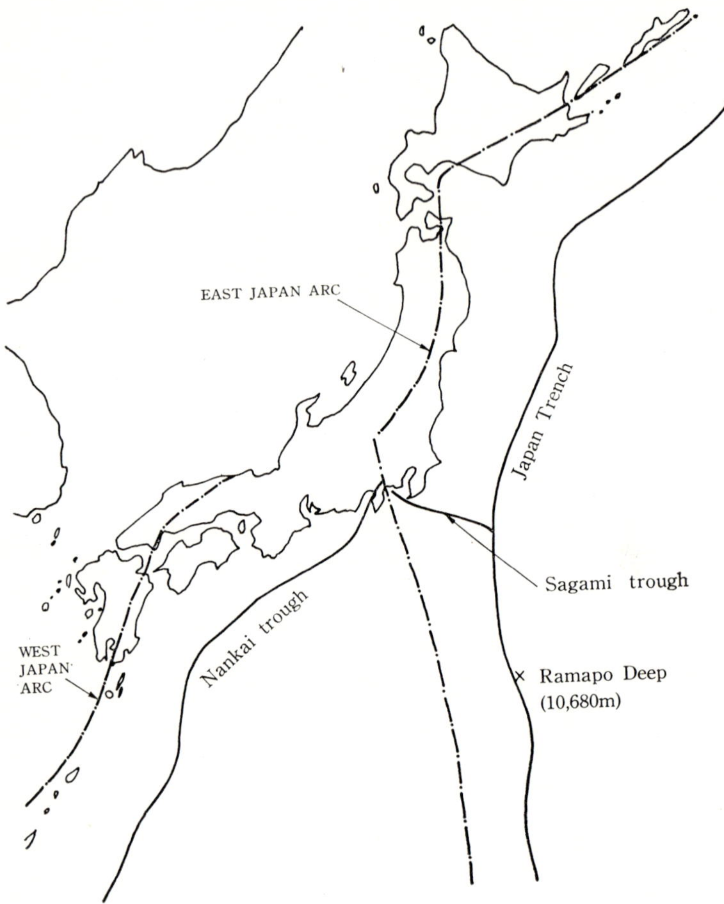

FIG. 2. Map showing the West Japan and East Japan arcs at the Eurasian/Pacific plate boundary (after Sugimura, 1972).

source is the Northern Equatorial Current, which originates in Micronesia, flows westwards to the Philippines and then, after touching their eastern seaboard, turns northwards to pass between Formosa and Iriomote Is. (the westernmost limit of the Ryukyus). From there, it flows through the East China Sea in a northeasterly direction, washing the shores of the northern Ryukyus, and passes through the Tokara Straits (between the Yaku-Tane island group and Amami-Oshima Is.) where it again enters the Pacific. The current then flows to the south of Kyushu, Shikoku and central Honshu (between the Kii and Boso Peninsulas), maintaining the same northeasterly

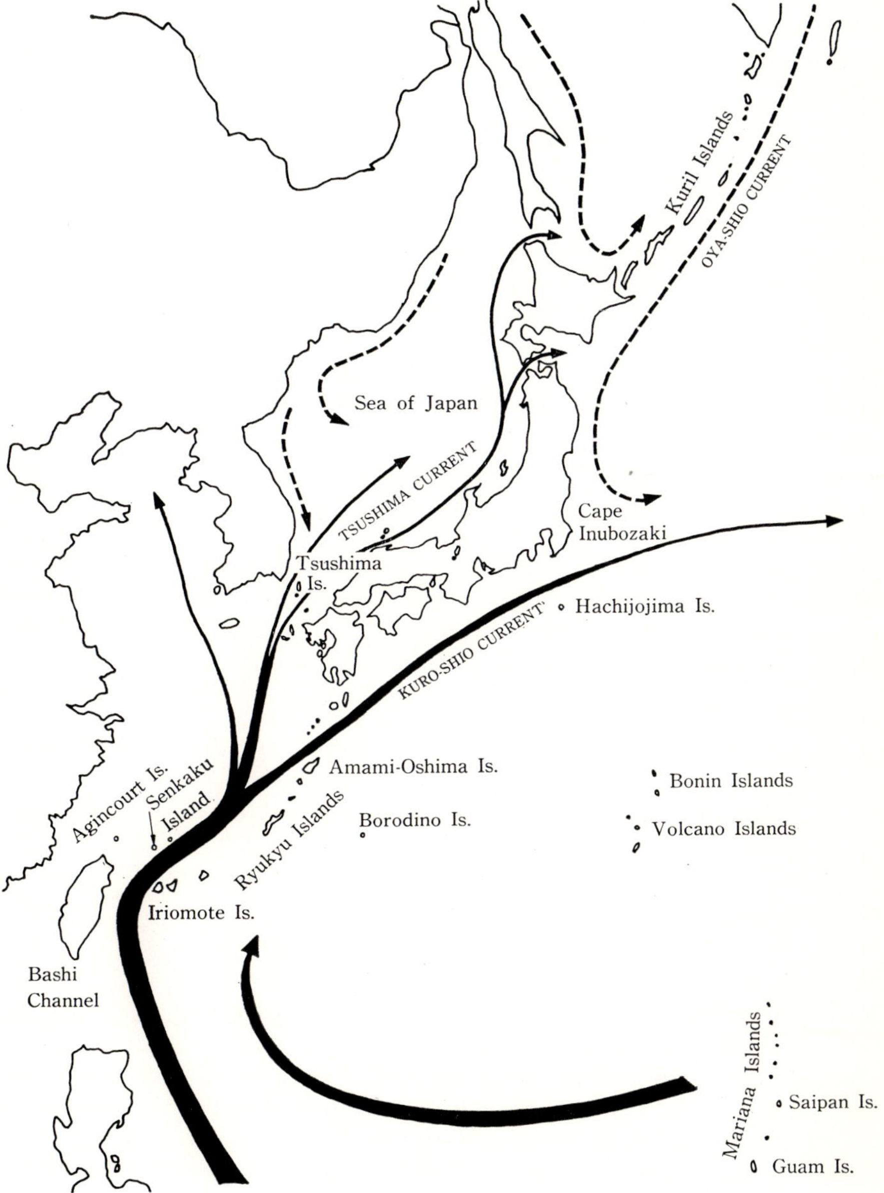

FIG. 3. Map showing the Japan Current (*Kuro-shio*) and other principal currents in the northwest Pacific region.

course, and crosses the Izu-Shichito archipelago just at the northern end of Hachijojima Is. Here, its colour is very deep, appearing as a blackish indigo (hence, the Japanese name *Kuro-shio*, or "black current"). It then leaves the Japanese region, flowing eastwards, the critical point of intersection with the cold current from the north being at Cape Inubozaki, near Choshi, Chiba Pref.

One powerful branch of the Japan Current is the so-called Tsushima Warm Current, which leaves the main stream to the west of Amami-Oshima Is. It flows northwards along the western seaboard of Kyushu, and then turns northeast off-shore from the southern end of the Korean Peninsula, passing through the Korean Straits and Tsushima Straits (from which its name derives). It then continues a northeast course through the Sea of Japan running along the Honshu and western Hokkaido coasts as far as the Soya Straits between Hokkaido and Sakhalin, where it peters out.

The subtropical and warm-temperate climates on Pacific-facing slopes of the Japanese islands owe much to the influence of the Japan Current, whereas the rather high summer temperatures, in contrast to the heavy winter snows, on the Sea of Japan-facing slopes of Honshu can be related to the Tsushima Current. These two currents thus show strong secondary effects on the vegetation and flora of eastern and western Japan, respectively, although their influence is not felt in northern areas.

The chief cold current is the so-called Chishima Current (or *Oya-shio* in Japanese), which flows along the western border of the Japan Trench, southwards from the eastern off-shore region of the Kuril Islands, along the eastern shores of Hokkaido and northern Honshu, to Cape Inubozaki (as mentioned above). In summer, this current brings severe foggy weather to eastern parts of Hokkaido and to the Tohoku region of Honshu, often resulting in heavy damage to the rice crop which, being tropical in nature, requires constant, adequate sunlight. Also, the interaction between the Chishima and Japan Currents in the sea to the south of the Kanto region is so variable from year to year that the climate and crops of the whole area show marked fluctuations due to the unstable conditions that generally prevail.

Another cold current, the East Sakhalin Current, washes the northeastern shores of Hokkaido, flowing southwards from Sakhalin. As a result, the seawater in coastal areas facing the Sea of Okhotsk usually freezes, and severe temperature drops often occur on the inland plateau of Kamikawa. The lowest recorded temperature is −40°C (on January 25, 1902, at Asahikawa).

In western Japan, there is a large inland sea known as the Seto Inland Sea (or *Seto-naikai* in Japanese). It stretches approximately east-west, having a total length of about 380 km and a breadth of 20–60 km. Numerous small islands are scattered throughout this sea, contributing to the very picturesque

scenery, and several straits connect it to the outer oceans. These straits are famous for their tidal vortices, which oscillate with a comparatively short period according to the ebb and flow of the tide. The Seto Inland Sea is thought to be very young, geologically speaking, since several fossil elephants (*Stegodon akashiensis* and/or *Elephas naumanni*) have been collected from the sea-bottom, usually caught up in fishing nets. Also, these same fossils have been found in the diluvial formations exposed on land. Based on such an age, the influence of the Seto Inland Sea in the overall differentiation of Japan's flora should be considered as rather minimal.

As indicated above, the elongate island chain of Japan is spread over a wide range of latitudes, but is in general restricted as regards longitude. It is almost entirely mountainous, with several principal ranges (commonly cut transversely) forming its backbone. The average height of the mountain ranges is not very great, but nevertheless sufficient for them to constitute an effective barrier to low altitude cloud-banks, etc. For these reasons, there is a clear climatic difference between the inner (Sea of Japan) and outer (Pacific) sides of Honshu. The climate of the inner zone (or *Ura-Nihon* in Japanese) exhibits rather high summer temperatures and high precipitation in winter (largely as snow, which often exceeds an overall depth of 2 m). The climate of the outer zone (or *Omote-Nihon*) is characterized by abundant rain in early summer, but winter conditions often approach draught. This climatic difference has had a direct effect on the character of the flora and vegetation of different regions, as well as on the living customs and habits of the people.

The number of open plains is very small, there being only four extensive ones, viz. the Kanto Plain (the largest, with an area of *ca.* 16,000 km^2), and the Echigo, Nobi and Ishikari Plains. All rivers are very short, the longest being the Shinano River (estimated total length, 369 km), which does not even exceed the Thames. These rivers often flow very swiftly, cutting narrow valleys and gorges with rocky cliffs and steep slopes. They have thus, due to their rapid erosive effects, yielded the uneven and steeply angular surface features characteristic of much of the Japanese islands, except perhaps Hokkaido. Nevertheless, the high precipitation and comparatively mild, humid climate enable rich forests to thrive on (and so cover) much of this rugged surface, apart from the mountaintops. The coastline, where the steep mountain-sides meet the sea, is characteristically made up of a highly indented, alternating sequence of protected sandy beaches and precipitous headlands.

Japan is at the same time subject to frequent volcanic activity and earthquakes, although these occur for the most part only in fairly well defined provinces (see Fig. 4). The chief volcanoes are found in the following areas:

FIG. 4. Map showing the location of the principal volcanoes in Japan.

(1) Central Hokkaido, continuing eastwards to the Kuril Islands.
(2) Southern Hokkaido, in the vicinity of Uchiura Bay (which is often referred to as "Eruption Bay").
(3) Median districts of Honshu from the northern end of the Tohoku region to the northern half of the Kanto region.
(4) Central Honshu, and the southward extension to the Izu-Shichito archipelago.
(5) The median sector of Kyushu.
(6) Southern Kyushu and adjacent islands of the Nansei archipelago to the south.

These regions include several active volcanoes. Also, there are a number of large calderas, some of which contain lakes (such as Lake Toya and Lake Mashu in Hokkaido, and Lake Towada in the Tohoku region of Honshu). In Kyushu, Mt. Aso (the largest caldera in the world) occupies the central part of the island, while the Aira caldera forms the northernmost part of Kagoshima Bay. Several of the coniform volcanoes show distinct zonation in their vegetation, and there are many excellent examples of cycles of damage and revival of the flora associated with periodic eruptions (e.g. the Asama volcano of central Honshu, the Mihara volcano of Izu-Oshima Is., the Komagatake volcano in Hokkaido, the Ontake volcano of Sakurajima Is., Kyushu, etc.).

Other regional differences in geographical and topographical characteristics, etc. of the islands of Japan will next be discussed. In central Honshu, a distinct structural zone (the so-called Shizuoka-Itoigawa Line) stretches north-south across the island (see Fig. 5). It is made up of a series of large-scale faults, initiated principally in the Miocene but also accompanied by many later, less well developed faults. To the east of this line lies the so-called "Fossa Magna" region, which is occupied by folded marine sediments of the later Miocene. It is also the site of several large volcanoes, formed in the Pleistocene and Holocene, such as Mt. Myoko (2448 m), Mt. Asama (2542 m), Mt. Yatsugatake (2899 m), Mt. Fuji (3773 m), Mt. Ashitaka (1188 m), Mt. Hakone (1439 m), Mt. Amagi (1407 m), etc. Among them, Mt. Fuji is the tallest; it also has the most extensive basal slopes. Its summit area has the most recent history, being built up in the Holocene. The Izu-Shichito archipelago, as mentioned above, constitutes the southerly extension of this volcanic zone, and includes several well-known active volcanoes such as Mt. Mihara (755 m) on Izu-Oshima Is. and Mt. Oyama (814 m) on Miyakejima Is. The more southerly situated Hachijojima Is. and Torishima Is., together with the Sulphur Islands, etc., are aligned in the same direction. However, the Bonin Islands (i.e. the three groups of small islets of the Mukojima, Chichijima, and Hahajima Islands), although they appear superfi-

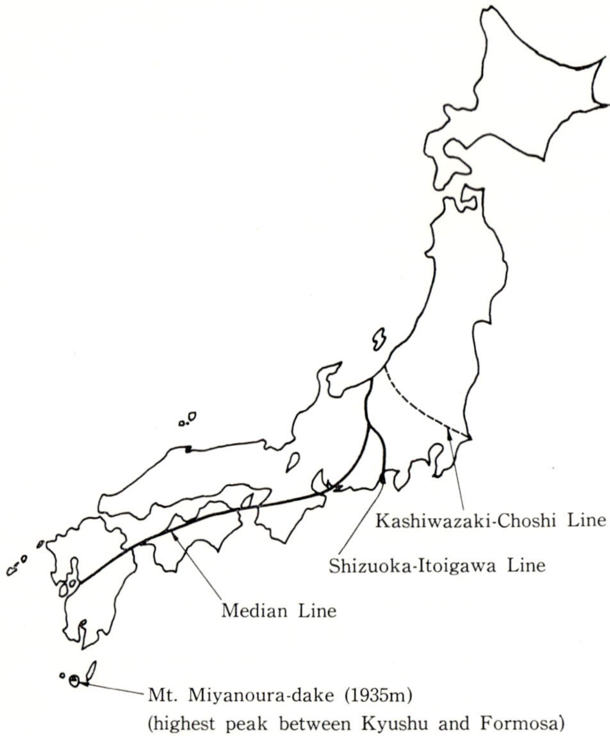

FIG. 5. General map of Japan, showing the Median Line and other major structural features.

cially to be part of this same belt, are in fact distinct from it, occupying a position on the edge of an ancient continental remnant block formed of Palaeogene sediments. These islands are famous for their highly endemic flora, which is clearly Asiatic in type.

Concerning the location of the critical eastern limit of the Fossa Magna region there is still some dispute, although increasing acceptance is being gained for the idea that it is buried under the Kanto plateau. Based on this assumption, the Chichibu mountain group should probably be regarded as an older block or massif retained in the central part of the Fossa Magna region during and since the Miocene. In fact, its general geological features and structure bear a strong resemblance to the Akaishi mountain range, which is located just outside the facing western border of the Fossa Magna region, i.e. to the west of the Shizuoka-Itoigawa Line opposite the Chichibu massif.

This western border itself represents the broadest part of Honshu, and contains three high mountain ranges (often exceeding 3000 m a.s.l.) which together constitute the so-called Japanese Alps. They run obliquely parallel to each other and are, from north to south, the Hida, Kiso and (just mentioned) Akaishi ranges. Associated with them in the north is a series of volcanoes, some of which are still active such as Mt. Tateyama (3015 m) and Mt. Yakedake (2458 m), while others are dormant, such as Mt. Norikura (3026 m) and Mt. Ontake (3063 m). These three ranges all show distinct signs of mountain glaciation, some having U-shaped valleys which face principally to the east. Rich alpine meadows, a direct consequence of the Ice Age, often occur near the summits; those of Mt. Shirouma (2933 m) and neighbouring Mt. Asahi (2418 m) in the Hida range, and of Mt. Senjo (3033 m) and Mt. Kitadake (3192 m) are the most famous, and include several rare and/or endemic plant species.

In western Japan there is another distinct structural line, the so-called "Median Line", which runs parallel to the axis of the island arc (see Fig. 5). At its eastern end it is deflected gradually northwards, making contact with the Shizuoka-Itoigawa Line just to the north of the Akaishi range. To the west, it runs across the entrance to Ise Bay, across the broadest part of the Kii Peninsula, and then passes just to the south of Awaji Is. at the eastern end of the Seto Inland Sea, so entering northern Shikoku. After traversing the complete length of Shikoku, it enters the western part of the Seto Inland Sea close to Matsuyama, reappearing again in Kyushu, where it runs southwesterly across the median part of the island from Oita to Yatsushiro.

Regions to the south of the Median Line exhibit a clear stratigraphic and structural parallelism with the line. That is to say, there is a broad north-south sequence of sedimentary rocks in approximately the order of the geological time-scale, ranging from Palaeozoic (Carboniferous to Permian) in the north, i.e. closest to the line, to Eocene in more southerly areas. Usually the Palaeozoic rocks are highly metamorphosed, although the region as a whole is remarkable for its lack of volcanoes. At the eastern end, as mentioned above, the land rises above 3000 m a.s.l. in the Akaishi range. However, to the west there is a gradual drop in the altitude of the tallest peaks to below 2000 m, such as Mt. Bukkyo-dake (1915 m) in Nara Pref. and Mt. Ishizuchi (1981 m) in Shikoku. As a result of this profile, the region as a whole barely reaches above the coniferous forest zone (except in the Akaishi range, which lies partly in the alpine zone); and in Kyushu, due to the low altitudes (highest point, Mt. Sobo, 1758 m), the vegetation does not even reach above the *Fagus* (cool-temperate) zone. Many of the rivers flow parallel to the general strike of the strata, although there are others which cut across it forming steep valleys and gorges, very often with limestone walls. In addi-

tion, the rather heavy rainfall in the region throughout the year yields a humid climate. Areas of low altitude are thus covered by rich broad-leaved forests, both deciduous and evergreen. The deciduous forests are distinct for their many endemic species, and also for certain characteristic genera whose distribution often extends to the Himalayas and Western China. The evergreen forests show a strong influence from subtropical elements, deriving from the Ryukyus, Formosa, and Malayasian floristic region.

The northern half of western Japan, i.e. that situated to the north of the Median Line, has several low-lying plateaus and eroded mountain ranges with an elevation of *ca.* 800–1300 m a.s.l. The exposed rocks are mostly granitic, while the small plains situated towards the coast, and the narrow valleys leading to them, are filled with sand and gravel derived from this granite. In general, the annual precipitation is comparatively low (among the lowest in Japan), and the vegetation is of a type characteristic of dry areas. The central part of the region, which covers both Okayama Pref. and Hiroshima Pref., has abundant limestone outcrops, and is famous for plant species which are closely related to the flora of the Korean Peninsula. Some of these plants have in fact evolved into species that are endemic to Japan.

Northern Kyushu is built up mainly of Palaeogene strata, and particularly in the western sector has sunk with respect to sea-level, resulting in the formation of numerous islets and peninsulas. At one time, these Palaeogene sediments contained several rich coal-bearing units, often with sufficient fossil plant material to give some idea of the former flora and vegetation of Japan. However, many of the workable seams have now been exhausted, and most of the coal-fields have been closed. Further to the west, are the islands of the Goto archipelago, and in the Tsushima Straits, the Tsushima and Iki Islands. These islands are largely mountainous and, as mentioned above, have a clear role as stepping stones for continental floral elements in their apparently rather recent invasion of Japan. Examples of invading forms are *Chloranthus Fortunei*, *Meliosma Oldhami*, *Chrysanthemum sibiricum*, etc.

The central part of Kyushu is covered with large volumes of lava and ash ejected from the Aso volcano. These ejecta are built up into gentle, undulating slopes and rises (referred to collectively in Japanese as *Namino*, or "wavy plateaus"), which support several continental prairie elements such as *Campanula glomerata*, etc. In southern Kyushu, many of the low hills contain a different type of volcanic ash (referred to in Japanese as *Shirasu*, or "white sand"), which has a different flora. This material is compact and solid when dry, but becomes very slippery and more plastic when wet, often leading to landslips.

The islands of the Yaku-Tane island group, which is situated further south on the opposite side of the Osumi Straits, show very divergent geographical

characters. Tanegashima is a very flat plateau-like island, formed chiefly of Tertiary strata, whereas Yakushima is a mountain with precipitous slopes, the highest point being Mt. Miyanoura-dake (1935 m). In fact, this peak represents the highest point in the entire region between Kyushu and the island of Formosa. Furthermore, it has an extremely high annual precipitation, the recorded maximum being 7354 mm (the fourth highest in the world) in the Kosugidani valley. Indeed, among people on the island there is a famous saying that "on this island, we have 366 rainy days a year". These climatic and topographical features have yielded dense and varied forests over most of the island. The lower slopes are covered by subtropical broad-leaved forests, followed by mixed deciduous forests (composed largely of *Cryptomeria japonica*), and, on higher slopes, a rich moss forest. The latter consists of large *Cryptomeria* trees whose branches are covered by dense moss colonies and several other epiphytic plants. The summit of the island resembles an alpine meadow, the dominant vegetation being a dwarf-bamboo association (*Pseudosasa Owatarii*) containing several alpine plants. This situation is undoubtedly a result of the so-called "mountaintop effect" (or *Spitzenerhebung*). The evergreen forests on the island contain many Malaysian subtropical and/or southern continental elements, their existence here representing in most cases the northernmost limit of their occurrence. The *Cryptomeria* forests are characterized by many temperate elements of Japan's typical flora, here existing near their southernmost limit of occurrence; there are also abundant endemic species.

The Ryukyus, which consist of more than 100 islands and islets, stretch southwestwards from the Yaku-Tane island group in the form of a broad curve *ca.* 1100 km in length. They constitute two sets of islands, fundamentally different from each other in geological characters (see Fig. 6).

The main set can be divided into four groups, which are (from north to south) the Amami, Okinawa, Miyako and Yaeyama archipelagos. These islands have a basement of Palaeozoic rocks, and in the northern sector (i.e. the Amami archipelago and northern half of Okinawa Is.) the mountains are composed principally of these rocks, averaging *ca.* 400 m in altitude. The tallest is Mt. Yuwan (694 m) on Amami-Oshima Is. In the remainder of the region, Tertiary beds cover the basement rocks, forming a low, hilly topography. Furthermore, in many places there is a covering of Ryukyu Limestone, which gives rise to many narrow plateaus and terraces: the Miyako archipelago is a good example. The smaller islets are generally the uplifted summits of coral reefs, and the shores of other islands are still often bounded by living coral reefs. On the island of Iriomote (the Yaeyama archipelago), there are also many drowned valleys and sunken estuaries in which broad areas of mangrove forest have developed.

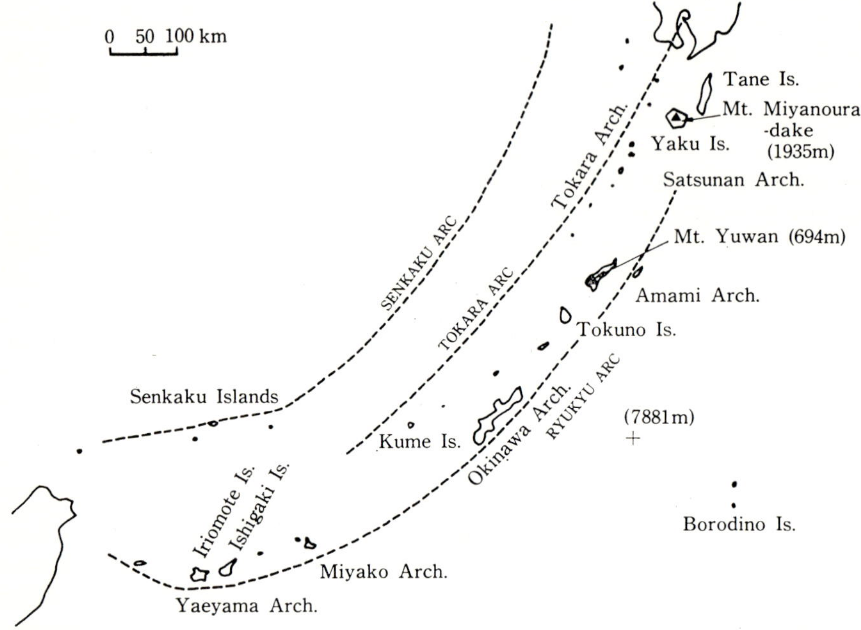

FIG. 6. Structural map of the Ryukyus.

The second set is known simply as the Ryukyu volcanic arc and runs approximately parallel to the main arc, to the north and along its western flank. It stretches southwards from the volcanic island of Sakurajima in Kyushu throughout the length of the Tokara archipelago, and includes many small volcanic islets, several of which are known to be active.

Some authors recognize a further set of islands to the west of the Ryukyus, known as the Senkaku arc. Their position is on the border of the Tun-hai continental shelf, and they are separated from the Ryukyus by the so-called Ryukyu trough. This arc stretches northeastwards to the Danjo archipelago and Goto Islands of Kyushu, and westwards to Agincourt Is., northeast of Formosa. The islands are composed mainly of pre-Miocene and/or Miocene rocks, and there are several "old-type" plants on them, such as *Heterotropa* sp. on Senkaku Is. However, full details of the flora are undetermined due to the lack of resources and general remoteness of the islands. Finally, to the east of the Ryukyu region, there is a submarine mountain range known as the Daito range. This lies to the east of the Ryukyu Trench, and its summits (covered with old, uplifted coral reefs) appear above sea-level as the Borodino (or Oagari) Islands.

Returning again to Honshu, a brief description will next be given of the northeastern part of the island (i.e. that to the east of the Fossa Magna region). The Kanto Plain (which includes the area occupied by the city of Tokyo) is very young geologically speaking, and constitutes a low, hilly region formed of recent diluvial deposits, eroded later into shallow valleys with an alluvial bottom. This plain is surrounded by mountains on all sides except the east, where it faces the Pacific Ocean directly. Its surface is widely covered by thin layers of aeolian volcanic ash (the last of which was deposited in the Ice Age), referred to collectively as "Kanto Loam". Also, the lower reaches of the plain formerly abounded in marshland rich in northern floral elements such as *Habenaria linearifolia*, *Gymnadenia conopsea*, *Viola Raddeana*, *Primula Sieboldi*, etc. However, this has now been almost completely destroyed, due mainly to the expansion of the cities and their suburbs, to the consequent pollution of the environment, and partly to the destructive activities of enthusiastic plant collectors, etc.

From this plain, Honshu curves almost due northwards, terminating in the two peninsulas of Shimokita and Tsugaru. These both face to the north and are separated from Hokkaido by the rather shallow Tsugaru Straits. This region of Honshu is known as the Tohoku district, and can be divided roughly into five topographically distinct strips or units, which together form three mountain ranges and two narrow rows of separating plains and basins (see Fig. 7).

The easternmost strip is mountainous, and built up chiefly of older rocks heavily eroded into plateau-like highlands. It comprises two distinct areas: the Kitakami massif in the north, and the Abukuma massif in the south. The eastern border of the northern mountains drops sharply to the sea, forming a highly indented coastline with precipitous cliffs. This area is famous for its scenic beauty, and represents the northern limit of many of Japan's warm-temperate floral elements. It is, at the same time, the site of considerable coastal damage, due to the earthquake-induced tidal waves that originate from the Japan Trench to the east, and frequently hit the many small harbours located along the narrow coastal strip. The highest peak in the Kitakami massif is Mt. Hayachine (1914 m), which is noted for its alpine meadows with many endemic plant species, some of which are also conspicuous for their relationship to the flora of the Hidaka and Yubari ranges of Hokkaido. The southern, Abukuma massif is similar in extent to the Kitakami massif, but is in general of somewhat lower elevation.

The second strip consists essentially of a series of plains, which are narrow by normal standards but nevertheless rather broad and long when compared with most plains in Japan. The two principal components are the Kitakami valley plain to the north and the Abukuma valley plain to the south.

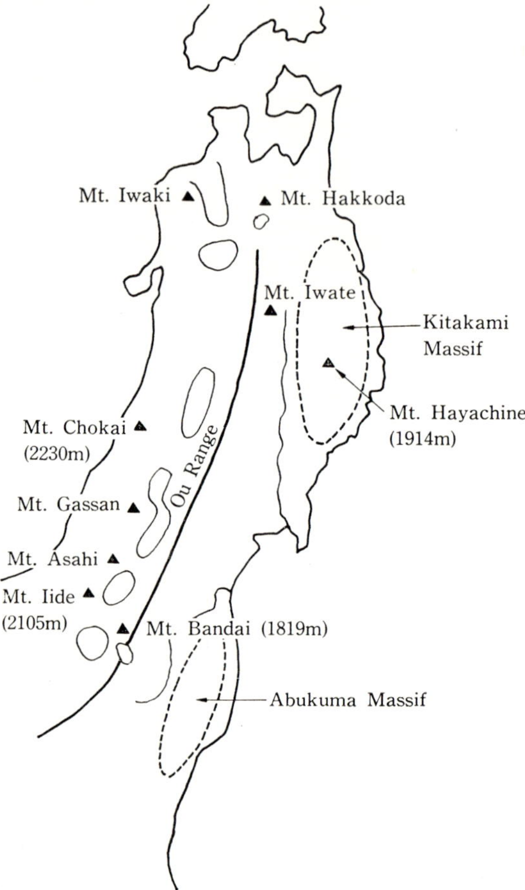

FIG. 7. General geographic map of the Tohoku district.

The third strip is also mountainous and contains many volcanoes formed during the Diluvial epoch. It is known as the Ou range. Some of the volcanoes are still active and others often have a history of recent eruption. For example, Mt. Bandai (1819 m) has undergone a serious recent eruption which caused the break up of almost half of the mountain's original mass, and has resulted in the damming up or otherwise closing of many valleys and the production of several small lakes. The Ou range also forms the principal barrier to cloud-banks passing across northern Honshu from the west, and so marks the approximate climatic demarcation line between east and west Japan mentioned above. It thus also represents the eastern limit of many Sea of Japan-side floral elements.

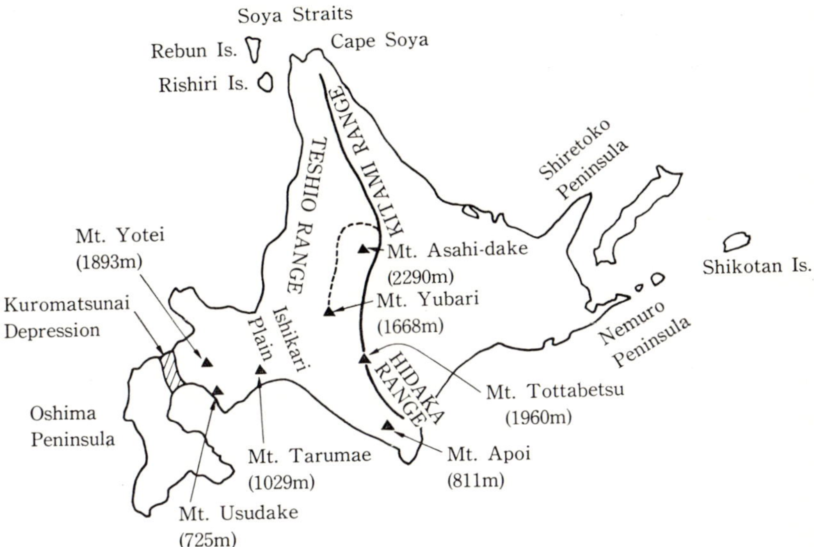

FIG. 8. General geographic map of Hokkaido.

The fourth strip constitutes a chain of small basins, bordered to the west by the fifth, mountainous strip. This last strip is essentially a combination of low mountain ranges and large volcanoes, the latter having numerous snowfields, even in summer months, and consequently several endemic alpine plants. The southernmost mountains (the Iide range; highest point, 2105 m) are not volcanic, but nevertheless are also rich in alpine species due to their rather high elevation.

Across the Tsugaru Straits, as mentioned above, is the northernmost of Japan's principal islands, Hokkaido (see Fig. 8). Its total extent is second only to Honshu (see Table 1), and it is often referred to as "Yezo", both in the literature and in botanical nomenclature. Its outline resembles that of a devilfish swimming eastwards, but with a whale-like tail. The island terminates to the north at Cape Soya (45°31′N) and is separated from Sakhalin, where similar geographical and geological trends are continued, by the Soya Straits. The southern, tail-like part of the island is known as the Oshima Peninsula, which curves round in a broad semicircle to enclose a gulf on its Pacific side. This peninsula is largely mountainous, but has a narrow area of lowland at its neck (the so-called Kuromatsunai depression) which marks the northernmost point of occurrence of many of Japan's warmer floral elements. Hokkaido's largest plain is known as the Ishikari Plain (or Ishikari depression in geological terminology), which is separated from the Kuromatsunai

depression by a mountainous area rich in volcanoes. Some of the latter are still active, such as Mt. Tarumae (1029 m), Mt. Usudake (725 m), etc. while Mt. Yotei (or Makkarinupuri in the Ainu language) is dormant but shows a typical cone-shaped form with the summit rising to 1893 m a.s.l.

The central backbone of Hokkaido is comprised of a central highland area plus two ranges of mountains, all of which are situated on the broad north-south axis of the island. The mountains are composed mainly of Mesozoic and/or Palaeozoic rocks, while the central plateau region is covered by large areas of recent volcanic rocks and volcanoes. The southern mountains are known as the Hidaka range and include many peaks with an elevation of the order of 1900 m a.s.l. Some of them, such as Mt. Tottabetsu (1960 m), show clear evidence of past glaciation near their summits. On the western side of this range there are two other mountain groups, which are significant for their endemic floral elements. One of them, the Apoi mountain group, is located near central Hokkaido's southernmost point (Cape Erimo) and has a rich alpine flora in spite of its low elevation (maximum, 811 m a.s.l.), due mainly to the "mountaintop effect" and to the influence of the cold sea currents in the area. The second mountain group is the so-called Yubari mountain range, which lies between the Ishikari depression and the Hidaka range. It is built up in part of serpentinitic rocks, rising to a maximum of 1668 m a.s.l., and shows many endemic species, perhaps arising under the influence of edaphic factors. The northern range, i.e. that on the opposite side of the central plateau, is best regarded as two mountain groups, the Kitami and Teshio ranges. These do not in general rise much above the 900 m level, but several of their valleys with areas of serpentinitic rocks exhibit an excellent endemic flora.

The central highland area is, as mentioned, largely volcanic in origin. Mt. Asahi-dake, its highest point, reaches a maximum altitude of 2290 m a.s.l., and supports many broad, rich alpine meadows. Some of the volcanoes are currently active, and the distinct volcanic belt originating in the area extends faraway to the northeast to the Kamchatka Peninsula via the Shiretoko Peninsula and Kuril Islands, which divide the Sea of Okhotsk from the Pacific Ocean.

At the eastern end of Hokkaido, to the south of the Shiretoko Peninsula, is a second large promontory, known as the Nemuro Peninsula. Its geographical trends extend eastwards as far as Shikotan Is., although it is in fact geologically distinct from the Central Highland–Shiretoko–Kuril volcanic chain, being made up principally of Mesozoic strata. Between these two geologically distinct zones are located two plains of Quaternary origin, which still contain many areas of relatively undisturbed marshland. Also, there are several lagoons along the coast of the Sea of Okhotsk, which often display the char-

acteristic red colour of *Saricornia* colonies in the summer and autumn.

Finally, on the Sea of Japan side of Hokkaido, to the west of Cape Soya in the north, are two small islands, Rebun and Rishiri. These are famous for their excellent alpine flora, although each is quite different as regards its geology, the former being a hilly lowland built up of Tertiary sediments, and the latter a cone-shaped volcano of the Diluvial epoch rising to a peak of 1718 m a.s.l.

2. Climate and the Distribution of Vegetation Zones

As has been discussed in general terms above, the islands of Japan lie approximately between 24° and 46°N latitude and between 122° and 148°E longitude, and so display a very wide range of climatic conditions, especially as regards temperature. One characteristic feature is the high precipitation throughout the year. In summer, the prevailing SE winds carry abundant rain to the area, including typhoons with strong winds and very heavy rainfall, whereas in winter NW winds bring considerable snowfall. The mean annual precipitation is *ca.* 1740 mm (max.>3000 mm; min. 800 mm). There are thus no arid regions, although the precipitation is comparatively low in certain parts of the Seto Inland Sea, the Chubu inland area and Kitakami Valley of Honshu, and the Ishikari Plain of Hokkaido.

Generally speaking, four principal climatic zones can be distinguished in Japan; viz. subtropical, warm-temperate, cool-temperate, and subarctic. These range from south to north along the island arc, or (with elevation) from the coastal strip, plains, and highlands, to the inland subalpine and alpine regions of the islands. As mentioned above, there is in general ample precipitation for the growth of plants, and so temperature is the dominant factor determining both the climatic and vegetational zones. Table 2 illustrates typical temperatures and precipitation levels for the four climatic zones.

These zones have formed (and still form) a useful working basis for forestry studies and general research on the ecology of Japanese plants. In 1912, Honda[1)] distinguished four principal vegetational types using values for mean annual temperatures: these he termed the "tropical" (banyan tree), "warm" (evergreen oak), "temperate" (beech), and "arctic" (fir tree) types. Mayr[2)] further separated the temperate type into two smaller divisions, the "chestnut tree zone" and the "beech tree zone". The former can be distinguished in certain areas at the junction between typical warm and temperate vegetational types. For example, on low hills and in the northern sector of the

Kanto Plain, forests consisting of broad-leaved deciduous trees such as the chestnut, oak, and *Zelkova serrata* are found. Some plant ecologists believed that these were in fact secondary forests, arising under the disruptive influence of human agricultural activity. However, Morishita and Kira[3] have shown that they are in fact distributed in parts of the warm zone that are subject to cool winter conditions.

In 1937, Imanishi[4] investigated the distribution of 17 specific, representative forest tree species which constitute the major components of climatic climax forests at various elevations in the Japanese Alps (central Honshu). From the results (see Fig. 2, Chapter 6), it can be seen that in the subalpine zone *Abies Veitchii* is the representative species on the Pacific Ocean side of Honshu, and *Abies Mariesii* the representative species on the Sea of Japan side. In the temperate zone, *Fagus crenata* is distributed widely on both flanks of the mountains, but on the Pacific side it occurs in association with *Abies homolepis*. On the Sea of Japan side the *Fagus crenata* zone borders directly onto evergreen oak forests, whereas on the Pacific side there is a transition zone with *Abies firma*.

In 1949, Kira[5] found that accumulated temperatures (expressed as warmth and cold indices) showed a remarkably good correlation with the distribution of forest types, and were in fact more suitable for explaining the actual distribution of forest zones than simple mean annual temperatures. He took a monthly mean temperature of +5°C as the standard, above which green plants would normally flourish. The "warmth index" was defined as the annual sum (in °C) of all positive differences between monthly means and +5°C; the "cold index" was the corresponding negative sum for monthly means below +5°C (see example in Table 3). He also drew up isotherm maps (Fig. 9) based on these warmth and cold indices. General relationships are apparent between the principal tree species and the warmth index (as illustrated in Fig. 10), and between the vertical distribution of forest zones and the warmth and cold indices (see Fig. 1, Chapter 6). Table 4 summarizes the major forest zones that can be defined in this way, together with their index limits.

As mentioned above, there are many distinct contrasts between life-forms and vegetation on the Sea of Japan and Pacific Ocean sides of the Japanese island arc. These can be attributed largely to the heavier snowfall on the Sea of Japan side. The precipitation is high (see Table 2), and in winter it is largely snowfall. The forestland becomes thickly covered with snow, which remains on the surface for *ca.* 4–6 months during the winter period. This snow-cover of course has a strong influence on the underlying vegetation, the most obvious effects arising from the incumbent snow pressure. The growing period of plants is effectively shortened, although the covering layer of snow

TABLE 2. Climatic conditions at various localities in Japan†
(T) = average temperature (°C); (P) = precipitation (mm)

		Month												Annual mean /total	Climatic zone
		J	F	M	A	M	J	J	A	S	O	N	D		
Naha	(T)	16.0	16.4	18.1	20.8	23.8	26.0	28.2	27.8	27.1	24.1	21.4	18.1	22.3	Sub-tropical
	(P)	122	116	154	142	244	320	174	253	152	149	151	140	2118	
Nase	(T)	14.2	14.5	16.5	19.6	22.4	25.3	28.1	27.8	26.6	23.0	19.8	16.2	21.2	
	(P)	178	170	191	204	368	452	215	325	311	237	219	169	3039	
Miyazaki	(T)	6.7	7.9	11.0	15.5	19.2	22.4	26.6	26.9	24.1	18.4	13.8	8.8	16.8	Warm-temperate
	(P)	56	97	143	216	289	440	343	302	296	207	139	67	2594	
Kochi	(T)	5.2	6.4	9.8	14.9	18.9	22.1	25.9	26.8	23.9	18.2	13.0	7.8	16.1	
	(P)	62	83	166	273	291	389	353	329	344	170	116	69	2645	
Ashizuri	(T)	8.1	8.8	11.8	16.3	19.8	22.5	26.1	27.3	25.1	20.6	16.0	10.9	17.8	
	(P)	80	133	160	236	280	371	269	214	293	254	154	102	2545	
Okayama	(T)	3.3	4.0	7.1	13.0	17.5	21.4	25.8	27.0	22.7	16.2	10.7	5.7	14.5	
	(P)	39	48	79	107	131	185	183	96	165	98	58	35	1218	
Hamada	(T)	5.5	5.4	7.9	12.6	16.7	20.4	25.0	26.2	22.2	16.5	12.3	8.1	14.9	
	(P)	105	99	106	123	132	200	238	105	247	125	110	113	1702	
Kanazawa	(T)	2.6	2.8	5.7	11.5	16.5	20.4	24.8	26.2	21.9	15.8	10.6	5.7	13.7	
	(P)	327	199	172	153	149	201	244	171	267	214	220	344	2662	
Shionomisaki	(T)	7.5	7.9	10.6	15.2	18.7	21.6	25.3	26.5	24.1	19.2	15.0	10.2	16.8	
	(P)	104	103	180	252	279	386	317	294	313	264	185	108	2785	
Shizuoka	(T)	5.7	6.3	9.2	14.0	18.0	21.4	25.2	26.4	23.3	17.9	13.2	8.1	15.7	
	(P)	76	96	180	242	239	345	244	275	255	193	131	79	2355	
Choshi	(T)	5.7	6.0	8.6	13.1	16.7	19.4	22.9	25.0	22.8	18.1	13.5	8.5	15.1	
	(P)	86	105	135	131	166	188	112	130	177	251	153	84	1729	
Tokyo	(T)	4.1	4.8	7.9	13.5	18.0	21.3	25.2	26.7	23.0	16.9	11.7	6.6	15.0	
	(P)	49	65	98	122	145	192	140	153	182	203	96	58	1503	
Hachijojima	(T)	10.2	10.2	12.1	16.1	19.2	21.9	25.3	26.6	25.0	20.9	17.2	12.9	18.1	
	(P)	182	203	243	236	289	350	181	231	336	484	373	174	3284	
Onahama	(T)	3.0	3.1	5.7	10.7	14.9	18.3	21.9	24.1	20.9	15.3	10.3	5.6	12.8	
	(P)	43	56	101	114	139	172	140	128	173	187	91	53	1396	
Takada	(T)	1.8	1.7	4.3	10.6	15.9	20.1	24.5	25.9	21.4	15.2	9.6	4.8	13.0	
	(P)	484	312	211	111	103	140	202	157	245	221	340	514	3038	

Station		J	F	M	A	M	J	J	A	S	O	N	D	Annual	Climate
Matsumoto	(T)	−1.2	−1.0	3.1	9.7	15.1	19.0	23.4	24.1	19.4	12.4	6.6	1.7	11.0	Cool-temperate
	(P)	36	45	65	84	96	164	128	107	138	109	53	33	1058	
Yamagata	(T)	−1.2	−0.8	2.4	9.2	15.1	19.1	23.2	24.5	19.5	12.7	6.8	1.7	11.0	
	(P)	98	75	72	68	66	107	174	127	124	106	79	116	1210	
Miyako	(T)	−1.0	0.0	2.7	8.6	13.2	16.2	20.3	22.5	18.6	12.8	7.6	2.5	10.4	
	(P)	63	63	88	80	100	129	127	134	195	161	80	69	1289	
Akita	(T)	−0.7	−0.6	2.4	8.5	13.9	18.3	22.6	24.3	19.4	13.0	7.1	1.9	10.9	
	(P)	135	100	108	137	115	135	196	161	196	168	183	173	1807	
Sapporo	(T)	−5.1	−4.4	−0.6	6.1	11.8	15.7	20.2	21.7	16.9	10.4	3.7	−2.3	7.8	Sub-arctic
	(P)	118	83	75	64	59	73	90	112	150	104	104	111	1141	
Asahikawa	(T)	−8.5	−7.7	−3.0	4.5	11.4	16.1	20.4	20.9	15.3	8.5	1.3	−4.9	6.2	
	(P)	81	66	61	58	79	75	119	160	138	105	117	99	1159	
Wakkanai	(T)	−5.8	−5.6	−1.7	4.2	8.7	12.4	16.7	19.2	16.4	10.6	3.1	−2.8	6.3	
	(P)	104	68	66	64	80	71	119	111	151	130	128	121	1212	
Abashiri	(T)	−6.6	−7.1	−2.9	3.9	9.3	12.6	17.2	19.0	15.7	9.9	3.0	−3.0	5.9	
	(P)	61	43	56	47	66	77	87	101	108	83	65	53	848	
Nemuro	(T)	−4.8	−5.6	−2.2	3.0	7.1	10.0	14.3	17.1	15.4	10.6	4.6	−1.2	5.7	
	(P)	48	41	70	77	98	107	97	116	149	124	87	65	1077	

† Data from *Rika Nenpyo* (Japanese) (ed. Tokyo Astronomical Observatory, University of Tokyo), Maruzen, 1973; based on figures for the period 1941–70.

TABLE 3. Warmth and cold indices for Sapporo City, Hokkaido

	Month												Annual mean (°C)
	J	F	M	A	M	J	J	A	S	O	N	D	
Monthly mean temp. (°C)	−5.5	−4.7	−1.0	5.7	11.3	15.5	20.0	21.7	16.8	10.4	3.6	−2.6	7.6
Difference between monthly mean and +5°C (°C)	−10.5	−9.7	−6.0	+0.7	+6.3	+10.5	+15.0	+16.7	+11.8	+5.4	−1.4	−7.6	
WARMTH INDEX				0.7	+6.3	+10.5	+15.0	+16.7	+11.8	+5.4			= +**66.4**
COLD INDEX	−10.5	−9.7	−6.0								−1.4	−7.6	= −**35.2**

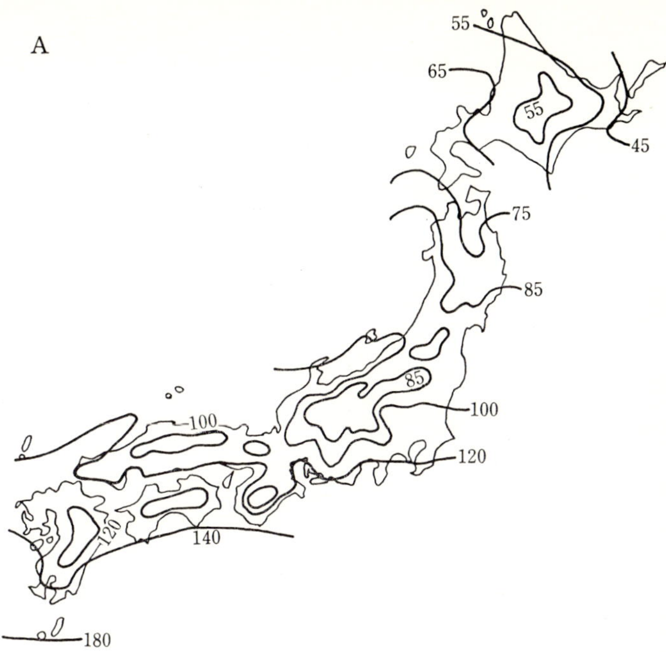

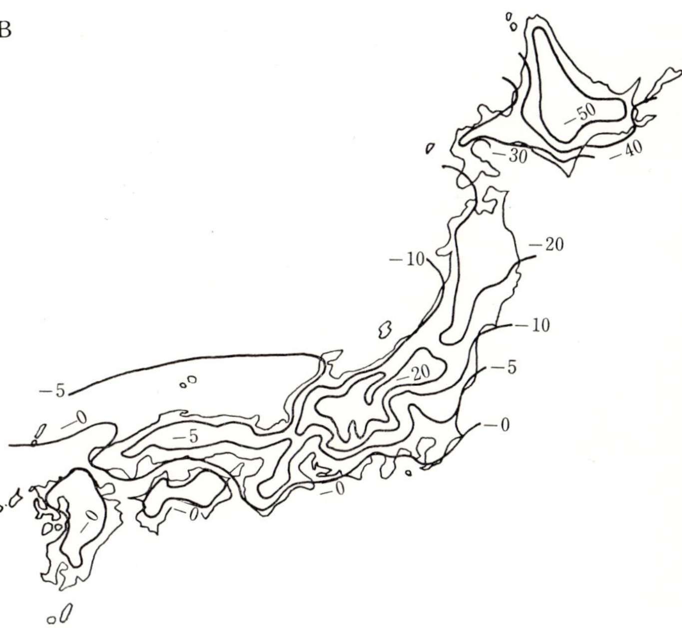

FIG. 9. Isotherm maps showing the warmth (A) and cold (B) indices for the Japanese islands (after Kira).

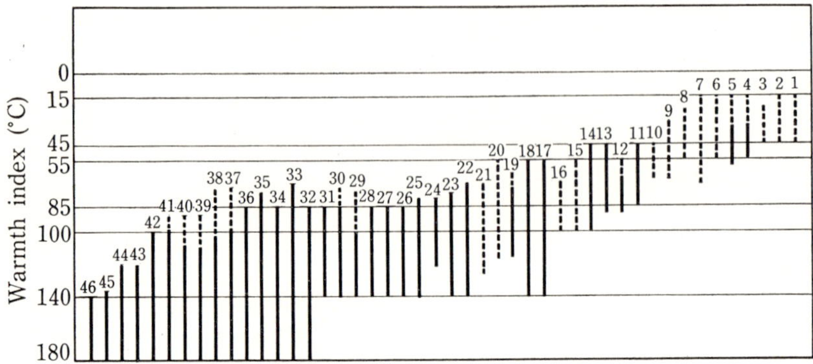

FIG. 10. Relationship between the warmth index and principal tree species. (Solid lines show horizontal (sea-level) ranges of warmth index for the given trees; broken lines show vertical (all altitude) ranges of warmth index for the given trees.)

1. *Abies Veitchii*
2. *Abies Mariesii*
3. *Picea jezoensis* var. *hondoensis*
4. *Picea jezoensis*
5. *Abies sachalinensis*
6. *Betula Ermanii*
7. *Larix leptolepis*
8. *Tsuga diversifolia*
9. *Betula platyphylla*
10. *Abies homolepis*
11. *Fagus crenata*
12. *Fagus japonica*
13. *Quercus crispula*
14. *Quercus dentata*
15. *Carpinus cordata*
16. *Carpinus japonica*
17. *Carpinus laxiflora*
18. *Carpinus Tschonoskii*
19. *Castanea crenata*
20. *Quercus serrata*
21. *Zelkova serrata*
22. *Tsuga Sieboldii*
23. *Abies firma*
24. *Thujopsis dolabrata*
25. *Cryptomeria japonica*
26. *Chamaecyparis pisifera*
27. *Chamaecyparis obtusa*
28. *Torreya nucifera*
29. *Quercus variabilis*
30. *Quercus acutissima*
31. *Cyclobalanopsis myrsinaefolia*
32. *Cyclobalanopsis stenophylla*
33. *Ilex integra*
34. *Celtis sinensis*
35. *Pinus densiflora*
36. *Pinus Thunbergii*
37. *Camellia japonica*
38. *Machilus Thunbergii*
39. *Quercus acuta*
40. *Cyclobalanopsis gilva*
41. *Castanopsis cuspidata*
42. *Podocarpus macrophylla*
43. *Cinnamomum Camphora*
44. *Podocarpus Nagi*
45. *Ficus Wightiana*
46. *Cycas revoluta*

TABLE 4. Major forest zones as defined by the warmth and cold indices

Forest zone	Warmth index (°C)	Cold index (°C)
Subtropical evergreen forest	>180	
Warm-temperate evergreen forest	85–180	>−10
(deciduous forest	85–180	<−10)
Cool-temperate deciduous forest	45–85	
Subarctic/subalpine coniferous forest	15–45	
Arctic/alpine meadow	<15	

does insulate them from the direct influence of the prevailing, severe climatic conditions. The upper limits of the forest zones thus tend to drop down towards the coast on the Sea of Japan side, and the plant communities differ in many respects from those on the Pacific side. These differences are of course also apparent among undergrowth tree species and shrubs, particularly in the cool-temperate zone. In northern districts, on the Sea of Japan side, several evergreen tree species vary from tall trees to creeping shrub forms, or different species, and these occur on the snow-covered mountain slopes (see Table 5).

The greatest recorded snow depths occur in the region around Mt. Chokai, Dewa Height and the Joetsu mountain range. For example, in mid-February 1950 a snow depth of 12 m was measured in a snow-field on the lee-side of Mt. Gassan at an elevation of 2000 m a.s.l. In such regions, for the greater

TABLE 5. Special distribution of trees and shrubs between districts with and without snow

	Genus	Districts with heavy snow	Districts without snow
Evergreen conifers	*Cryptomeria*	*C. japonica* var. *radicans*	*C. japonica*
	Torreya	*T. nucifera* var. *fruticosa*	*T. nucifera*
	Cephalotaxus	*C. Harringtonia* var. *nana*	*C. Harringtonia*
	Taxus	*T. cuspidata* var. *nana*	*T. cuspidata*
Evergreen broad-leaved trees and shrubs	*Aucuba*	*A. japonica* var. *borealis*	*A. japonica*
	Daphniphyllum	*D. macropodum* var. *humile*	*D. macropodum*
	Ilex	*I. leucoclada*	*I. integra*
	Ilex	*I. crenata* var. *paludosa*	*I. crenata*
	Camellia	*C. japonica* var. *procumbens*	*C. japonica*
Dwarf bamboo	*Sasa*	*S. kurilensis*	*S. nipponica*
	Sasamorpha /*Sasa*	*Sasa palmata*	*Sasamorpha purpurascens*

part, there is no subalpine coniferous zone but instead there is a zone of shrub-like deciduous oak bushes. This zone has thus been termed "pseudo-subalpine", and the absence of tall conifers such as *Abies Mariesii* is attributed to the deep snow-cover and consequent high incumbent pressure on the land surface.

3. Soil Types and Their Plant Productivity

In relation to the wide variety of climatic and geographic conditions in Japan, there is a corresponding variety of soil types. These can be classified in general into four broad groups; viz. black soil (*ando* soil), red soil, podsol, and brown forest soil.

Also, as described in section 1, the Japanese island arc has several longitudinal and transverse volcanic mountain chains running through it. Present volcanoes include *ca.* 60 that are either active or dormant, and *ca.* 200 that are extinct; the severest recent activity probably occurred in the late Tertiary. These volcanoes have scattered a very large volume of ash over the entire island arc, so influencing strongly the properties and composition of the surface soil layer.

The typical black soils formed or derived from volcanic ash occur widely in Japan, e.g. in volcanic piedmont regions and lava plateaus, in river basins and associated depressions, old river terraces, etc., covering a total area of $>$3,000,000 ha from northern Hokkaido to southern Kyushu. The vegetational types supported by such black soils are typically grassland or Japanese red pine (*Pinus densiflora*) forest, or sometimes artificial larch, *Cryptomeria* or *Chamaecyparis* forest. Areas previously covered by deposits of volcanic ash (and so now with the typical black soil) have thus probably had their previous vegetation reduced largely to grassland owing to the lower productivity of this soil type. In general, it is very compact, wet soil, lacking in porosity, and showing a high phosphorus requirement.

Red soils are produced under very hot, humid conditions where rock weathering and decomposition are intense. They owe their red colour to iron oxidation, and, dependent on their precise composition, show a colour range from dark red to orange-yellow. They are typically rich in iron and aluminium, but lack most of the essential substances required for high plant productivity. Present climatic conditions in Japan are unsuitable for the further formation of such lateritic soils, and in fact those that do occur (mostly in lowland areas with low erosion rates) are Pleistocene in age. These "fossil" soils are nevertheless distributed widely over the island arc, especially in southwest districts, occurring mostly in restricted patches at elevations of $<$2000 m a.s.l. Their typical vegetation is *Pinus densiflora* forest, although in

some areas they are cultivated as orange groves or bamboo woods. Besides the typical red soils, other related reddish-brown soils occur, originating from the weathering of limestone or serpentinitic rocks, or by thermal metamorphism of existing soils through volcanic activity. These soils, although of very limited distribution, also clearly show a very low plant productivity.

Podsol is the characteristic soil of subarctic/subalpine coniferous forests and alpine meadows in Japan. Its general characteristics are as follows: a comparatively thick A_0-layer, a greyish A-layer, and a chocolate, reddish-brown or orange (iron-stained) B-layer, sometimes including pans. The acidity of the A_0- and A-layers is high, giving rise to strong leaching activity.

In Japan, podsols can be divided into four broad types: P_D, $P_{W(h)}$, $P_{W(1)}$, and P_G. P_D podsol exhibits strong F and H strata in the A_0-layer, and the A-layer has a crumbled or granular structure (see Fig. 11). Three sub-types (P_{DI}, P_{DII}, P_{DIII}) are distinguished among P_D podsols based on the degree of leaching in the A-layer (i.e. "very strong", "weak or spot-like" or "oblique" leaching, respectively). These P_D podsols typically occur on well-drained ridges, and support natural stands of *Chamaecyparis*, *Sciadopitys*, *Thujopsis*, *Tsuga* (two species) and *Pinus pumila* throughout the subalpine coniferous and alpine zones.

$P_{W(h)}$ podsol exhibits an especially well-developed H stratum, and has a crumbled or wax-like structure. The A-layer is rich in humus and blackish in colour. The principal leaching stratum can be identified at the upper part of this layer, and is underlain by a reddish-brown accumulation stratum which is sometimes hard and compact. The $P_{W(h)}$ podsol type is generally found in coniferous zones with *Abies Veitchii* or *Abies Mariesii* forests.

$P_{W(1)}$ podsol also has a well-developed H stratum (Fig. 11). The A-layer is composed of grey, clayey soil, and the B-layer is represented by a light orange

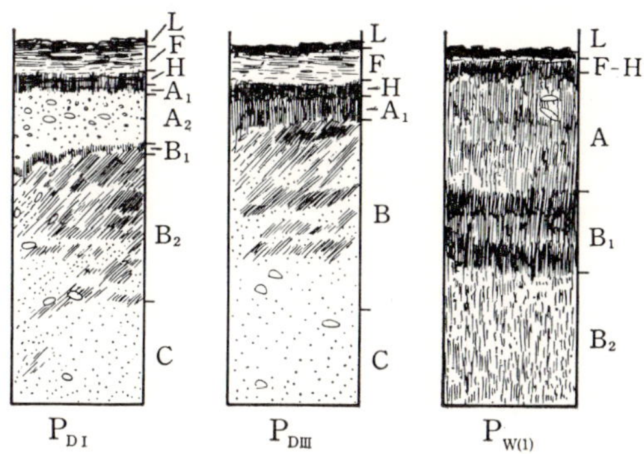

FIG. 11. Profiles of P_{DI}, P_{DIII} and $P_{W(1)}$ podsols.

iron-rich accumulation, occasionally with a pan. This type of podsol occurs on wet clayey plateaus with abundant stagnant water, and consequently is low in oxygen content (a typical characteristic). $P_{W(1)}$ podsols are also found in districts where the dominant underlying rock is quartz porphyry, or where there is abundant snow. Here, *Chamaecyparis*, *Fagus* or *Cryptomeria* forests develop in the warm-temperate zone; however, owing to the low oxygen levels in the surface soil, such areas are unsuitable for artificial reforestation and even the natural trees tend to grow in patches or around old stumps.

P_G podsol is found in association with P_W type podsols, and occurs in flat river basins. The upper soil layer is represented by normal podsol elements, but the lower layer consists of the so-called G (or Grey)-layer.

Brown forest soil is the typical dark brown soil found in cool-temperate areas with broad-leaved deciduous forests, where the humidity is only moderate. There is no leaching or excess accumulation of iron or aluminium. The humus in the surface layer is gradually transported to deeper layers, and in profile the soil displays appreciable porosity throughout. In Japan, brown forest soils always show an acidic reaction, and have been classified by

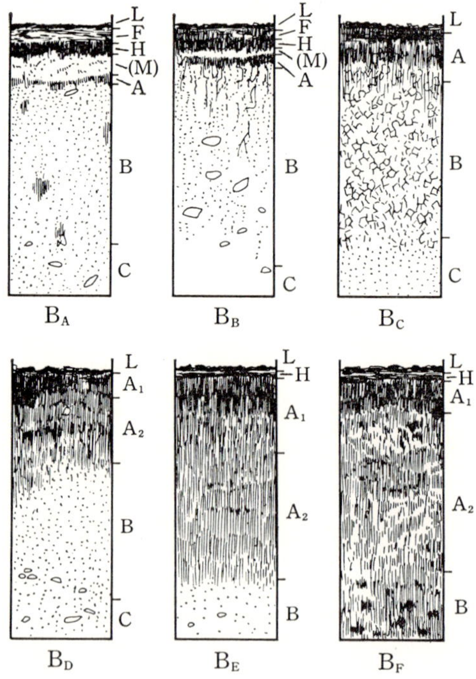

FIG. 12. Profiles of B_A–B_F brown forest soils.

Oomasa[6] into six types (B_A to B_F) on the basis of their moisture content. B_A and B_B are very dry brown soils, B_C is dry, B_D moderately moist, B_E moist, and B_F damp soil. The nature and structure of the soil profile varies according to the moisture content. B_A and B_B have clear boundaries between the A- and B-layers, and the B-layer shows a finely granular or massive structure. These types are found on ridges and on the upper slopes of hills and mountains. The B_D type usually occurs below the B_A or B_B types, sometimes interspersed by the B_C type, i.e. typically at the foot and on the lower slopes of hills and mountains, but sometimes on higher, gently sloping land. The B_E and B_F types are restricted to the flanks of valleys and other flat or low-lying areas. Fig. 12 illustrates typical profiles of the six brown forest soils, and Fig. 13 shows their general relationships to topography.

In northern districts of Japan, broad-leaved deciduous (mainly *Fagus*), *Thujopsis* and *Pinus densiflora* forests occur naturally on B_A and B_B type brown soils. In central Japan, broad-leaved, deciduous *Pinus densiflora*, *Abies homolepis* and *Abies firma* forests grow on the B_A and B_B type soils, but in southern districts only the broad-leaved and *Pinus densiflora* forests are found on these soils. In the case of B_C and B_D type brown soils, almost all kinds of forests can thrive, such as the well-developed *Fagus* and *Cryptomeria* forests of the Tohoku region. B_F and B_E type soils are usually dominated by mixed broad-leaved forests.

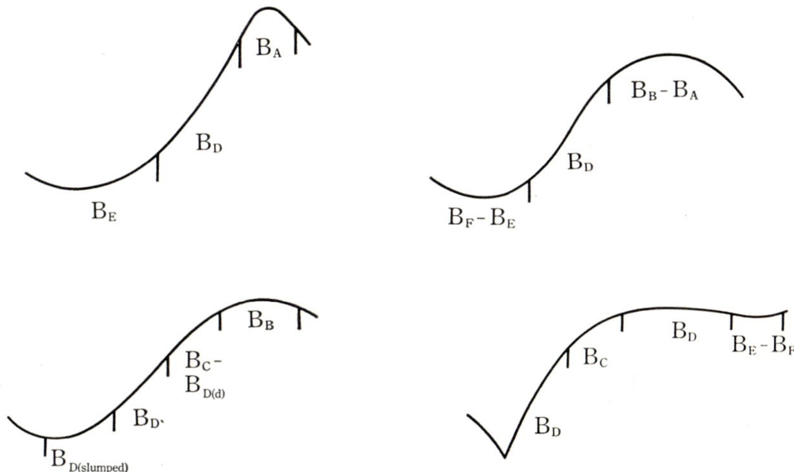

FIG. 13. Relationships between brown forest soil types and surface topography.

References

1. S. Honda, *Studies on Forest Zones in Japan* (Japanese), Tokyo, 1922.
2. H. Mayr, Waldbau auf naturgesetzlicher Grundlage 2, Aufl., 1925.
3. M. Morishita and T. Kira, *Nature: Ecological Studies* (Japanese), Tokyo, 1967.
4. K. Imanishi, *Natural Scientific Studies on Japanese High Mountains* (Japanese), Tokyo, 1969.
5. T. Kira, Forest zones in Japan, *Ringyo Kaisetsu Series* (Japanese), no. 17, Nippon Ringyo Gijutsu Kyokai, Tokyo, 1949.
6. M. Oomasa, Studies on beech-forest soil types, *Rept. Forest Soil Survey 1* (Japanese), 1951.

2

Origin and Characteristics of Japan's Flora

Fumio MAEKAWA*

1. The Geohistory and Past Flora of Japan as a Key to the Present Flora
2. The Present Flora: Its General Features and Regional Divisions
    2.1. Vegetation of the warm-temperate zone
    2.2. Vegetation of the cool-temperate zone
    2.3. Vegetation of the subarctic coniferous zone
    2.4. Vegetation of the alpine zone
    2.5. Local floristic regions in Japan
        2.5.1. Yezo-Mutsu region
        2.5.2. Kanto region
        2.5.3. Sea of Japan region
        2.5.4. Fossa Magna region
        2.5.5. Sohayaki region
        2.5.6. Atetsu region
        2.5.7. Mino-Mikawa region
        2.5.8. Bonin region
        2.5.9. Ryukyu region

* *Emeritus Professor of the University of Tokyo, 1-13-12 Shimizu, Suginami-ku, Tokyo 167, Japan*

1. The Geohistory and Past Flora of Japan as a Key to the Present Flora

It is clear that the flora and vegetation of different regions are determined chiefly by local geographic factors, including the soil properties. The flora should therefore be regarded as unstable in the sense that it will readily undergo modification when changes are induced in the surrounding environment. The present or recent flora and vegetation are the cumulative product of many changes that have occurred in the past. To gain a better understanding of the contemporary nature and distribution of Japan's flora and vegetation, it is therefore necessary to review briefly the ancient flora, so far as it is known, together with the geological history of the region, i.e. insofar as they form the background to the present flora.

As an example, Fig. 1 shows a map of the Kanto district (central Japan)

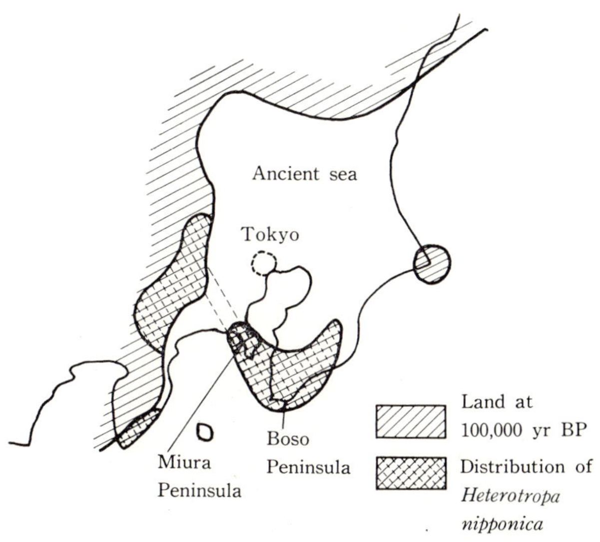

FIG. 1. Map showing the distribution of land and sea in the Kanto district about 100,000 yr BP, together with the distribution of *Heterotropa nipponica* in the Kanto district.

at the latest stage of the Diluvial epoch, about 100,000 yr BP. As can be seen, the greater part of the region was at that time covered by sea, which was open to the east and continuous with the Pacific Ocean. However, there was a large peninsula/island to the south, at the site of the two present peninsulas of Miura and Boso, which were then joined together in an east-west direction as a single terrace. It is reasonable to suppose also that most members of the land flora were influenced by this equilibrium between land and sea, taking on a distribution pattern that corresponded to the general form of the land. However, even after the relatively short time-lapse involved, it cannot be expected in most cases that clear remnants of this past distribution are still preserved at present.

One exception is *Heterotropa nipponica* (see also Fig. 1, and Fig. 2). Its present distribution covers foothills in the western mountainous area of the southern Kanto district, together with the Miura Peninsula and southern parts of the Boso Peninsula (mostly on hills). It is entirely absent from the neighbouring alluvial plains. Moreover, there is a very distinct disjunction in the distribution of *H. nipponica* between the Tama Hills and northern part of the Miura Peninsula, even though ecological conditions in this region are similar to those normal for growth of the species, and certainly adequate to support it. It thus seems reasonable to suppose that the anomalous distribution is a direct result of the geohistory of the area.

The disjunction coincides with the presumed neck of the old Miura-Boso peninsula, when it was connected to the Tama Hills region. It is thus believed that the formation of a sea-channel across the old peninsula neck separated the distribution area of *H. nipponica* into two parts, and that insufficient time has since elapsed for any relinking to occur. Moreover, even though the two

FIG. 2. *Heterotropa nipponica* from Kamakura (Kanagawa Pref.).

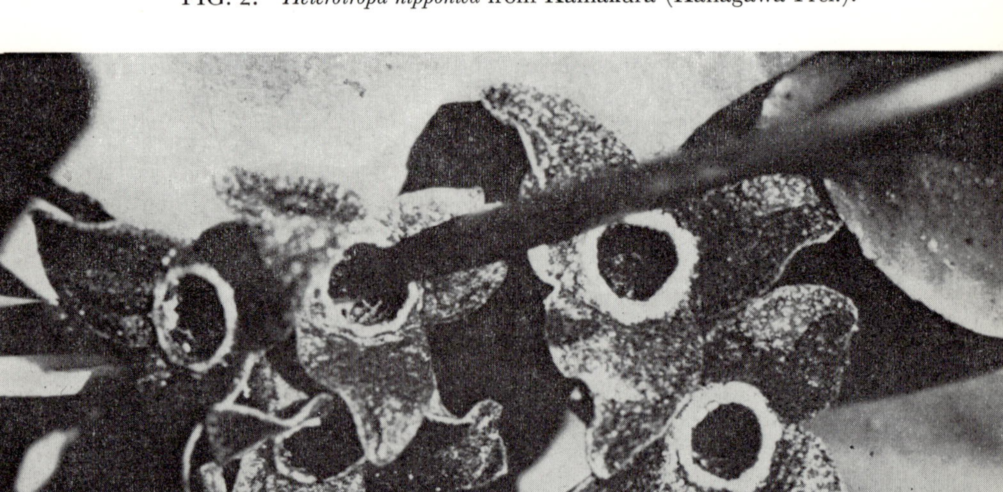

peninsulas of Miura and Boso are now separated by a channel (the *Uraga-suido*), and in spite of *Heterotropa*'s inability to cross the sea-surface, this species shows clear evidence of a continuous distribution across the two peninsulas. In each case, it occupies hills composed of Pliocene sediments, growing vigorously in shaded habitats. This distribution pattern was therefore presumably established prior to the formation of the *Uraga-suido*, and has remained essentially intact to the present. These observations are all consistent with the fact that *H. nipponica* is exceedingly slow in invading new areas, even when they are hospitable to it. Its distribution speed has been calculated at only about 1–3 km/10^4 yr.

Generally speaking, most species of ferns and flowering plants which make up the flora and vegetation, show a relatively high distribution speed. This circumstance thus limits to a great extent any detailed analysis of the past floral composition in Japan. Nevertheless, if we consider a long span of time on a more global scale, it is possible in some cases to reconstruct the past floral successions, making use of geohistorical data on the distribution of land and sea. The next paragraphs therefore summarize briefly the geologic development and characteristics of the Japanese islands.

The site of Japan was formerly just at, or very close to, the point of junction between the Tethys Sea and Pacific Ocean, i.e. in the marginal portion of the old Asian continent. The oldest known group of sedimentary rocks extends from the Silurian (400 m.y. BP)* to the Permian (230 m.y. BP), and, according to currently accepted theory, represents deposits forming in a large geosyncline, often termed the Honshu geosyncline. It is thought that this geosyncline was composed of several parallel axes, rather like the present floor of the Philippine Sea, reaching a maximum depth of over 10,000 m. A general uplifting of the land surface creating numerous islands occurred in the early Jurassic (180 m.y. BP), and such an age can thus be safely ascribed as the latest at which the direct ancestral elements of Japan's present flora were established in the region. The series of orogenic movements (the Honshu orogeny in its broad sense) which led to the formation of these islands, took place on a global scale, and is thought to be intimately related to the similar movements occurring across Malaya, Borneo, etc. in Southeast Asia.

A discussion of the plate tectonics of the Japanese region is also essential to understanding the geohistory fully. One typical example of the geological processes involved is seen in the Tohoku district (i.e. Northeast Japan, covering the entire area of Honshu northeast of the Fossa Magna). About 200 km to the east of this district lies the Japan Trench, which is itself *ca.* 800 km in length. The West Pacific plate, which is moving slowly westwards at a

* m.y. BP=10^6 years before present.

calculated speed of *ca.* 9 cm/yr, turns down as it reaches the trench, passing under the Japanese island arc (i.e. the peripheral part of the Asian continental plate) at an angle of about 20°. Movement of this ocean plate is of the Cordilleran type, i.e. a counterpart of the collision type, which is distinct in the case of the Caledonian and Alpine orogenics. It has caused a high and characteristic gravity anomaly in the area. When the Pacific plate material has travelled about 130 km downwards from the trench region, it experiences high temperatures sufficient to induce the molten state, and then escapes upwards as liquid magma. A chain of surface volcanoes has been formed as a result, passing through the central part of the Tohoku district. This volcanic "front-line" stretches approximately north-south, running parallel to the Japan Trench and *ca.* 300 km from it. Along and to the west of the line, there are many Quaternary volcanoes and volcanic rocks, the chemical and mineralogical properties of which show a clear zonation parallel to the "front-line" (see Fig. 3), as a tholeiite zone, a high-alumina basalt zone, and an alkali basalt zone, respectively westwards. These zones are an important key in the analysis of Cordilleran-type tectonic processes in general. Moreover, the many ore beds in the region, formed largely as a result of hydrothermal activity associated with Miocene volcanoes, show a similar pattern of zonation. They include deposits of native sulphur, so-called "black ore", etc. Based on available evidence, it also

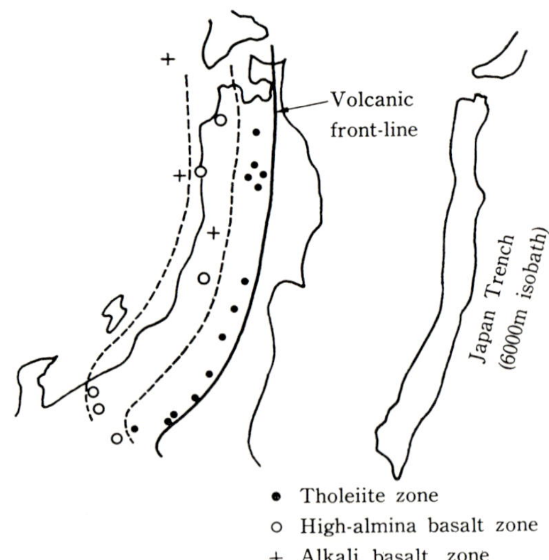

FIG. 3. Volcanic front-line of Northeast Japan (adapted from Horikoshi, 1972).

appears that the volcanic front-line has not shifted its position significantly since the Miocene (i.e. for at least 10 m.y.). This leaves a distinct possibility that the basic floral stock of Japan has remained essentially intact for this rather long span of time, although it was no doubt frequently disturbed and partially damaged by periods of intense volcanic activity.

The geological history of Northeast Japan prior to the Miocene is highly complex and difficult to analyze accurately. Therefore, the structure and history of Western Japan will next be considered, since this region has developed and retained a rather distinct zonal configuration that is capable of broad analysis. The overall geohistory can be summarized according to the postulated series of orogenic cycles of T. Kobayashi (1941), which have been revised more recently by his followers according to the concept of plate tectonics. The essential details are shown in Table 1.

TABLE 1. The principal orogenic cycles of Western Japan

Age	Name of orogeny	Metamorphism
1) Early Carboniferous to Early Permian	Akiyoshi orogenic cycle	Hida-Sangun metamorphics
2) Middle Carboniferous (?) to Cretaceous or Palaeogene	Sakawa orogenic cycle	Ryoke-Sambagawa metamorphics
3) Late Permian (?) to Middle Miocene	Shimanto orogenic cycle	—
4) Early Miocene–	Mizuho orogeny (revival of Shimanto orogeny)	—

Typically, each orogenic cycle commences with the formation of a trench, which is deepened as the oceanic plate turns obliquely down and passes beneath the colliding continental plate. After about 3 m.y., the forward edge of the ocean plate reaches the zone of high temperatures where it becomes molten as magma. As described above, this magma then rises through the continental plate material, leading to the formation of a series of volcanoes. The oblique intrusion of the oceanic plate and associated volcanic activity continue for a variable span of time up to the climax of the cycle, after which there is a gradual cessation of both. At this time, deep-seated igneous activity occurs, causing severe metamorphism of the surrounding crustal material. The rocks towards the trench are subjected to rather low-temperature but high-pressure metamorphism, whereas those away from the trench undergo high-temperature but rather low-pressure metamorphism. The different grades of metamorphism are clearly preserved in the structural and mineralogical properties of the resultant metamorphic rocks. The trench is uplifted

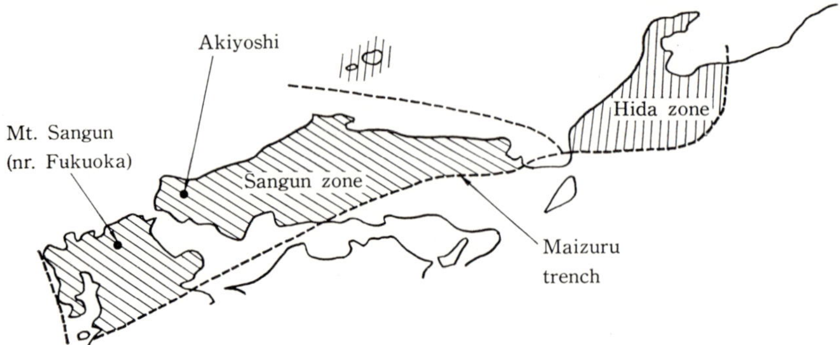

FIG. 4. Map showing the structural components of the Akiyoshi orogenic zone (adapted from Horikoshi, 1972).

under compression, often creating new mountain or island chains, which mark the final stage of the completed orogenic cycle.

In the Akiyoshi orogenic cycle, the oldest established one in the Japanese region, two distinct metamorphic provinces can be identified: the Sangun high-pressure metamorphic zone, and the Hida high-temperature metamorphic zone (see Fig. 4). The so-called Maizuru zone represents the old, fossilized trench. It stretches from Wakasa Bay to the western part of the Seto Inland Sea, cutting obliquely across Okayama Pref. It contains the famous Yanahara mine, whose ore bed is thought to have originated in the central (ridge) part of the Pacific region and travelled slowly westwards to reach its present site prior to the cessation of activity in the ancient trench. No non-volcanic regions at present occur between the Maizuru and Sangun zones, although they might be expected. Instead, it is conjectured that they may have been dragged down to a deeper level with the intrusion of the ancient ocean plate material, i.e. in a manner similar to that proposed for other examples of older orogenic cycles. In the case of the southern boundary of the Hida zone, no clear junction with a non-volcanic zone, the Sangun metamorphics, or even with the Maizuru trench can be distinguished, and it is proposed instead that these have been crushed and amalgamated into a complex peripheral structural belt. No evidence of any land flora has yet been found in the rocks of this region, and it is estimated that plate motion ceased in the mid-Permian (*ca.* 250 m.y. BP).

The Sakawa orogenic complex lies to the south of the Akiyoshi region and shows a rather clear zonation (see Fig. 5). The different components, from north to south, are the Mino-Tamba zone, the Ryoke metamorphic zone, the Sambagawa metamorphic zone, the Chichibu serial zone, and the Kurosegawa zone (the old, fossilized trench), respectively.

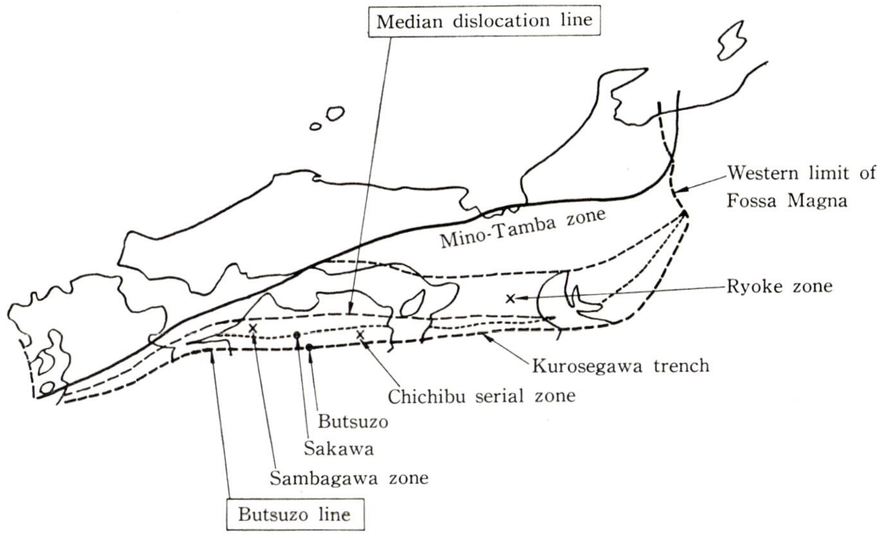

FIG. 5. Map showing the structural components of the Sakawa orogenic zone (adapted from Horikoshi, 1972).

The northernmost, Mino-Tamba zone is the broadest in extent, and essentially unmetamorphosed. It received extensive intrusions of igneous rocks during the Cretaceous.

The Ryoke metamorphic zone stretches southwards from a rather indistinct boundary with the Mino-Tamba zone that passes approximately east-west through the city of Kobe. Its southern limit, on the other hand, which forms the boundary with the Sambagawa metamorphics, is quite distinct and known as the Median dislocation line (or "Median Line"). The Ryoke zone is characterized by its high-temperature, low-pressure metamorphism, and the Median Line originated at least at the end of the Early Cretaceous. In certain parts of Shikoku, the Median Line is still active, occasionally causing appreciable damage to subterranean cables and mountain bridges.

The Sambagawa metamorphic zone stretches from the western Kanto district to the eastern side of Kyushu. Its total length is thus about 1000 km; however, it forms an extremely narrow belt, measuring only *ca.* 30 km across at the broadest point in Shikoku. The entire zone shows a remarkable low-temperature, high-pressure metamorphism, the original rocks generally being Early to Middle Cretaceous in age. Together with the Ryoke zone, it forms one of the two metamorphic belts which accompany a typical Cordilleran-type collision between two crustal plates (in this case, that of the Sakawa orogeny).

The Chichibu serial zone of Western Japan is somewhat similar to the Sambagawa zone in extent. However, it is quite distinct from the latter in that it has not undergone any major metamorphism.

Further south, the Kurosegawa zone occurs disjunctly forming a number of rather small lens-shaped outcrops. It derives its name from a locality in Ehime Pref. where the structural relationships have been determined. The zone is considered to represent a remnant of an ancient trough which existed from the Middle Carboniferous to Early Jurassic. In the Late Jurassic, this trough underwent gradual burial, which was completed by the Late Cretaceous, in parallel with the intrusion of igneous rocks in the Mino-Tamba zone and the metamorphism of the Ryoke and Sambagawa zones.

The southern boundary of the Chichibu serial zone is a distinct dislocation line, the Butsuzo line, which stretches east-west through the southern suburbs of the city of Kochi in Shikoku, and marks the northern boundary of the so-called Shimanto geological area (see below). Butsuzo is the name of a small village near Sakawa, Kochi Pref., through which the line passes, and Sakawa is the famous geological type-locality which suggested the name for the entire orogenic cycle. It should be noted also that the Sakawa district has a special importance for Japanese botany, since it was here that T. Makino (1862–1957), who made many of the basic investigations that led to the establishment of systematic and floristic botany in Japan in the early 1900's, was born. The son of a local wine brewer, he spent much of his younger life collecting botanical specimens, which were later used critically as type specimens for his studies and designation of new species. These specimens are now housed in the University Museum of the University of Tokyo, and in the Makino Herbarium of Tokyo Metropolitan University.

Radio-isotopic age determinations have been made on metamorphic rocks of the Samabagawa zone. They give an age of 100–82 m.y. BP, or Middle to Late Cretaceous, when the conflicting plates of the Sakawa orogeny are believed to have ceased their active movement. The intrusion of igneous rocks which ensued is calculated to extend from the Middle Cretaceous to Middle Palaeogene. On examining the distribution of these igneous rocks, which are largely granitic, a sporadic but broad occurrence in Southwest Japan, northwards from the Median Line and in both the Akiyoshi and Sakawa orogenic belts, can be distinguished. In the neighbouring Asian continental region, there are also several remarkable features in the distribution pattern of granitic rocks (see Fig. 6). The northern limit of Mesozoic granites encloses part of the maritime province of the U.S.S.R., crosses the northwest corner of the Sea of Japan, and then continues obliquely across the central part of the Korean Peninsula. There is thus a distinct gap in the distribution of Mesozoic granites in the continental part of the northwest

FIG. 6. Reconstruction of the dislocation and former site of Southwest Japan based on a comparison between its Cretaceous igneous rocks and those of the Korean Peninsula and maritime province of the U.S.S.R. (adapted from Horikoshi, 1972). Solid areas indicate the major areas of granite; the arrow indicates the direction and degree of movement of Southwest Japan.

sector of the Sea of Japan. Based on the general configuration and pattern of granitic intrusions in Southwest Japan, a convenient fit can be obtained between the three regions, i.e. southern Korea, Japan, and maritime U.S.S.R. This fit has recently been developed into a new hypothesis regarding the origin of the Sea of Japan. It is suggested that Southwest Japan moved southwards from an original site on the border of the Asian continent, and that the site now forms the northwest sector of the Sea of Japan. If correct, the new theory would replace the earlier idea of a relatively stable location for the Japanese islands, and development of the Sea of Japan through subsidence of former continental terraces. Following the new hypothesis, and utilizing other geological data, it is further suggested that such movement of Southwest Japan could not have occurred before the end of the Cretaceous, i.e. at least until the cycle of igneous intrusions associated with the Sakawa orogeny had completely ceased. Moreover, considering the wide occurrence of green tuffs of Miocene age around the present shoreline of the Sea of Japan, the movement may be younger than Palaeogene in age. This being so, the origin of Japan's flora, or at least the major part of it, should be sought in the Asian

continental region with which there would have been a direct connection up to Palaeogene times. Indeed, there is already some evidence to suggest a former relationship between the ancient floras of the Korean Peninsula (especially northern and central areas) and Japan.*

One of the older terrestrial floras known in Japan occurs in the fossil vascular plant bed of Nariwa, Okayama Pref. Its age is Upper Triassic, and the deposits are thought to be of freshwater, inland type. The principal components are ferns and gymnosperms (there are no angiosperms), and the total number of species exceeds 100. Most of the genera have a world-wide distribution pattern, with *Hausmannia*, *Nilssonia*, *Ginkgoites*, etc. all comprising a rather large number of species. *Oishia* (Dipteridaceae) is the only endemic genus. The flora is noteworthy in that it includes several distinct tropical families, the Marattiaceae, Dipteridaceae and Cycadaceae. The principal species are *Marattiopteris asiatica* (Marattiaceae); *Clathropteris meniscoides*, *Dictyophyllum Nilssoni* and *Hausmannia nariwaensis* (Dipteridaceae); and *Nilssonia orientalis* and *Ctenis japonica* (Cycadaceae). Based on the common occurrence of these tropical plants, it appears reasonable to suppose that the Upper Triassic climate was tropical-subtropical over a rather broad area of Japan. Certain elements of the flora are maintained in the so-called Tetori flora of the Hida region, which is earlier Jurassic in age. Also, in the maritime provinces of the U.S.S.R., the species *Nilssonia orientalis* is known from the Lower Jurassic, while in the Upper Jurassic, *Hausmannia ussuriensis* and *Ctenis latiloba* (which are very similar to the vicarious Japanese species *H. nariwaensis* and *C. japonica*, respectively) occur, suggesting that tropical-subtropical climatic conditions extended to northern parts of the region, and that a delicate difference in speciation between the continent and Japan had already occurred in these genera.

Two other problems, however, must be considered before an understanding of the general situation regarding the origin of Japan's flora can be gained. The first is the species difference between Japan and the continent. The Japanese flora, which was virtually identical to the continental flora in the Triassic, has been evolving to form a distinct floristic region through a slow but clear tendency towards species differentiation. The exact cause, or causes, of this change are not yet confirmed. However, the present author is inclined to believe that the progressive isolation of the flora may be a direct result of the geotectonic dislocation of the Southwest Japan area which, in the Upper Jurassic, is suggested to have begun drifting southwards from an original site in close proximity to the continent.

* It should be noted that this relationship is one quite distinct from the later (Plio-Pleistocene) migrations of species to and from the Korean Peninsula contemporaneously with the so-called Naumann elephant.

The second problem concerns the past orientation of climatic zones. According to the author's present hypothesis for the palaeoquatorial distribution of plants, which is a synthesis of many independent lines of evidence, mostly palaeomagnetic and biological, the ancient climatic and vegetational zones are arranged parallel to the old equator. The biological evidence derives mainly from the past and present distribution patterns of the flora. For example, *Coriaria* (Coriariaceae), *Nothofagus* (Fagaceae), *Oreomyrrhis* (Umbelliferae), etc. show a relict of a past distribution pattern which was oriented parallel to a line that is not consistent with the present equator (see Fig. 7). From the available data, the former site of the equator can be reconstructed. In the East Asia region, it passed westwards of present-day Japan, oriented obliquely northwest-southeast, running approximately from western Siberia and the Tien Shan mountains of western China, across China proper, between Formosa and Hainan Is. to the Philippine Sea, then through the northeastern part of New Guinea, ultimately to New Zealand in the far south. The similarity of tropical climatic conditions in Southwest Japan and maritime provinces of the U.S.S.R. would thus fit satisfactorily with this ancient palaeoequatorial orientation, and also the proposed connection between the two regions and subsequent southward drift of Japan would follow naturally (see Fig. 8).

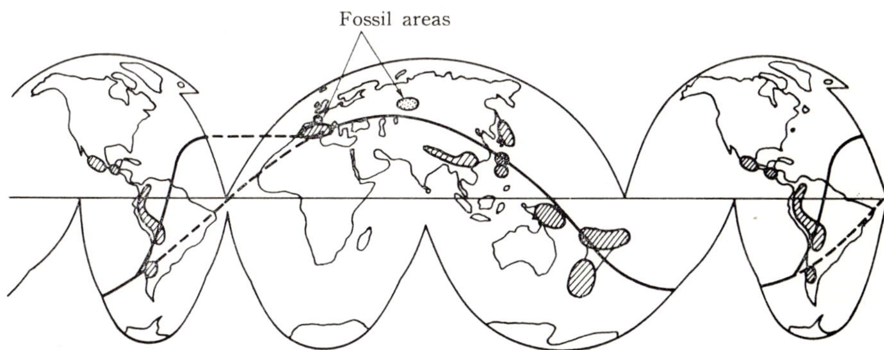

FIG. 7. Palaeoequatorial distribution of *Coriaria*. Shaded parts show present areas; dotted parts show fossil areas.

FIG. 8. Mesozoic floras of Japan viewed in the context of the author's hypothesis for the palaeoequatorial distribution of plants.

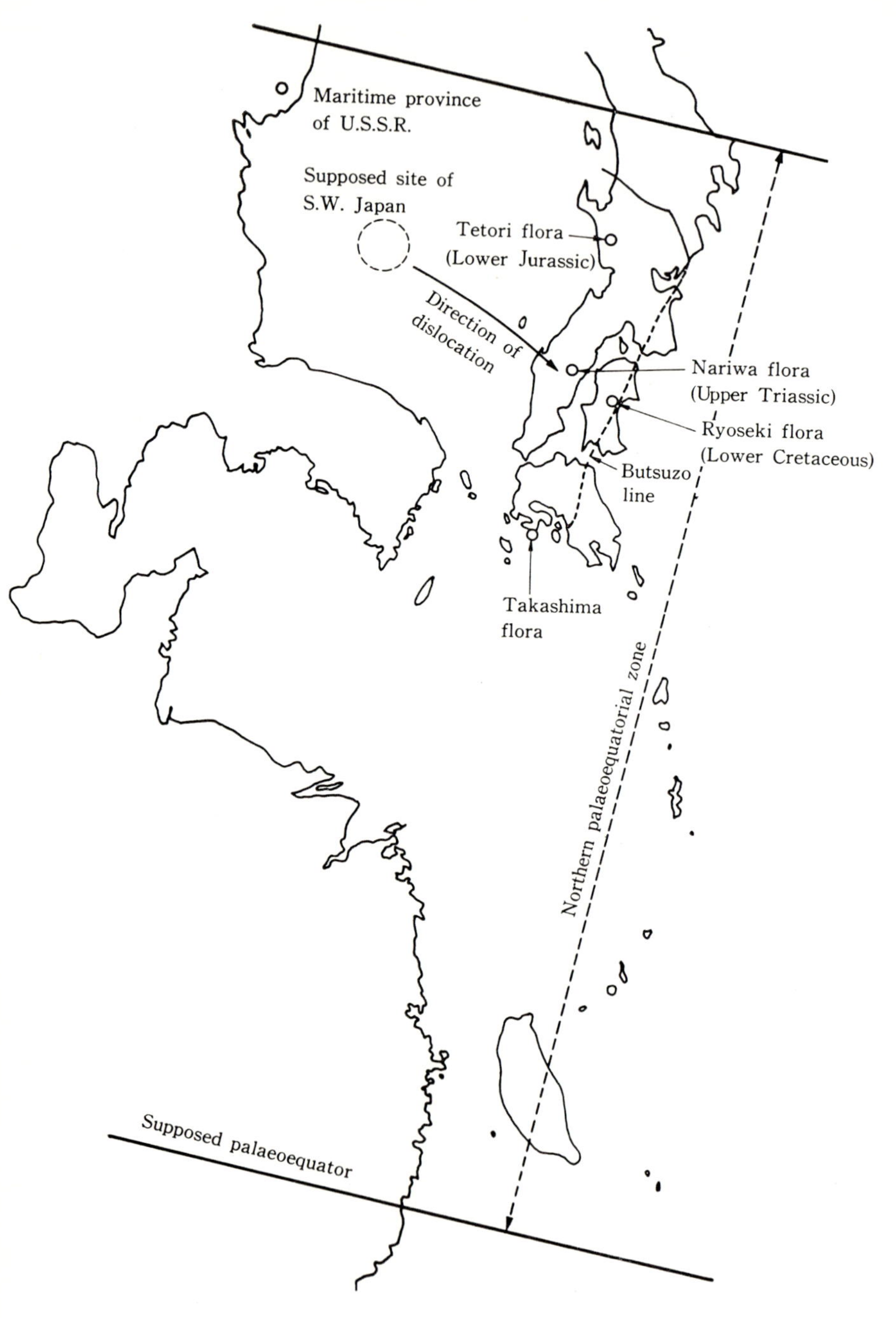

Maritime province
of U.S.S.R.
Supposed site of
S.W. Japan
Tetori flora
(Lower Jurassic)
Direction of
dislocation
Nariwa flora
(Upper Triassic)
Ryoseki flora
(Lower Cretaceous)
Butsuzo
line
Takashima
flora
Northern palaeoequatorial zone
Supposed palaeoequator

The above conclusions are further supported by the following data. *Coniopteris hymenophylloides* (Cyatheaceae) and *Ginkgoites digitata* (Ginkgoaceae) occur both in the Tetori flora and in maritime U.S.S.R. Also, in Kochi Pref. there is a Lower Cretaceous plant bed, containing the so-called Ryoseki flora, which indicates the end of the long-continued dominance of a mesophytic flora in Japan. One distinct component of the flora, *Nathorstia Oishii* (Gleicheniaceae), has a strong phylogenetic relationship with *N. alata*, a species discovered in the Patagonian Lower Cretaceous sediments of Argentina. This relationship suggests a long-range affinity between the two species, i.e. in accordance with a similar position relative to the palaeoequator.

Although the above-mentioned fossil plants have no direct ancestral relationship to the components of Japan's recent flora, they certainly mark an adaptation to modified climatic and physical conditions. In this sense, therefore, they represent a precursor or pioneer community which preceded, and so influenced indirectly, the recent flora. Moreover, from the Upper Cretaceous onwards, fossil remains of angiosperms are known, although regrettably they are rather rare and fragmental and the state of preservation in the Upper Cretaceous sediments of Japan is poor.

From the Butsuzo line southwards, several parallel, fault-contacted blocks of strata run with an east-west strike across Southwest Japan. The northern series is known as the Shimanto group and belongs to the Cretaceous, whereas the southern series, the Muroto group, is of Eocene age. The geographical situation and age of these sediments, together with the present form of the Nankai trough (which lies about 180 km off Cape Muroto), suggest the former existence of a large trench system in the region (see Fig. 9). Concerning the age of the presumed activity of the old Nankai Trench, some evidence is provided from the northern boundary region, i.e. that with the Chichibu zone of the Sakawa orogenic belt. The boundary zone to the south is known as the Sanpozan zone, and is composed of three or four main stratal blocks of Middle Permian age. They run parallel to each other, with fault contacts, and are associated with many other faulted insertions of strata of very different ages. The oldest is Silurian and the youngest, later Cretaceous. The precise significance of this wide age-range is difficult to assess, and cannot be considered here. It is necessary therefore, for the present purpose, simply to select those rocks of the Upper Permian to Lower Jurassic which contain clear evidence of past volcanism. This volcanic material indicates the existence of a volcanic front line in the region that was closely associated with the old Nankai Trench. It marks the first stage of the so-called Shimanto orogenic cycle, i.e. in later Permian times. It is thus very significant that, based on these conclusions, the earlier half of the Shimanto orogenic cycle would represent a period when two trenches existed contemporaneously in

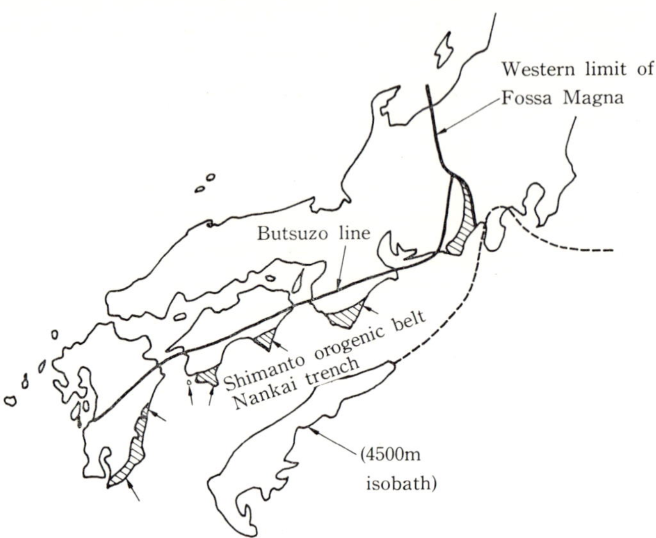

FIG. 9. Map showing the structural components of the Shimanto orogenic zone in Southwest Japan (adapted from Horikoshi, 1972). Shaded parts represent areas of dominant Eocene deposits. Arrows indicate locations of Miocene intrusive rocks.

Southwest Japan, i.e. the Kurosegawa Trench (initiated in the Middle Carboniferous) and the old Nankai Trench.

Granitic intrusions occur in the Eocene beds of the southern half of the Shimanto orogenic zone. These are most extensive in the Osuzu mountains of Kyushu, on Okinoshima Is., southwest Shikoku, and in the Kumano district of the Kii Peninsula. Isotopic age determinations have given a Miocene age (calculated value, 14 m.y. BP). Furthermore, both the Osuzu and Kumano rocks show some clear evidence of subaerial extrusion on a terrestrial surface. Based on this fact, it appears justified to assume that the region of Southwest Japan northwards of the Butsuzo line had already been uplifted above sea-level before the Miocene, and so supported a terrestrial flora. In Shikoku, the Tertiary strata south of the Butsuzo line are almost all of Eocene age, the Oligocene is limited to small fault-controlled blocks in the west, and Miocene and Pliocene strata are extremely limited in extent. These facts, considered together with the data on igneous rocks, suggest that the establishment of a land flora in the region may have been by the later Oligocene. Also, in northern and northwestern parts of Kyushu, as well as at the western end of Honshu, several large Oligocene coalfields are known, although they have now mostly been exhausted. A broad gulf, with a southwest-northeast extension, is thought to have existed between Japan and the old

Korean Peninsula, and an extensive peninsula, with a comparatively rich flora, is thought to have occupied the Southwest Japan region during the Oligocene. The present author, in 1957, proposed that this plant assemblage represents the starting point of Japan's recent flora, and named it the "Makinoesia flora" in honour of Dr. T. Makino.

So far as details of the Palaeogene plants are concerned, the so-called Takashima, Sakito and Kishima fossil floras are known from Southwest Japan, and the Kitaaiki flora from central Honshu. However, they are insufficiently well determined for a broad synthesis of the general floristic pattern across the region to be made with certainty. Although a general tendency towards lower temperatures has been established through the Eocene and Oligocene, some doubt remains as to whether the Takashima flora was actually tropical or not. Tropical genera such as *Acrostichum*, *Musophyllum*, *Sabalites*, *Geonomites*, *Mangifera*, etc. have been reported, but the first three represent old determinations which are in need of reexamination, *Geonomites* was found in a core of different lithology obtained from 800 m below sea-level, and *Mangifera* was unfortunately misidentified. The other principal floral components of the Takashima beds are *Glyptostrobus europaeus*, *Machilus Nathorsti*, *Parabenzoin eotrilobum*, *Platanus Chaneyi*, *Disanthus eocercidifolia*, *Viburnum Awabukioides*, etc. The seindicate a subtropical and/or warm-temperate climate, and certainly not a tropical one. In fact, Japan has no known pure tropical plants equivalent to the Early Ravenian flora of Alaska (reported by Wolfe in 1969), which displays an Indo-Malaysian character and represents a paratropical climate. Nevertheless, as mentioned above, the Japanese floristic components do suggest a gradual decline in temperature from a probable subtropical climate (the Takashima flora, Early Eocene), through an intermediate climate (the Sakito and Kishima floras, Oligocene), to a colder climate (the Kitaaiki flora, at the end of the Palaeogene). This sequence was followed by the Sasebo flora of northern Kyushu (earliest Miocene), which indicates a slow recovery of higher temperatures leading up to the common, mild to somewhat subtropical floral elements of the mid-Miocene.

Palaeogene plant fossils are also known from the coalfields of Hokkaido. They include conifers with deciduous branches (*Glyptostrobus*, *Metasequoia* and *Taxodium*) and many deciduous forest trees such as *Carya*, *Alnus*, *Carpinus*, *Corylus*, *Ulmus*, *Zelkova*, *Cercidiphyllum*, *Corylopsis*, *Disanthus*, *Liquidambar*, *Platanus*, *Acer*, *Alangium*, etc. Certain of these are commonly intermingled with the subtropical fossil flora of Kyushu, and they form part of the so-called Arcto-Tertiary floral elements which continued even until the Miocene. The name "Arcto-Tertiary" was applied since the occurrence of such temperate and/or subtropical genera in circum-polar districts at high latitudes (espe-

cially in Greenland and Spitzbergen) was unexpected, i.e. inasmuch as no living temperate forest could occur in such severe, low-temperature climatic conditions as now prevail.

Under the conventional concept of an Arcto-Tertiary flora, it is generally considered that even in the arctic region a rather mild climate prevailed, allowing the growth of forests of deciduous trees and large conifers, and that the tropical zone covered a much larger proportion of the globe than at present. However, it must be remembered that this concept was established at a time when the available fossil data were somewhat insufficient and the known fossil localities were comparatively few in number. We now know in more detail the precise sequence of climatic changes, even including the Palaeogene to some extent (as described above). Furthermore, the concepts of polar wandering and continental drift have, largely through the study of palaeomagnetism, gained a wide acceptance. It is clear that polar wandering will directly coincide with a shift in the position of the equator, and so in a translocation of climatic zones which in general remain parallel to the equator.

A reexamination of the facts will now be given in the light of the author's own palaeoequatorial distribution hypothesis. The wide occurrence of fossil remains of forest trees at high latitudes, such as in Greenland and Spitzbergen, indicates that their past position was closer to the equator, i.e. closer to the early Palaeogene equator. These areas are now widely separated from each other and also far from the present equator, a situation which is considered chiefly the result of polar wandering and drift that was completed in the Palaeogene, at least before the end of the Oligocene. On this basis, the corresponding tropical locality farthest south from the present equator would have been the South Pacific region, including New Zealand and part of the Antarctic continent. The author considers that comparatively low temperatures always prevailed in arctic regions, rather than any warming up of the kind envisaged in the conventional concept of an Arcto-Tertiary flora, but that there has been considerable transmigration of the arctic regions, which perhaps continues to the present. It is therefore reasonable to assume that the equatorial-tropical, and also temperate zones have moved gradually (and presumably often rather quickly) across the surface of the globe, and that such movement still continues at present.

Returning to the special features of Mesozoic Japan discussed above (see the equator locus in Fig. 8), it can be said that the subtropical environment of the Takashima flora in the earlier Palaeogene represented a maintenance of the palaeoequator in a similar position to that presumed for the Jurassic and Cretaceous. The subsequent temperature decline in the Palaeogene, finally reaching the cold conditions of the Kitaaiki flora of central Honshu, was thus

a direct result of the translocation of the palaeoequator to a new position, approximately coincident with the present equator. The climatic changes experienced in Southwest Japan would therefore be just one small part of a widespread, global event, including the establishment of cold climatic conditions in Siberia, the Antarctic and southern Australasia region, etc. Moreover, the present author considers that these widespread climatic changes were one of the principal factors involved in the explosive development and differentiation of angiospermic plants, accompanied by a lesser but nevertheless vast development of gymnosperms, especially of conifers in the northern hemisphere.

One other important point is that although large changes in climate took place locally, resulting in major changes in local floras, the absolute size of each wandering climatic zone as it moved slowly or rapidly over the global surface probably did not vary appreciably. Also, theoretically, two areas should have existed where the old and new equatorial zones were coincident. Here, little or no change in climate would have occurred, so allowing the maintenance of a relatively unchanged flora. The two areas are estimated to coincide with points west of the Pacific Ocean and west of the Atlantic Ocean, i.e. Malaysia and Guiana/North Brazil, respectively.

The great changes in local climate occurring with movement of the palaeoequator presumably spread from at least the Cretaceous, through the Eocene, terminating before the end of the Oligocene. In Japan, the fossil flora of the earliest Palaeogene exhibits subtropical features, a fact which (at least superficially) appears to contradict the proposed movement pattern of the palaeoequator. However, as suggested above, it is probably that the rather long maintenance of subtropical conditions in the area was a result of the southward movement of Japan (Fig. 8), which to some extent cancelled out the direct effects of the southwestward retreat of the palaeoequator. By reason of its special geohistory, i.e. proximity to the crossing point of the old and present equators and a generally very humid climate due to the proximity of the Pacific Ocean, Japan is thought not to have experienced any sudden or drastic climatic change, but rather to have had a predominantly mild climate. It is therefore very natural to expect on overall uniformity of character in the fossil floras, excluding differences that of course result from vegetational zonation, and it can be said that known species constitute an integral standard floral assemblage. Floral zonation in the early Palaeogene would have been arranged in a northwest-southeast direction, and would then have gradually shifted to run east-west, the present situation. It can be expected that relict elements have survived, and in fact, as mentioned above, several examples are known in the present-day flora.

One additional point should be made concerning the special situation of

the Early Ravenian tropical flora of Alaska. Based on the author's own hypothesis, which has been expressed in another publication, this can be summarized as follows. The contemporary palaeoequator of the flora ran through a point in western Siberia at *ca.* 60°N, and the arctic polar region thus received a paratropical climate (see Fig. 10). The location of Alaska in the Early Eocene was just in this polar position (i.e. in the Arctic Ocean,

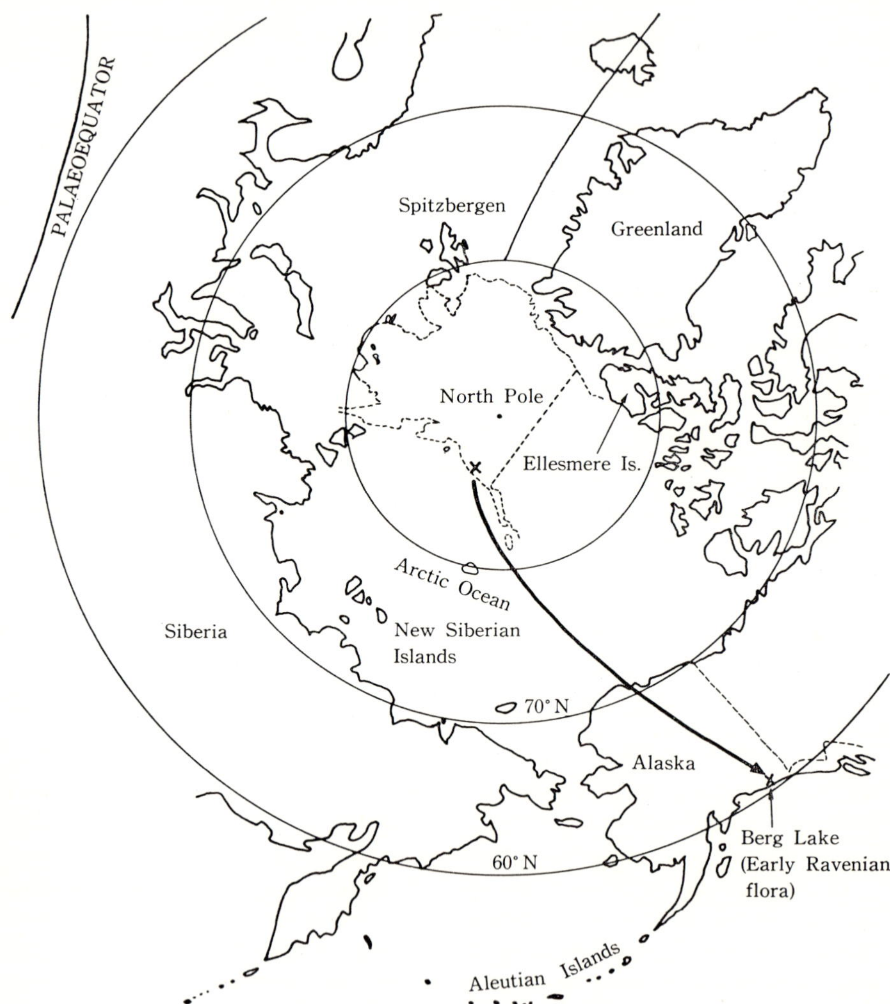

FIG. 10. Map showing the proposed drift of Alaska, viewed in the context of the author's hypothesis for the palaeoequatorial distribution of plants.

FIG. 11. Palaeogeography of Japan in the Middle Miocene (adapted from Tanai, 1972).

north of Siberia), and it so had a prevalent tropical climate. However, at least by the Middle Eocene, Alaska had started to drift southwards, reaching its present position in the Oligocene.

In the Miocene of Japan (23 m.y. BP), activity was again restored to the Nankai Trench in western districts, as part of the so-called Mizuho orogeny (see Table 1) which had begun in Eastern Japan. However, in general there were no very major geographical changes in most land areas under the somewhat mild tectonic conditions that prevailed. The flora was essentially a continuation of that from the Eocene, with the result that a rich plant assemblage with many relict genera and species was retained.

In Eastern Japan, the main stage of the Mizuho orogeny was represented by the so-called "green-tuff" metamorphism. At that time, several important events occurred, such as the initiation of the Fossa Magna across the median part of Honshu, the separation of land areas into many small blocks and invasion of the sea into the intervening regions (see Fig. 11), and the deposition of green-tuff sediments. These conditions can be expected to have influenced the floral composition, but a full analysis of the facts is yet to be made.

Two distinct fossil floras have been described from Eastern Japan, i.e. the so-called Aniai type and Daijima type, respectively. The former preceded the latter, and evidently developed in a distinctly cool-temperate climate. The Daijima type exhibits subtropical to warm-temperate tendencies. This difference was, at first, interpreted as pure, local floristic variation; however, it is now generally accepted that it corresponds to the shifting zonation of vegetation of the region. Initially in the Miocene, Japan was covered mostly by temperate vegetation which had continued from the preceding Oligocene, but by the Middle Miocene, conditions had become warm-temperate to subtropical, allowing the Daijima-type vegetation to spread as far north as northern Honshu (see Fig. 12). A decline in temperature then followed into

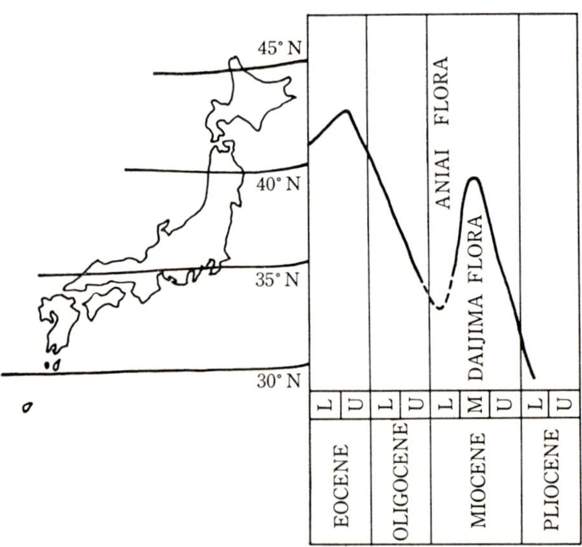

FIG. 12. Fluctuations between the Aniai and Daijima type vegetational zones across Japan during the Eocene–Pliocene (adapted from Tanai, 1972).

the Pliocene. These variations in climate and vegetation are clearly indicated in the broad fossil sequence at any one locality (see the example in Table 2).

TABLE 2. Vegetation sequence in the Aniai district, Akita Pref.

Vegetational type	Fossil flora	Age
Cool-temperate Aniai type	Aniai flora	Early Miocene
Warm-temperate Daijima type	Utto flora	Middle Miocene
Cool-temperate type with dominant *Fagus palaeo-crenata*	Miyata flora	Late Miocene

The following list shows the principal genera of the Aniai and Daijima floral types which are important in discussing the floristic relationships between Japan and adjacent areas. Genera marked with an asterisk (*) are representatives of the Daijima type (i.e. warm-temperate to subtropical elements), and unmarked genera belong to the Aniai type (i.e. warm and/or temperate elements), but are also accompanied by others from the subtropical to cool-temperate zones (indicated by a triangular mark (△)). It is very important to note that *Cryptomeria* did not yet exist in the Miocene of Japan.

Cephalotaxus△, *Picea*, *Pseudotsuga*, *Pseudolarix*, *Tsuga*, *Pinus*△, *Keteleeria*, *Sequoia*, *Taxodium*, *Metasequoia*, *Glyptostrobus*, *Cunninghamia**, *Taiwania**, *Thuja*, *Thujopsis*, *Fokienia**, *Comptonia**, *Juglans*, *Carya*, *Pterocarya*, *Platycarya*, *Engelhardtia**, *Alnus*△, *Betula*, *Carpinus*, *Corylus*, *Ostrya*, *Castanea*, *Castanopsis**, *Quercus*△, *Cyclobalanopsis**, *Fagus*, *Celtis*△, *Ulmus*, *Zelkova*, *Eucommia*, *Trochodendron*△, *Cercidiphyllum*, *Liriodendron*, *Magnolia*△, *Schisandra*, *Cinnamomum**, *Actinodaphne**, *Machilus**, *Parabenzoin*, *Lindera*△, *Sassafras*, *Berberis*△, *Mahonia**, *Hydrangea*△, *Schizophragma*, *Pittosporum**, *Disanthus*, *Corylopsis*, *Parrotia**, *Hamamelis*, *Liquidambar*△, *Platanus*, *Prunus*△, *Sorbus*, *Cercis*, *Gleditschia*, *Cladrastis*, *Maackia*, *Wisteria*, *Sophora*△, *Robinia*, *Ailanthus**, *Mallotus**, *Cedrela**, *Sapium**, *Buxus**, *Rhus*, *Euonymus*, *Ilex*△, *Acer*, *Aesculus*, *Dodonaea**, *Sapindus**, *Koelreuteria**, *Paliurus**, *Zizyphus**, *Tilia*, *Camellia**, *Stewartia*, *Ternstroemia**, *Alangium*, *Hemitrapa*, *Aralia*△, *Kalopanax*△, *Cornus*, *Nyssa**, *Clethra*, *Rhododendron*△, *Tripetaleia*, *Maesa**, *Diospyros*△, *Symplocos*△, *Meliodendron**, *Styrax*△, *Chionanthus*, *Forsythia*, *Osmanthus**, *Fraxinus*, *Ehretia**, *Catalpa*, *Viburnum*△.

Many of these genera still occur naturally in Japan. In particular, *Picea*, *Pinus*, *Alnus*, *Quercus*, *Cyclobalanopsis*, *Magnolia*, *Prunus*, *Euonymus*, *Ilex*, *Acer*, *Rhododendron*, *Symplocos*, *Viburnum*, etc. are in a flourishing condition, having differentiated into many species and/or varieties and including a vast number

of native individuals. The majority may be said now to represent an oligospecies stage, and, viewed from the standpoint of their worldwide distribution, often contain a small number of endemic groups which are clearly the result of disjunction of the broader distribution areas.

The following 27 genera are all known to be extinct from the Japanese area: *Pseudolarix, Keteleeria, Sequoia, Taxodium, Metasequoia, Glyptostrobus, Cunninghamia, Taiwania, Fokienia, Comptonia, Carya, Engelhardtia, Liriodendron, Sassafras, Parrotia, Liquidambar, Eucommia, Platanus, Cercis, Robinia, Cedrela, Dodonaea, Zizyphus, Hemitrapa, Nyssa, Meliodendron, Catalpa.*

This assemblage is often termed, in the palaeobotanical sense, the "*Metasequoia* fossil group". Many of the genera disappeared from Japan during the Pliocene; however, some of them (for example, *Metasequoia, Glyptostrobus, Cunninghamia, Liquidambar*, etc.) survived up to the Middle Pleistocene, at the latest, when all remaining members of the *Metasequoia* group were entirely lost.

Recently, much information on the Plio-Pleistocene floras of Japan has been accumulated, especially in the Kansai and Chubu districts. About 20 stages, extending from the later Pliocene ($\pm$ 4 m.y. BP) to the latest Holocene (10,000 yr BP), have been identified and reconstructed in their geological time sequence (see Fig. 13). Much of the basic data was obtained from macrofossil and pollen analysis carried out in conjunction with stratigraphical studies on the so-called Osaka Formation, which was deposited in and around the Osaka area over a time-span from the Late Pliocene to Middle Pleistocene. Concurrent investigations of the evolutionary development of the Japanese fossil elephants, *Stegodon* and *Elephas*, were also carried out, and as a result the whole floral sequence has been widely accepted, at least for the present.

The oldest stage is represented by the so-called *Pinus trifolia* fossil flora discovered in the vicinity of the city of Seto, which is situated about 20 km eastwards of Nagoya, central Honshu, and famous since ancient times for its earthenware industry. In fact, the collective Japanese word for earthenware is "*Setomono*" (lit. produce of Seto). Vast argillaceous deposits were laid down in the area during the Late Pliocene, on the floor of an inland sea or lake, including the special clay now used for manufacturing *Setomono* and numerous fossil plant beds. The *Pinus trifolia* flora contains 118 known species belonging to 90 different genera. Of these, 68 genera are common with Japan's recent flora, including 38 modern species and 57 extinct species; 13 genera are now absent from the Japanese area, and 9 genera are totally extinct. The main forest trees, which are direct successors from the Miocene, are evergreen conifers (*Sequoia* and *Cunninghamia*), deciduous conifers (*Pseudolarix, Metasequoia* and *Glyptostrobus*) and broad-leaved evergreen trees (*Cyclobalanopsis*),

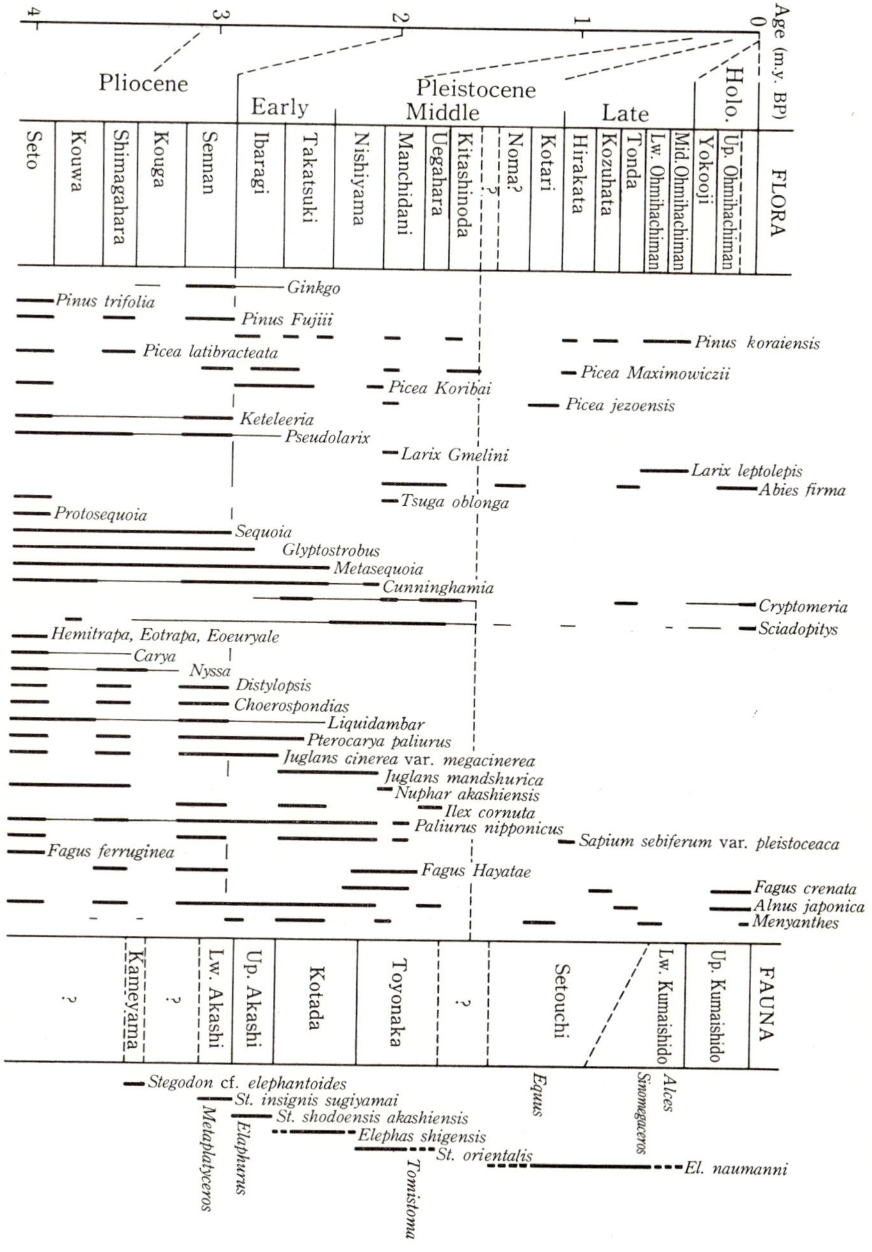
Age (m.y. BP)
4
3
2
1
0
Pliocene
Pleistocene
Early
Middle
Late
Holo.
FLORA
Seto
Kouwa
Shimagahara
Kouga
Sennan
Ibaragi
Takatsuki
Nishiyama
Manchidani
Uegahara
Kitashinoda
?
Noma?
Kotari
Hirakata
Kozuhata
Tonda
Lw. Ohmihachiman
Mid. Ohmihachiman
Yokooji
Up. Ohmihachiman
Ginkgo
Pinus trifolia
Pinus Fujiii
Pinus koraiensis
Picea latibracteata
Picea Maximowiczii
Picea Koribai
Picea jezoensis
Keteleeria
Pseudolarix
Larix Gmelini
Larix leptolepis
Abies firma
Tsuga oblonga
Protosequoia
Sequoia
Glyptostrobus
Metasequoia
Cunninghamia
Cryptomeria
Sciadopitys
Hemitrapa, Eotrapa, Eoeuryale
Carya
Nyssa
Distylopsis
Choerospondias
Liquidambar
Pterocarya paliurus
Juglans cinerea var. megacinerea
Juglans mandshurica
Nuphar akashiensis
Ilex cornuta
Paliurus nipponicus
Sapium sebiferum var. pleistoceaca
Fagus ferruginea
Fagus Hayatae
Fagus crenata
Alnus japonica
Menyanthes
FAUNA
?
Kameyama
?
Lw. Akashi
Up. Akashi
Kotada
Toyonaka
?
Setouchi
Lw. Kumaishido
Up. Kumaishido
Stegodon cf. elephantoides
St. insignis sugiyamai
St. shodoensis akashiensis
Elephas shigensis
St. orientalis
El. naumanni
Metaplatyceros
Elaphurus
Tomistoma
Equus
Sinomegaceros
Alces

of which the latter may be the dominant genus. However, this dominance is based largely on pollen data (*Cyclobalanopsis* shows a *ca.* 75% occurrence in pollen diagrams), possibly giving a distorted impression of the true situation since pollen analysis is normally restricted to the counting of wind-carried pollen grains. *Nyssa, Carya, Eucommia* and *Liquidambar* are intermingled with these genera, together with certain other genera not yet reported from the Miocene, i.e. *Keteleeria* (an evergreen conifer), *Fortunearia* and *Rehderodendron* (broad-leaved semi-deciduous trees) and *Tetrastigma* (a woody vine). All of these trees (except *Cyclobalanopsis*) belong to the *Metasequoia* fossil plant group and are so now extinct from Japan. On the other hand, the following other species, which are either identical to or almost indistinguishable from recent forms, are found. (Those indicated by a dagger (†) are restricted in their distribution to the western or westernmost parts of Japan.)

Broad-leaved evergreen trees: *Pasania glabra, Camellia Sasanqua*†.
Broad-leaved deciduous trees: *Carpinus japonica, Alnus japonica, Lindera umbellata. L. citriodora*†, *Fagara ailanthoides, Melia Azedarach*†, *Acer rubrum* var. *pycnanthum, A. diabolicum, A. palmatum, Meliosma rigida, Cornus controversa, Cynoxylon japonica, Fraxinus japonica.*
Evergreen shrubs: *Buxus japonica*†, *Pieris japonica, Dicalyx myrtacea*†, *D. lancifolia*†, *Viburnum japonicum*†.
Deciduous shrubs: *Magnolia salicifolia, Euscaphis japonica.*
Evergreen vines: *Sabia japonica*†.
Deciduous vines: *Sinomenium acutum, Berchemia racemosa.*

Thus, in general, the central part of Honshu in the Late Pliocene had a rich colour of Sohayaki elements (see below) within its flora, and so belonged to the higher part of the warm-temperate climatic zone. Moreover, Miki has identified extinct examples of the following additional fossil genera: *Protosequoia, Distylopsis, Palaeodavidia, Meliodendron, Bambusoides* and *Lissopepo*, none of which (except *Meliodendron*) are known from any other stratal division of the world or from any other period. Seen in this light, the *Pinus trifolia* flora appears as a very unique assemblage, and quite distinct from others, although this impression may arise partly from the fact that the extensive studies of Miki contrast strongly with the rather imperfect work on other fossil floras.

Judging from the characteristics of the other floras, however, it can be said that the climate in other areas was in general similar to that of central Hon-

FIG. 13. Sequence of 20 fossil floras of the Plio-Pleistocene in central Japan (after Nasu, 1972). Thin lines indicate pollen data; thick lines indicate the presence of macrofossils.

shu, i.e. warm-temperate with moderate precipitation. Also, it is clear that these floras were a continuation of those extant in the Miocene, with the gradual addition of more numerous recent species. In most cases, plant remains are comparatively rare, and the total number of species present in any one fossil flora probably represents less than one percent of the total original species. However, it is clear that the general characters of these fossils are reliable in several respects as an indicator of the essential features of the past living floral assemblage. On this basis, the *Pinus trifolia* and other floras can be assigned to the same general floristic region as their contemporary warm-temperate counterparts on the Asian continent. This regional grouping is still true today, and all plants undoubtedly formed part of the characteristic temperate flora of the northern hemisphere.

From the Late Pliocene onwards, the Japanese region experienced several colder, and some hotter periods. These included, for example, the so-called Gokenya cold period (*ca.* 1 m.y. BP) and Manchidani cold period (*ca.* 0.3 m.y. BP), which prevailed in the low plains of central Honshu and were characterized by *Larix Gmelini* forests of the type that now occur in Sakhalin. Following this, a luxuriant flora (the so-called Uegahara flora) developed in the same localities under rather hot climatic conditions. This flora was characterized by *Syzygium* (*S. buxifolium*), the indicator genus of the northern limit of the subtropical zone.

These changes were part of a longer alternation of glacial and interglacial periods in Japan. However, even though there is evidence of former glacial scarring in the Japanese Alps, Hidaka mountain range of Hokkaido, etc., there was apparently no occasion at which the temperate flora was entirely swept away and replaced by an arctic flora (in periods of cold) or a tropical flora (in periods of heat). Instead, a north-south oscillation of temperate floras probably occurred, ensuring that the greater part of Japan retained an essentially temperate flora at all times. As a result, the floral continuation from the Eocene has in general been maintained to the present. This is in fact the main reason for the richness of relict elements. Nevertheless, Japan does have two distinct types of habitats displaying younger speciation, i.e. those developing in volcanic areas and those developing under heavy snow cover. These are discussed in later parts of this book.

2. The Present Flora: Its General Features and Regional Divisions

As described above, Japan's flora has been maintained in a relatively unchanged state since the Eocene, although it has often been modified by

climatic and palaeogeographic changes, and disturbed in part by volcanic activity. Concerning the recent flora, the standard manual is Ohwi's *Flora of Japan* (Japanese revised edition, 1965; English edition published by the Smithsonian Institute, Washington DC, 1965).

From Ohwi's book, the present author has selected all reliably native, spontaneously growing species, and excluded the many plants naturalized recently or in historic time, together with those which are presumed to have been prehistorically naturalized. The latter were introduced to Japan from the Indo-Malaysian floristic region, following the ancient introduction of rice cultivation, and also from continental Asia, following the introduction of wheat-barley agriculture in prehistoric times. Furthermore, in this discussion, use is made of the present author's evolutionary studies to emend certain generic concepts. Some are divided into smaller, evolutionarily independent units; for example, *Polygonum* (*sensu stricto*), *Reynoutria* and *Tovara* replace the broader *Polygonum* concept. Others are amalgamated into a single genus; for example, *Gymnaster* and *Kalimeris* are both included under *Aster* since they merely represent the end products of degeneration of the pappus. In order to gain a better understanding of the richness of Japan's flora, a numerical comparison is made with temperate northeastern North America and New Zealand (see Table 3). As can be seen, Japan has approximately 3900 species, compared with 2900 for northeastern North America and 1900 for New Zealand. Clearly, geomorphological and climatological differences, both past and present, play a principal role in deciding the richness or otherwise of the recent flora, and also the preservation or otherwise of relict species.

From the worldwide viewpoint, Japan's vegetation belongs to the Holoarctic region and its general character is a temperate one. The casual visitor from overseas, arriving through the airports of Haneda (Tokyo) or Itami (Osaka), or through the seaports of Yokohama or Kobe, will soon come in contact with signs of the vegetation which is characteristic of the upper parts of the warm-temperate floristic zone. Examples can be seen in public parks, the precincts of Buddhist temples or Shinto shrines, and in private gardens. The low mountains and highlands of the immediate hinterland, i.e. away from the ports and cities, are covered by a green carpet which corresponds to a transitional zone between typical warm-temperate and cool-temperate vegetation. However, due to recent urban and industrial expansion, this zone now unfortunately shows frequent signs of damage or destruction.

In parks and gardens, the tree canopy is commonly composed of broad-leaved evergreen species, such as the "shii" (*Castanopsis cuspidata* var. *Sieboldii*) and several oaks (or "kashi") which are often treated together as a distinct genus, *Cyclobalanopsis* (instead of *Quercus*). The principal species are *Cyclo-*

TABLE 3. Comparison of Japan's flora with that of temperate North America and of New Zealand

Plants / Region	Ferns (gen./sp.)	Gymnosperms (gen./sp.)	Dicotyledons (gen./sp.)	Monocotyledons (gen./sp.)	TOTAL (gen./sp.)	Latitude
Japan†1	81/401	17/39	737/2353	275/1064	1110/3857	30–45.5°N
Northeastern North America†2	32/108	10/26	438/1727	178/974	658/2835	36.5–48°N
New Zealand†3	47/164	5/20	233/1249	115/438†4	400/1871	34–47.5°S

†1 From Ohwi, *Flora of Japan*, 1965; excludes the Ryukyus and Bonin Islands.

†2 From Gleason and Cronquist, *Manual of Vascular Plants of Northeastern United States and Adjacent Canada*, 1963.

†3 From Allan, *Flora of New Zealand*, vol. 1, 1961; Moore and Edgar, *ibid.*, vol. 2, 1970.

†4 Since information on the Gramineae of New Zealand was not available, an attempt was made to derive reasonable figures for the gramineous genera and species, as follows. For Japan, the monocotyledons included 97/210 (gen./sp.) gramineous plants, while the corresponding figures for northeastern North America were 69/295. Expressed as percentages of the total monocotyledons, these figures represented 32.7/19.7 (%) and 38.7/30.2 (%), respectively. Taking an approximate mean of 35/25 (%), the gramineous genera and species of New Zealand were calculated as 40/109.

balanopsis glauca ("arakashi"; lit. vigorous oak) and *C. myrsinaefolia* ("shira-kashi"; lit. white oak) which has a glaucous undersurface to its leaves. Conifers such as *Pinus densiflora* ("akamatsu"; lit. red pine, derived from the bark colour), *P. Thunbergii* ("kuromatsu"; lit. black pine) and *Chamaecyparis pisifera* ("sawara", origin unknown) also occur. The lower trees and/or undergrowth evergreen shrubs are *Machilus Thunbergii, Ilex integra, I. crenata, Ternstroemia gymnanthera, Camellia japonica, C. Sasanqua, Ligustrum japonicum, Viburnum Awabuki, Photinia glabra, Osmanthus aurantiaca, Euonymus japonica, Pieris japonica, Aucuba japonica, Fatsia japonica,* etc. Of these, the last species has large, shiny-green palmate leaves and forms distinct bushes and undershrubs. Its generic name, *Fatsia*, derives from a misapplied pronunciation, Hattshu, of the Chinese characters that make up its name; the correct Japanese pronunciation is "yatsude".

In early spring, the sweet-smelling *Daphne odora* can be seen in full bloom, while in the late spring, several species of "tsutsuji" (the so-called Azalea) make up a picturesque array of bushes in many areas. The latter correctly belong to the genus *Rhododendron*, subgenus *Anthodendron*, and are as follows: *Rhododendron mucronatum* (purple-red), *R. hortense* (white) and *R. pulchrum* (deep purple-red). All are supposed to be cultivars from *R. macrosepalum* and/or hybrids between the latter and *R. scabrum* of the Ryukyus. Subsequently, another Azalea with laterite-red flowers, *Rhododendron indicum*, can be seen in bloom. This species, despite its scientific name, occurs in rock crevices along river-banks in southern Japan, and its Japanese name ("satsuki") is an elegant appellation of the month of May (based on the lunar calendar), during which it flowers. It is at present a very popular plant for pot-cultivation (*Bonsai*) due to its interesting colour variegation. In winter, the bright red fruits of *Aucuba japonica*, which was also introduced into Europe about 100 yr ago, are conspicuous within the characteristic green foliage of adult plants of this species. Its generic name was proposed by the Swedish botanist, C.P. Thunberg (one of the pioneering investigators of Japan's flora), based on the Japanese name, "aoki" or "aokiba" (lit. green tree or foliage).

Most of the evergreen trees and shrubs mentioned above are of strictly Japanese origin, and together they show clearly that the cities of Tokyo, Osaka, etc. are situated within the warm-temperate vegetational zone. Based on the dominant tree genera, this zone is commonly known as the Shii–Tabu zone (or *Castanopsis–Machilus* zone). However, recently, another name has been proposed, i.e. the Camellietea japonicae zone, based on the ubiquity of *Camellia* as the undershrub representative of the Laurisilvae in southern Japan. Among these evergreen thickets in the cities, several deciduous trees also occur. The largest is the maiden-hair tree, *Ginkgo biloba*. This species was introduced from China at an unknown time in the past; however, it is also

present in Tertiary sediments as a natural species of the Japanese area. In autumn, the maiden-hair tree is decorated with fan-shaped, golden-yellow leaves which, when they fall, create one of the most picturesque landscapes of the Japanese autumn. The tallest of the other trees is *Zelkova serrata* (or "keyaki"), which is an important timber tree of the Ulmaceae. It forms a very attractive canopy when covered with new, green foliage. Its buds open in a characteristic way, i.e. rather sporadically, proceeding in groups here and there all over the tree and even along a single branch.

Cherry trees are of course commonly planted in towns and cities. The species usually met with is the so-called "someiyoshino", or *Prunus yedoensis*, named after Yedo (the old name for Tokyo) where it was first identified under cultivated conditions. It has recently been suggested that this cherry tree was probably brought from the Izu Peninsula, where it had been produced by natural hybridization between *P. pendula* var. *ascendens* and *P. Lannesiana* var. *speciosa*, of which the latter is the ancestor of the decorative many-petaled garden cherry. In countryside areas, the most common cherry is the mountain cherry, *Prunus Jamasakura* (a name proposed by Siebold based on the Japanese). This species now occurs in a very variable condition, especially as regards the delicate form of the flower and colour of the young leaves. The latter open simultaneously with the flowers, leading to considerable variety in the general appearance of the trees at full bloom. In sharp contrast, *Prunus yedoensis* is normally praecox in flowering, and about two weeks later (in mid-April), *P. Lannesiana* (the so-called "satozakura") decorates the towns and cities with its fine, many-petaled pink flowers which are globose in form and large in size. The latter species originated as a cultivar from *P. Lannesiana* var. *speciosa*, which is itself a native of the Izu Peninsula and Archipelago and has single-petaled, pale-white flowers. It was developed under the skillful horticultural techniques practiced in the gardens of Yedo during the latter half of the Tokugawa Era (18–19th century).

Appearing before the cherry blossoms, are the white flowers of *Magnolia Kobus* and *M. denudata** and the yellow flowers of *Forsythia suspensa** and *F. viridissima**. Blooming in succession after the cherry, i.e. up to the spring climax in May when the evergreen Azaleas are in full blossom, are the white bell-shaped flowers of the "dodan-tsutsuji" (*Enkianthus perulatus*), the red-and-white flowers of the "hakone-utsugi" (*Weigela coraeensis*, which is a pure native of Japan despite its name) and the long, pendulous racemes of the "fuji" (*Wisteria floribunda*). These decorative deciduous shrubs and trees are also representative species of the temperate zone, and were introduced to urban areas from native localities within Japan and/or continental China (the latter being indicated by an asterisk (*)).

A visit to suburban districts reveals the occurrence of deciduous bushy

TABLE 4. Major vegetational zones of Japan

Principal northern boundary on the Pacific side	Zonal name based on:			Name applied by ecologists	Main upper altitudinal limit in central Japan (m a.s.l.)	Principal vegetational type
	latitude	altitude	vegetation			
(45°N)†	arctic (frigid) zone	alpine zone	*Pinus pumila* zone	Vaccinio–Picetea zone		alpine meadow
43–43.5°N	subarctic (sub-boreal) zone	subalpine zone	*Picea–Abies* zone		± 2500	evergreen coniferous forest
42–43°N	cool-temperate zone	montane zone	*Fagus* zone	Fagetea crenatae zone	± 1500	broad-leaved deciduous forest
37.5–38°N	warm-temperate zone	submontane (hilly) zone	*Castanopsis–Machilus* zone	Camellietea japonicae zone	± 500	broad-leaved evergreen forest
30°N	subtropical zone	lowland zone	*Pandanus–Cycas* zone			

† Figure for the Sea of Japan side of Hokkaido.

forests of *Quercus serrata* (or "ko-nara") and *Q. variabilis* (or "kunugi"), intermingled with *Carpinus Tschonoskii*, *Styrax japonica*, *Viburnum dilatatum*, *Mallotus japonicus*, *Albizzia Julibrissin*, etc. These forests are man-made, having been formed through repetitive cutting at an interval of 20–30 yr for at least 300 yr, the wood formerly being used for charcoal manufacture. (This practice was discontinued at the end of World War II as a result of the fuel revolution.) On the forest floor, certain evergreen species occur as young plants. They include *Cyclobalanopsis myrsinaefolia*, *Neolitsea Sieboldii*, *Camellia japonica* *Eurya japonica*, *Machilus Thunbergii*, *Aucuba japonica*, etc., which mark a transitional stage of the plant succession corresponding to the northern limit of the warm-temperate zone.

Apart from the temperate vegetation described above, several other major vegetational zones can be distinguished in Japan. These are developed sequentially, both with increasing longitude south-north across Japan, and with increasing elevation in mountainous areas. Table 4 summarizes the essential features of this zonation (see also Chapters 1, 3 and 6), and the following sections give detailed descriptions of specific zones and of floristic regions.

2.1. Vegetation of the warm-temperate zone

As mentioned above, the dominant members of the forests of this zone are *Castanopsis cuspidata* var. *Sieboldii* and *Machilus Thunbergii*. In southern parts, *Cinnamomum Camphora*, *Cyclobalanopsis gilva* and *C. acuta* are often intermingled with them, while close to the northern boundaries, other evergreen oak species (*Cyclobalanopsis glauca*, *C. salicina*, *C. myrsinaefolia* and *C. sessilifolia*) are dominant. The following other species are also frequent: *Cephalotaxus Harringtonia* (as conifers), *Camellia japonica*, *Aucuba japonica*, *Eurya japonica* and *Neolitsea sericea* (as chief lower trees or large shrubs), and *Kadsura japonica*, *Trachelospermum asiaticum* and *Hedera rhombea* (as evergreen lianas). On the forest floor, the liliaceous plants *Liriope* and *Ophiopogon*, the orchidaceous plants *Cymbidium Goeringii*, *Cremastra appendiculata* and *Calanthe discolor*, the myrsinaceous *Bladhia japonica*, and tufts of the evergreen ferns *Polystichum* and *Dryopteris erythrosora*, are commonly found. All of these genera and species are East Asiatic elements and indicate a very close relationship between the flora of Japan and that of the Sino-Himalayan region.

Conifers: *Podocarpus Nagi*, *P. macrophylla* (mainly in coastal districts).

Broad-leaved evergreen trees and shrubs: *Myrica rubra*, *Lithocarpus edulis*, *L. glabra*, *Machilus japonica*, *Actinodaphne lancifolia*, *A. longifolia*, *Michelia compressa*, *Distylium racemosum*, *Prunus spinulosa*, *P. Zippeliana* (=*P. macrophylla*), *Photinia glabra*, *Cleyera ochnacea*, *Ternstroemia gymnanthera*, *Ilex integra*, *I. macrophylla*, *Daphniphyllum macropodum*, *Elae-*

ocarpus sylvestris var. *ellipticus*, *E. japonicus*, *Dicalyx* spp. (evergreen group of *Symplocos*, often called *Bobua*, and containing *D. glauca*, *D. lancifolia*, *D. lucida*, *D. myrtacea*, *D. prunifolia*), *Bladhia Sieboldi* (=*Ardisia Sieboldi*), *Athruphyllum neriifolia* (=*Myrsine Seguinii*), *Maesa japonica*, *Ligustrum japonicum*, *Viburnum Awabuki*, *V. japonicum*.

Other species of coastal slopes and littoral areas: *Livistona subglobosa* (southernmost areas), *Quercus phillyraeoides* (dry thickets), *Cinnamomum japonicum* (formerly *C. pedunculosa*), *Fiwa japonica* (=*Litsea japonica*), *Raphiolepis umbellata* var. *integerrima*, *Pittosporum Tobira*, *Eurya emarginata*, *Euonymus japonica*, *Hydrangea macrophylla*, *Daphniphyllum Teijsmanni*, *Fatsia japonica*.

Evergreen lianas: *Piper Kadzura*, *Stauntonia hexaphylla*, *Rubus Sieboldi*, *Elaeagnus macrophylla*, *E. glabra*, *Uncaria rhynchophylla*.

Evergreen ferns and lower shrubs: *Arachniodes aristata* and *A. pseudoaristata* (formerly under *Rumohra*; both very common ferns, forming large communities in shaded places), *Dicranopteris linearis* and *Gleichenia japonica* (usually *G. glauca*) (ferns of rather dry open spaces), *Rubus trifidus*, *Bladhia crenata*, *B. crispa*, *B. pusilla*, *Damnacanthus indicus*.

Dwarf bamboos: *Pleioblastus* (often treated as *Arundinaria* but different in having many branches from each node). This genus occurs as many local species, with corresponding variants. In western Japan, *P. communis* is common; in eastern areas, *P. Chino* is common; and *P. Simonii* often occurs on rather open riversides.

Broad-leaved deciduous trees: several deciduous trees are found in the evergreen forests. *Quercus serrata–Q. variabilis* forests are frequent, especially near towns and cities. In higher or more northerly areas, the following species occur: *Zelkova serrata*, *Aphananthe aspera*, *Carpinus Tschonoskii*, *Castanea crenata*, *Prunus Jamasakura*, *Acer crataegifolium*, *Styrax japonica*, *Clethra barbinervis*, *Cornus controversa*. In southern parts, the following species occur: *Debregesia edulis*, *Mallotus japonicus*, *Albizzia Julibrissin*, *Fagara ailanthoides*, *Acer palmatum*, *Rhus javanica*, *R. succedanea*, *R. silvestris*, *Euscaphis japonica*, *Stachyurus praecox*, *Idesia polycarpa*, *Cornus brachypoda*.

Deciduous shrubs: *Deutzia Sieboldiana*, *D. scabra*, *Lespedeza bicolor*, *L. Buergeri*, *L. homoloba*, *Rosa multiflora*, *Rubus palmatus*, *R. corchorifolius*, *Lyonia Neziki*, *Callicarpa japonica*, *C. mollis*, *Palura chinensis* (deciduous group of *Symplocos*), *Rhododendron reticulatum*, *R. dilatatum*, *Viburnum dilatatum*, *V. erosum*, *Weigela floribunda*, *Helwingia japonica*, *Pertya scandens*, *P. glabrescens*.

Deciduous or herbaceous lianas: *Parthenocissus tricuspidata*, *Ampelopsis brevipedunculata*, *Cayratia japonica*, *Aristolochia Kaempferi*, *Akebia quinata*,

Clematis Maximowicziana, Paederia chinensis, Pueraria lobata, Milletia japonica, Wisteria brachybotrys, W. floribunda, Rubia Akane, Humulus japonicus.

Distinct herbs: *Athyrium niponicum, Lemmaphyllum microphyllum, Pyrrosia lingua, Dryopteris erythrosora, D. varia*-group, *D. lacera, Thelypteris decursivepinnata, Cyclosorus acuminatus, Pteris multifida, P. cretica, Woodwardia orientalis, Houttuynia cordata, Fatoua villosa, Pellionia radicans, Urtica Thunbergiana, Pilea petiolaris, Achudemia japonica, Boehmeria Sieboldiana, Duchesnea indica, Geum japonicum, Boenninghausenia japonica, Viola grypoceras, V. Keiskei, Centella asiatica, Hydrocotyle sibthorpioides, Angelica pubescens, A. decursiva, Ostericum Sieboldii, Mitchella repens, Cacalia delphiniifolia, Ainsliaea apiculata, Macroclinidium robustum, Carex Morrowii, C. conica, C. brunnea*-group (=*C. Nakiri*), *Lycoris sanguinea, Iris japonica, Zingiber Mioga, Alpinia japonica, Cymbidium Goeringii, Cephalanthera falcata.* Certain genera are also rich in local species, such as *Athyrium, Polystichum, Diplazium, Heterotropa, Chrysosplenium, Sedum* sect. Sino-Japonicae, *Viola, Carex, Arisaema,* etc.

Among the above warm-temperate species, several different types can be distinguished on the basis of their distribution patterns. A large proportion of the species are endemic to Japan, but as a rule they also have their closest counterparts among continental and/or Formosan species. The remainder are conspecific with other members of the continental and/or Formosan flora. Furthermore, one or several other related species are often present on the southern slopes of the Himalayas. Examples are *Cardiocrinum, Aucuba* and *Helwingia,* in which at least three different species occur side-by-side in a west-east direction across the whole region south of 30°N; and, at the eastern limit (i.e. the Japanese archipelago), they extend much farther northwards. According to recent reports, *Aucuba* and *Helwingia* can even be distinguished on the basis of their chromosomal rank, i.e. quite apart from external morphological differences. This suggests strongly that the two genera had their own centre of differentiation in Southwest China (in particular the Si Kiang–Yunnan–Szechwan region), from which they simultaneously spread westwards to the Himalayas and eastwards to Japan. In this way, the different species in these areas were established, and the time of differentiation can be assigned fairly reliably to the Miocene. Apart from this, rather many species of ferns growing in shaded habitats of southern Japan, certain orchids (*Cremastra appendiculata* and *Cymbidium lancifolium*) and also several dicotyledonous herbs (*Houttuynia cordata, Swertia bimaculata, Geranium nepalense, Adenocaulon himalaicum*) form single species that have not yet undergone any regional differentiation. Based on the above evidence, it can be safely assumed that Japan belongs to a Sino-Japanese floristic region, extending in a large but

slender belt westwards as far as the Himalayas. The genera belonging to this category can be enumerated as follows: *Cephalotaxus, Houttuynia, Sarcandra, Cyclobalanopsis, Heterotropa, Nanocnide, Trochodendron, Stauntonia, Nandina, Euryale, Parabenzoin, Loropetalum, Platycrater, Hydrobryum, Skimmia, Boenninghausenia, Stachyurus, Hosiea, Hovenia, Aucuba, Helwingia, Monotropastrum, Eukianthus (s.s.), Gardenia, Macroclinidium, Chionographis, Heloniopsis, Liriope, Ophiopogon, Aspidistra, Cardiocrinum, Heterosmilax, Shibataea, Phaenosperma, Trachycarpus, Cremastra.*

The following genera are distributed further southwards in Malaysia, and in some cases even as far as Eastern Australia and/or the islands of the Pacific: *Aphananthe, Broussonetia, Kadsura, Michelia, Dunbaria, Cleyera, Camellia, Buxus, Peracarpa, Pollia, Livistona, Aulacolepis, Gastrodia, Microtis, Lecanorchis.*

Certain genera are distributed in the New World (i.e. North America and even Central and/or South America, the latter being indicated collectively with an asterisk (*)). The genera are as follows: *Torreya, Castanopsis, Magnolia**, *Illicium, Mitrastemon**, *Hydrangea**, *Wisteria, Lyonia, Pieris, Adenocaulon**, *Disporum, Calanthe**, *Liparis**, *Pogonia**. Among them, particular attention should be paid to those with an asterisk (in spite of their rather small number), since judged from the author's hypothesis on the palaeoequatorial distribution of plants, they form a remnant of an ancient distributional region that girdled the earth. This is thought to have been established at least before the Palaeogene, along the old equator, and subsequently disrupted by various geological events.

A sequence for the disjunction and deformation can be formulated as follows: *Coriaria* type distribution (original girdle distributed along the palaeoequator)–*Magnolia* type distribution (that enumerated in the above list as *Magnolia*, etc.)—Himalayo-Sino-Japonica type distribution (i.e. as listed under *Cephalotaxus, Houttuynia*, etc. above)—Japanese endemic elements. However, due to the relatively uniform subtropical/warm-temperate climate prevalent since the Palaeogene, widely disjunctive differentiation did not occur in general. In fact, only the four genera, *Apodicarpum, Pterygopleurum* (Umbelliferae), *Alectorurus* (Liliaceae) and *Neofinetia* (Orchidaceae), are strictly endemic to Japan.

2.2. Vegetation of the cool-temperate zone

The cool-temperate zone contains the primary or essential part of Japan's flora, and is rich in distinct endemic genera and endemic species.

A superficial glance reveals that the lowest part is composed of *Quercus serrata* and *Castanea crenata*, with a sporadic occurrence of the conifers *Abies firma* and *Tsuga Sieboldii*. However, strictly speaking, these trees form a transitional zone from the lower zone, i.e. the upper part of the warm-temperate

TABLE 5. List of the principal Japanese genera of the *Fagus* (cool-temperate) zone, together with lists of the principal members of the Himalayo-Sino-Japonica and *Magnolia* types

(NOTE: Horizontal alignment between genera in different columns does not necessarily imply that they are corresponding genera between the different types.)

Life-form	Family	Genera endemic to Japan	Genera of the Himalayo-Sino-Japonica type (* indicates those semi-endemic to Japan)	Genera of the *Magnolia* type
Conifer	Sciadopitydaceae	*Sciadopitys*		
	Taxodiaceae		*Cryptomeria**	
	Cupressaceae	*Thujopsis*		*Thuja* *Chamaecyparis*
	Pinaceae			*Tsuga* *Pseudotsuga*
	Taxaceae			*Torreya*
Tree	Salicaceae		*Chosenia* *Toisusu*	
	Magnoliaceae			*Magnolia*
	Eupteleaceae		*Euptelea*	
	Cercidiphyllaceae		*Cercidiphyllum*	
	Lauraceae		*Parabenzoin*	*Lindera*
	Hamamelidaceae		*Disanthus**	*Hamamelis*
	Theaceae			*Stewartia*
Shrub	Santalaceae			*Buckleya*
	Saxifragaceae		*Platycrater** *Cardiandra*	*Itea* *Deutzia*
	Hamamelidaceae		*Corylopsis*	
	Rosaceae		*Kerria* *Rhodotypos* *Stephanandra*	
	Leguminosae			*Lespedeza*
	Ericaceae	*Tripetaleia*	*Enkianthus* *Tritomodon*	*Leucothoe* *Menziesia* *Epigaea*
	Caprifoliaceae		*Weigela* *Macrodiervilla*	

TABLE 5—*Continued*

Liana	Schisandraceae			*Schisandra*
	Lardizabalaceae		*Akebia*	
	Menispermaceae		*Sinomenium*	
	Saxifragaceae		*Schizophragma*	
	Leguminosae			*Apios* *Wisteria*
	Actinidiaceae		*Actinidia*	
Herb (dicot.)	Saururaceae			*Saururus*
	Chloranthaceae		*Chloranthus*	
	Aristolochiaceae		*Asiasarum* *Heterotropa*	
	Polygonaceae		*Reynoutria*	*Tovara*
	Ranunculaceae	*Anemonopsis*	*Isopyrum*	*Trautvetteria*
	Podophyllaceae	*Ranzania* *Glaucidium*		
	Berberidaceae			*Diphylleia* *Caulophyllum* *Achlys*
	Papaveraceae		*Hylomecon* *Macleaya*	
	Crassulaceae		*Orostachys*	*Penthorum*
	Saxifragaceae	*Peltoboykinia*	*Kirengeshoma** *Deinanthe* *Rodgersia*	*Boykinia* *Astilbe* *Tiarella* *Mitella*
	Umbelliferae		*Spuriopimpinella* *Dystaenia*	
	Diapensiaceae	*Schizocodon*		*Shortia*
	Borraginaceae	*Ancistrocarya*		
	Labiatae		*Chelonopsis*	*Meehania*
	Phrymaceae			*Phryma*
	Campanulaceae		*Adenophora*	
	Compositae		*Ainsliaea* *Pertya*	
Herb (moncot.)	Liliaceae	*Brachycyrtis*	*Tricyrtis* *Hosta*	
	Stemonaceae			*Croomia*
	Gramineae	*Hakonechloa*		*Muhlenbergia* *Diarrhena*
	Orchidaceae		*Ponerorchis*	*Tipularia* *Aplectrum*

forests, and would be replaced by *Cyclobalanopsis* if there were no human interference.

The main part of the cool-temperate zone is covered by *Fagus crenata* forests, often admixed with another member, *F. Sieboldi*, especially in lower parts. Forests of *Quercus mongolica* var. *grosseserrata* commonly occur in flat diluvial plains, while *Euptelea polyandra* and *Acer carpinifolium*, with their horizontally spreading branches, fill many of the narrow mountain valleys. The latter species can be distinguished by its leaf-form, but this is easily mistaken with *Carpinus*, or more so with the "keyaki" (*Zelkova serrata*). However, the exact species can be determined on the basis of the opposite phyllotaxis of *Acer*, in contrast to the alternating leaf arrangement in *Carpinus* and *Zelkova*. The "katsura" (*Cercidiphyllum japonicum*) is often present in the lower parts of valleys, while in spring, the slopes are decorated with cherry blossoms (*Prunus Jamasakura*, *P. Sargentii*, *P. apetala*, *P. incisa*, *P. verecunda*). In autumn, the variegated red/yellow leaf colour of numerous maple species (*Acer japonicum*, *A. crataegifolium*, *A. mono*, etc.) can be seen widely in mountainous areas. The undershrubs are luxuriant and rich in relict genera, which include many species endemic to Japan and several endemic genera. The forest floor is covered with *Sasa*, which can be divided into many local species that usually belong to the section Crassinodi (named on the basis of its unique character of sudden inflation at each node of the stem*). Among other genera that occur, the following in particular show multiple speciation: *Aconitum*, *Chrysosplenium*, *Mitella*, *Viola*, *Isodon*, *Cacalia*, *Cirsium*, *Saussurea*, *Arisaema* and *Hosta*.

The principal Japanese-endemic and semi-endemic (*) genera of the cool-temperate zone are shown, together with Himalayo-Sino-Japonica and *Magnolia* type genera, in Table 5. Under the author's concept of palaeo-equatorial distribution, the endemic elements are thought to have been derived from the Himalayo-Sino-Japonica type assemblage through extinction of any opposing counterparts, and the Himalayo-Sino-Japonica elements are thought to have been established from the *Magnolia* type assemblage through the loss of their American partners, or through a superfluous extinction resulting from more advanced generic differentiation into two or three distinct genera different between the two sides. As examples, the following relationships can be given: *Houttuynia* and *Gymnotheca* (Asian) vs. *Anemiopsis* (American), among the Saururaceae; *Epimedium* vs. *Vancouveria*, among the Berberidaceae; *Macleaya* vs. *Bocconia*, among the Papaveraceae; *Fatsia* and *Boninofatsia* vs. *Oreopanax*, among the Araliaceae; *Tripetaleia* and *Botryostege* vs. *Elliotia*, among the Ericaceae. The existence of this disjunctive distribution of the *Magnolia* type assemblage was discovered and recorded by the

* Strictly speaking, the exact position of inflation is in the part just above the true node.

eminent American taxonomist, A. Gray, based on Siebold's data published in the earlier 1800's; however, the present author (in 1960) was the first to propose an integral explanation of the different distributions based on the worldwide organization. Among other genera, *Ranunculus* is often treated as cosmopolitan, *Chrysosplenium* as bipolar, *Hydrangea* as a *Magnolia* type, and *Coriaria* as special or peculiar. All have distinct representative(s) in the Andes or southern Chile. Based on these facts, the author considers that such genera represent the oldest distribution, more or less deformed and/or modified by later, secondary extension, etc. The ancestral distribution pattern is termed the *Coriaria* type, and is exemplified by the following genera: *Julgans*, *Alnus*, *Berberis*, *Ranunculus*, *Thalictrum*, *Ribes*, *Hydrangea*, *Chrysosplenium*, *Fragaria*, *Coriaria*, *Osmorrhiza* and *Clethra*.

The relict elements derived from the palaeoequatorial distribution type are by no means restricted to the region covering the southern Pacific. They are found in another disjunct distributional region across Japan and adjacent areas, Mediterranean Europe and/or the Near East. The representative genera from Japan, which grow mainly in the cool-temperate zone, are as follows: *Pterocarya* (Juglandaceae), *Zelkova* (Ulmaceae), *Pseudostellaria* (Caryophyllaceae), *Thelygonum* (= *Cynocrambe*) (Thelygonaceae), *Eranthis* (Ranunculaceae), *Epimedium* (Berberidaceae), *Syringa* and *Forsythia* (Oleaceae) and *Kengia* (Gramineae).

2.3. Vegetation of the subarctic coniferous zone

The subarctic coniferous zone is developed principally in mountains north of central Japan that are higher than 1400–1500 m a.s.l. in elevation, although there are small corresponding areas in the southern Kinki district and on the summits of the two highest mountains in Shikoku. The lower limit of the zone is 1000–1400 m a.s.l. in the Tohoku district, but it is often established at sea-level in Hokkaido. There is a clear difference in coniferous species between Hokkaido and Honshu, the former having *Abies Mayriana* and the latter, *A. Veitchii* as the vicarious species, together with *A. Mariesii* (which has no counterpart in Hokkaido). The genus *Picea* shows slighter differentiation at the varietal level, i.e. *P. jezoensis* vs. *P. jezoensis* var. *hondoensis*. In the case of the genus *Larix*, there is no natural representative in Hokkaido,* whereas *L. leptolepis* occurs widely in central Honshu. (In its natural distribution, this species is lacking from northern districts; however, it has now been planted so widely in northern Honshu, and also Hokkaido, that the original distribution pattern is no longer easily discernible.)

* Pollen analysis has revealed *Larix Gmelini* pollen on Shikotan Is., southern Kuril Islands, and a spontaneous, disjunctive distribution of this species, which forms luxuriant growths in Sakhalin, is also known there.

The above distributional differentiation of conifers between Hokkaido and Honshu appears at first sight to suggest that the Tsugaru Straits represent a significant floral demarcation line. However, in view of the fact that the straits are known to have disappeared and reappeared several times in the Pleistocene, and also that there is no detectable species difference among many broad-leaved trees and shrubs (and also herbaceous plants), the division at the Tsugaru Straits can be considered of lesser importance.

Many of the subarctic species are Eurasiatic or Aleutian elements, and the number of endemic genera is few. The latter are as follows: *Ranzania* (Podophyllaceae), *Pteridophyllum* (Papaveraceae), *Botryostege* (Eriaceae), *Kinugasa* (Liliaceae) and *Dactylostalix* (Orchidaceae).

Semi-endemic genera which are known to extend further northwards are *Brylkinia* (Gramineae) and *Ephippianthus* (Orchidaceae).

Relict genera include *Macropodium* (Cruciferae), which is of the so-called *Forsythia* type, and *Oplopanax* (Umbelliferae) and *Lysichiton* (Araceae), which are of the *Magnolia* type. Of the species, *Oplopanax japonicus* has large palmate leaves which are heavily armed with numerous strong needles, while *Lysichiton camtschatense* is famous for its snowy-white spathe in early spring and large *Musa*-like leaves in mid-summer. (The American partner has a yellow spathe.)

Sphagnum moors are often found in this zone, and their components (especially the sedges and eriaceous plants) are generally conspecific with the Eurasian or American arctic region. Among the plants, *Narthecium* (Liliaceae) is the sole *Magnolia* type genus, and *Fauria* (Gentianaceae) is a semi-endemic genus.

2.4. Vegetation of the alpine zone

The alpine zone in Japan is restricted to very small areas over 2500 m in altitude in central Honshu, over 1900–2000 m in the Tohoku district, and over 1400–1500 m in Hokkaido. It is often covered by dense carpets of *Pinus pumila*, which usually occurs in prostrate form with decumbent stems and branches, but is sometimes semi-erect at lower elevations close to the timberline. In its general character and appearance, this vegetation presents a very similar aspect to that of the alpine zone of Europe. It also contains many identical species, especially among the grasses and eriaceous members. The sole endemic genus is *Japonolirion* (Liliaceae), which occurs sparsely in serpentinitic areas, and *Callianthemum* is a relict genus of the *Forsythia* type. The latter is found as two different species on Mt. Kitadake (southern Japanese Alps) and Mt. Apoi (Hokkaido), respectively.

2.5. Local floristic regions in Japan

In 1949, the present author proposed a scheme for the floral division of the

FIG. 14. Map showing the local floristic regions in Japan proposed by the author.

Japanese region. Emendments were later made in 1961, and the re-examined results are illustrated in Fig. 14. The major divisions are as follows:

- Yezo-Mutsu region (Y)
- Kanto region (K)
- Sea of Japan region (J)
- Fossa Magna region (F)
- Sohayaki region (S)
- Atetsu region (A)
- Mino-Mikawa region (M)
- Bonin (or Ogasawara) region (B)
- Ryukyu region (R)

2.5.1. Yezo-Mutsu region (Y)

The principal part of the Yezo-Mutsu region lies in Hokkaido, and is covered by *Abies sachalinensis* forests. The northern and northeastern boundaries are Schmidt's line in Sakhalin and Miyabe's line in the southern Kuril Islands, respectively, of which the latter is to some extent a climatic boundary since it marks the point where the timber-line reaches sea-level at the Etorup Straits. The southern boundary is still not fixed precisely. However, the Kitakami mountain range is included in the Yezo-Mutsu region based on the rather recent finding there of *Picea Glehnii*, one of the representative species. The northern limit of the natural distribution of *Larix leptolepis* (on Mt. Zao, Miyagi Pref.) is a useful regional indicator; however, part of the Pacific side of the Tohoku district may be best regarded as an area of intermigration between the Y, J and K regions. Along sea-shores in northern areas, *Chrysanthemum yezoense* and *Carex macrocephala* are characteristic species.

2.5.2. Kanto region (K)

The Kanto region has two principal centres, the Chichibu and Nikko-Ashio mountainlands. The former, plus a third centre (the Abukuma range), is rich in Sohayaki elements, which are thought to have been present in the area before the formation of the Fossa Magna region and Kanto plateau.

The distinctive species are *Salix Shiraiana, S. japonica* (Salicaceae), *Heterotropa nipponica, H. albivenia* (Aristolochiaceae), *Tritomodon subsessilis* (Eriaceae), *Ajuga incisa* (Labiatae), *Arisaema limbatum, A. monophyllum* (Araceae), *Hosta longipes* (Liliaceae).

Other species were probably established first in this region, although they are now broadly distributed over Honshu. These are *Larix leptolepis, Abies Mariesii* (Pinaceae), and *Hydrangea involucrata* (Saxifragaceae).

2.5.3. Sea of Japan region (J)

This region extends from the Oshima Peninsula, Hokkaido, to the mountainlands at the western end of Honshu, and is characterized by its deep winter snows.

Four types can be distinguished among the floral components, as follows:

(1) *Ranzania* type. This is the most characteristic and contains certain relict elements unique to the region. Among the species, *Ranzania japonica, Diphylleia Grayi, Glaucidium diphyllum, Kinugasa japonica, Rhododendron nipponicum, Carex podogyna*, etc. are distinct representatives of Japan's flora.

(2) *Viola vaginata* type. In this case, there is always a vicarious counterpart on the Pacific Ocean side of Honshu. Examples are as follows: *Epimedium sempervirens* vs. *E. grandiflorum* (Berberidaceae), *Viola vaginata* vs. *V. Bisseti* (Violaceae), *Sasa kurilensis* vs. *Sasamorpha purpurascens* (Graminae), *Rhododendron nudipes* vs. *R. Wadanum* (Eriaceae), etc.

(3) The so-called *Mallotopus* type. This type has a very special distribution pattern in that although it is represented further southwards on the Pacific side (in the Kinki district, Shikoku, and even Kyushu), it is entirely absent from the Tokai and Kanto regions. The origin of this peculiar distribution has not yet been determined. Typical species are *Thujopsis dolabrata* (Cupressaceae), *Quercus aliena* (Fagaceae), *Magnolia salicifolia* (Magnoliaceae), *Tripterygium Regelii* (Celastraceae), *Ilex Sugeroki* (Aquifoliaceae), *Hugeria japonica* (Eriaceae), *Pertya verticillata, Mallotopus japonicus* (Compositae), etc.

(4) The fourth type constitutes a rather recently established varietal rank, clearly a result of the physical effects of deep snow and often characterized by forms with decumbent branches. The typical species are *Taxus cuspidata* var. *nana, Torreya nucifera* var. *radicans* (Taxaceae), *Ilex crenata* var. *radicans* (Aquifoliaceae), *Daphniphyllum macropodum* var. *humile* (Daphniphyllaceae), *Skimmia japonica* var. *repens* (Rutaceae), etc.

2.5.4. Fossa Magna region (F)

The so-called Fossa Magna region covers the Fuji-Hakone volcanic area, the Izu Peninsula, Izu-Shichito archipelago, and stretches further south to the Sulphur Islands. The latter, although often treated as part of the Bonin Islands, are geographically and floristically quite distinct in spite of the fact that there are common southern elements.

Much of the Fossa Magna region has surface volcanic rocks, including rather recent volcanic features, and it appears that these circumstances may have induced species differentiation along many distinct lines among the plants invading from adjacent areas. This differentiation is exemplified by *Weigela coraeensis* (Caprifoliaceae), *Hydrangea macrophylla* (Saxifragaceae),

Stephanandra Tanakae (Rosaceae), *Cacalia idzuensis* (Compositae), *Rhododendron Tanakae* (formerly treated as *Tsusiophyllum*, a monotypic genus), etc.

2.5.5. Sohayaki region (S)

This region comprises the greater part of Southwest Japan, and its eastern boundary with the Fossa Magna region is at the Akaishi mountain range. The Sohayaki flora is the richest in Japan, and many of its species have congeners in western China.

The name, Sohayaki, was proposed by Dr. G. Koidzumi in 1931, based on the initial characters of the names of three of the important localities in Southwest Japan (see Fig. 15). "So" is derived from "So-no-kuni" (lit. home of the ancient Kumaso tribe), which is located in southern Kyushu; "Haya" is derived from "Hayasui-no-seto", the ancient name for the Hoyo Straits between Kyushu and Shikoku; and "Ki" is the symbolic character of "Ki-no-kuni" (or Kii province) which is now represented by Wakayama Pref. in the southern Kinki district.

The type areas are the foothills of the Akaishi range (Shizuoka Pref.), Mt. Odaigahara and associated peaks of the Kii Peninsula, the backbone mountain range of Shikoku, and southeastern Kyushu (i.e. the Osumi Peninsula, southern parts of Miyazaki Pref., and Yakushima Is.).

The relict genus *Callianthemum* (species, *hondoense*) occurs in the alpine zone on Mt. Kitadake. Among the conifers, a delicate speciation is displayed in Shikoku, i.e. *Abies shikokiana* vs. the parent species *A. Veitchii* of highland Honshu. The region is famous for the native occurrence of the umbrella pine, *Sciadopitys verticillata*, whose fasciculated leaves at the tip of each branch are important from the viewpoint of their morphology, i.e. they are an example of perfect fusion between adjacent true leaves. *Pseudotsuga japonica* is also confined to the Sohayaki region.

Many endemic and semi-endemic genera and species occur which show a pecuriar affinity with the continental flora of southwest China. They are often met with in the *Fagus* zone and/or upper half of the *Castanopsis–Machilus* zone.

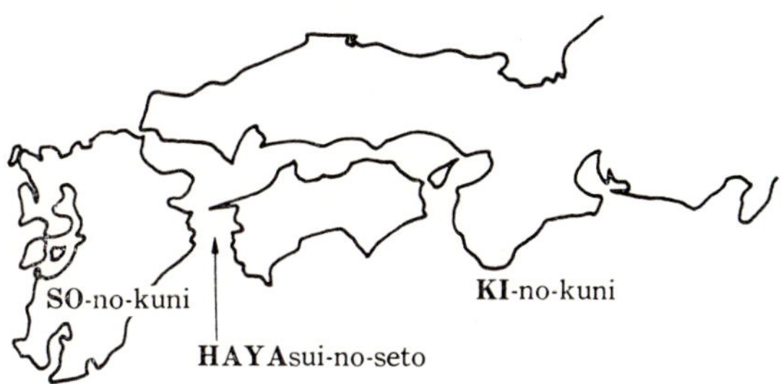

FIG. 15. Location and etymology of the "Sohayaki" region.

For example, even within the single family Saxifragaceae, the following distinct members can be enumerated: *Kirengeshoma palmata*, *Hydrangea shikokiana*, *Platycrater arguta*, *Deinanthe bifida*, *Peltoboykinia Watanabei*, *Saxifraga sendaica*, *Tanakaea radicans*, *Mitella stylosa* and its congeners. Also, *Parabenzoin trilobum* (Lauraceae), *Disanthus cercidifolia* (Hamamelidaceae), *Brachycyrtis macrantha*, *Chionographis japonica*, *C. Koidzumiana* (Liliaceae), *Gentiana shikokiana* (Gentianaceae), etc. are frequent in some areas. The genus *Heterotropa* (Aristolochiaceae) is differentiated into many local species, among which *H. asaroides*, *H. aspera*, *H. Minamitaniana*, *H. sakawana*, etc. are distinct for the peculiar form of their flowers.

The lower part of the *Castanopsis–Machilus* zone is very rich in ferns, which occur under climatic conditions very similar to those of mountains in southwest China and Formosa. Besides these, the southern half of the region is rich in members of a more southerly floral group, i.e. Malaysian elements. Typical examples occur in the Orchidaceae, as follows: *Liparis formosana*, *Calanthe furcata*, *C. gracilis*, *Gastrochilus japonicus*, *Luisia teres*, *Sarcanthus scolopendrifolius*, *Neofinetia falcata*, *Oberonia japonica*, etc. However, the most significant representative among genera of this type may be taken as *Mitrastemon Yamamotoi*, which is parasitic on roots of *Castanopsis* species and is distributed far southwards to New Guinea, and in Mexico/Guatemala.

2.5.6. Atetsu region (A)

This small region covers part of the Chugoku district, the western half of Okayama Pref., and the eastern half of Hiroshima Pref. The word, Atetsu, derives from the name of the county where all representative species have been detected. It seems probable that the Atetsu flora is a remnant flora of ancient Japan, derived directly from the luxuriant "Makinoesia flora" that thrived in the Oligocene–Miocene. Recent studies on the geological dislocation of Western Japan, i.e. across the Sea of Japan region in the Cretaceous, suggest that it exerted a strong influence on the initiation and development of the florula of the Atetsu region. Indeed, it seems natural that this region should represent the main remnant of the basal floristic assemblage existing before the Sohayaki florula.

The principal representative species of the Atetsu region are *Chloranthus Fortunei* (Chloranthaceae), *Morus tiliaefolia* (Moraceae), *Rhodotypos scandens* (Rosaceae), *Corylopsis coreana*, *Hamamelis bitchuensis* (Hamamelidaceae), *Elaeagnus Yoshinoi* (Elaeagnaceae), *Rhododendron yedoense* var. *poukhanense* (Ericaceae), *Forsythia japonica* (Oleaceae), *Gardneria multiflora* (Apocynaceae), *Viburnum Carlesii* var. *bitchiuense*, *Zabelia integrifolia* (Caprifoliaceae), *Paraixeris Yoshinoi* (Compositae), *Streptolirion volubile* (Commelinaceae), etc. Most of these species are conspecific with the continent.

2.5.7. Mino-Mikawa region (M)

The Mino-Mikawa region is similar to the Atetsu region in that it may represent a relict floristic block within the Sohayaki region. It covers the central part of the Tokai district, and takes its name from the provinces of Mino (southern Gifu Pref.) and Mikawa (eastern Aichi Pref.).

The characteristic species are as follows: the sylvatical plants, *Rhododendron Makinoi* (Ericaceae) and *Ainsliaea dissecta* (Compositae), open lowland forms such as *Eulalia speciosa* (Gramineae), and swampland species such as *Magnolia stellata* (Magnoliaceae) and *Eriocaulon nudicuspe* (Eriocaulaceae). Swamp trees are represented by *Acer rubrum* var. *pycnanthum*, and *Chionanthus retusa* is of *Magnolia* type in its sectional or generic rank.

2.5.8. Bonin region (B)

This region is comprised of a small island group, the Bonin (or Ogasawara) Islands, which represent old volcanic basement uplifted during the Miocene. The region retains in a modified form its old Miocene florula. The present climate is very similar to that of Miami (Florida), although the mean annual temperature is slightly (about 1.3°C) lower. The florula is rather small but rich in endemic forms. The tree and shrub species number 99, of which 67* are endemic. This endemic ratio (i.e. 67%) is the highest among the islands of the western Pacific Ocean, although it is surpassed by the Galapagos, Juan Fernandez and Hawaiian islands. In the case of herbaceous species, there is a total of 80 species, of which 40 are endemic.

Among the elements that occur as slightly differentiated endemic species (i.e. in comparison with other or counterpart species), the following genera can be enumerated: *Machilus, Neolitsea* (Lauraceae), *Raphiolepis* (Rosaceae), *Eurya, Schima* (Theaceae), *Euonymus* (Celastraceae), *Stachyurus* (Stachyuraceae), *Fatsia* (Araliaceae), *Vaccinium* (Eriaceae), *Osmanthus* (Oleaceae), *Gardenia* (Rubiaceae), *Livistona* (Palmae), etc. These are all members of the Asiatic subtropical or warm-temperate floral assemblage. Besides them, *Distylium* (Hamamelidaceae), *Osteomeles* (Rosaceae), *Trachelospermum* (Apocynaceae) and *Viburnum* (Caprifoliaceae) are the Asiatic representatives of the *Magnolia* type distribution. The genus *Pittosporum* (Pittosporaceae) includes four endemic species, while *Ilex* and *Callicarpa* include three each. They thus show a small-scale but distinct speciation across this small island group.

Other remarkable endemic species are *Rhododendron boninense* (Eriaceae), *Crepidiastrum ameristophyllum* and *C. linguafolium* (Compositae), which have

* This includes 14 species known in the Sulphur Islands but clearly distributed there in recent times.

clear affinities to their Asiatic congeners. On the other hand, *Santalum boninense* (Santalaceae), *Meterosideros boninensis* (Myrtaceae), *Lobelia boninensis* (Campanulaceae) and *Clinostigma Savoryana* (Palmae), all have a close kinship with Polynesian elements.

Formerly, the following other genera were considered endemic: *Boniniella* (Aspleniaceae), *Boninofatsia* (Araliaceae), *Platypholis* (Orobanchaceae), *Boninia* (Rutaceae) and *Dendrocacalia* (Compositae). However, the first has now been discovered on Hainan Is., southern China, and the Borodino Islands (east of Okinawa). The second and third have been reassigned to the genera *Fatsia* and *Orobanche*, respectively. *Boninia* has been shown to have a close relationship to the Australo-Asiatic *Acronychia* and *Melicope* and Hawaiian *Pelea*. *Dendrocacalia* is a very unique and isolated plant, and its true generic position is now considered problematic.

2.5.9. *Ryukyu region (R)*

In the Ryukyu region, tropical floral elements have invaded southwest-northeastwards along the island arc. Different genera show different distribution patterns and there are clear limit lines where numbers of representative elements suddenly decrease. The two most important lines are that between the Yaeyama archipelago and Okinawa main island, and that north of Amami-Oshima Is. These are illustrated in Fig. 16 on the basis of two rather distinct floral groups of the Ryukyu landscape, i.e. flowering plants that are submerged in sea-water and members of the mangrove forests, both of which are adequate as tropical symbols. However, the same general tendency is also apparent among all vascular plants in the Ryukyus (see Table 6).

Clearly, the two lines represent very sharp drops in tropical floristic elements. Also, it is noteworthy that among the 53 genera of the Yaeyama archipelago, 12 have no distribution in Formosa and another 8 have their locus only on Botel Tobago, a small islet off the southeastern coast of Formosa. The principal members are as follows: *Acrostichum* (Polypodiaceae), *Microgonium* (Hymenophyllaceae), *Illigera* (Hernandiaceae), *Dalbergia, Pterocarpus* (Leguminosae), *Intsia* (Connaraceae), *Nothapodytes* (Icacinaceae), *Sonneratia* (Sonneratiaceae), *Ochrosia* (Apocynaceae), *Cyrtandra* (Gesneria-

TABLE 6. Numbers of genera and species of vascular plants in the Ryukyu region† which have their northern limit of distribution at:

	Yaeyama archipelago	Okinawa Is.	Amami-Oshima Is.
Genera	53	27	41
Species	129	54	142

† Data from Hatusima, *Flora of the Ryukyus*, 1971.

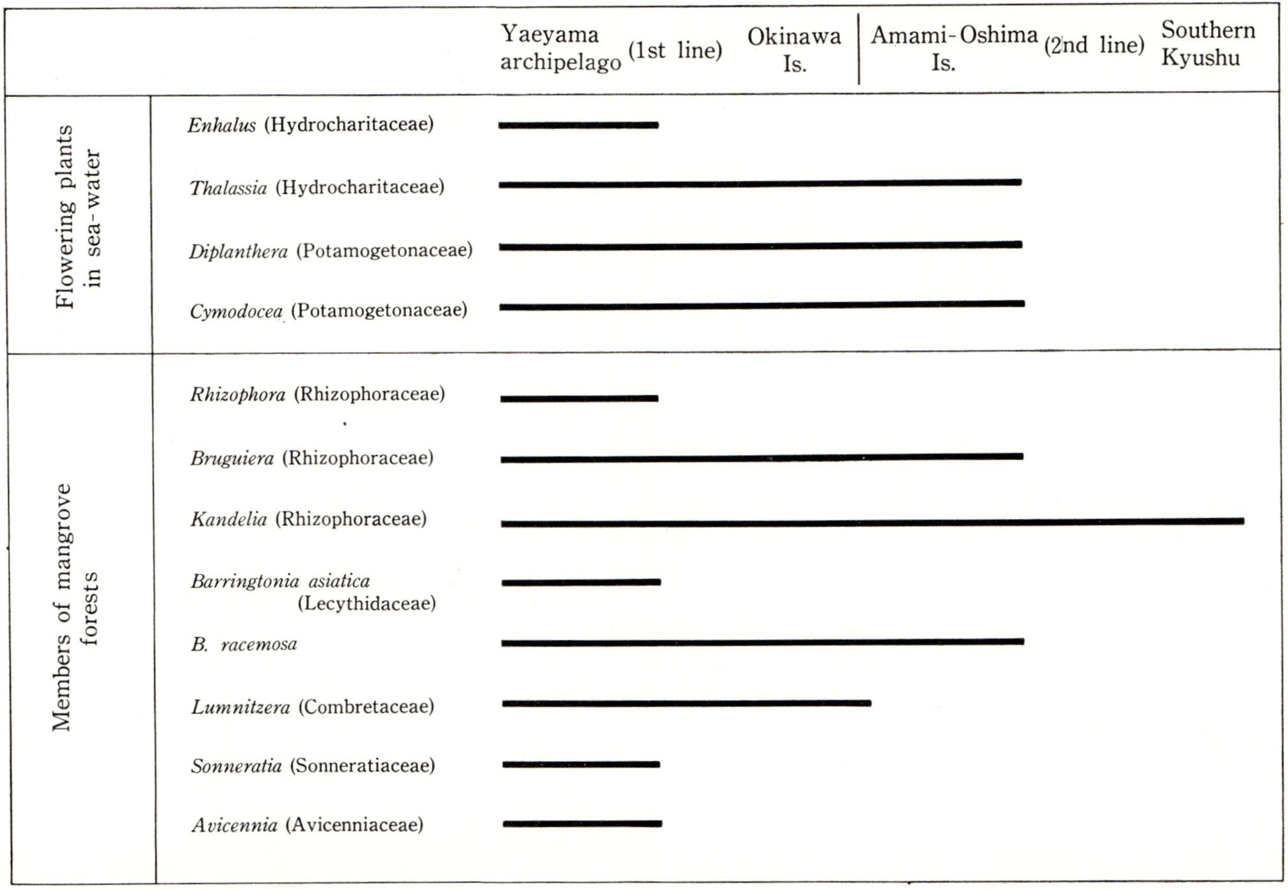

Yaeyama archipelago (1st line) Okinawa Is. | Amami-Oshima Is. (2nd line) Southern Kyushu

Flowering plants in sea-water

Enhalus (Hydrocharitaceae)

Thalassia (Hydrocharitaceae)

Diplanthera (Potamogetonaceae)

Cymodocea (Potamogetonaceae)

Members of mangrove forests

Rhizophora (Rhizophoraceae)

Bruguiera (Rhizophoraceae)

Kandelia (Rhizophoraceae)

Barringtonia asiatica (Lecythidaceae)

B. racemosa

Lumnitzera (Combretaceae)

Sonneratia (Sonneratiaceae)

Avicennia (Avicenniaceae)

ceae), *Argostemma* (Rubiaceae), *Lepistemon* (Convolvulaceae), *Enhalus* (Hydrocharitaceae), *Rhaphidophora* (Araceae), *Nipa* (Palmae), *Garnotia* (Gramineae), *Staurochilus, Stereosandra, Spathoglottis* and *Vanda* (Orchidaceae). Besides the above plants, other elements in the same category occur as the endemic genus *Satakentia* (Palmae), and also as several distinct species such as *Nervilia Aragoana* (Orchidaceae), etc. The overall floral demarcation can best be understood by accepting the so-called Kano line, proposed by T. Kano as an emended elongation of Merrill's line between mainland Formosa and the islet, Botel Tobago. Clearly also, the two fundamental characters of the flora of the southern Ryukyus are its membership to the Asiatic region but at the same time its reception of a strong Australo-Polynesian (rather than Formosan) influence under the regional control of the Northern Equatorial Current.

The following lists summarize certain of the distinct tropical elements according to their northernmost limit of occurrence:

Yaeyama archipelago: *Dipteris* (Dipteridaceae), *Lomariopsis* (Aspidiaceae), *Pithecelobium* (Leguminosae), *Briedelia* (Euphorbiaceae), *Kleinhovia* (Sterculiaceae), *Saurauia* (Actinidiaceae), *Garcinia* (Guttiferae; this tree is planted extensively in the Ryukyus as a wind shield, but its natural occurrence is in Yaeyama only), *Ecdysanthera* (Apocynaceae), *Avicennia* (Avicenniaceae), *Hemiboea, Titanotrichum* (Gesneriaceae), *Hemigraphis* (Acanthaceae), *Guettarda* (Rubiaceae), *Frecynetia* (Pandanaceae), *Corymborkis* (Orchidaceae).

Okinawa main island: *Aglaomorpha* (Polypodiaceae), *Hemigramma* (=*Anapausia*) (Aspidiaceae), *Schizaea* (Schizaeaceae), *Cleome* (Capparidaceae), *Atylosia* (Leguminosae), *Tristellateia* (Malpighiaceae), *Melanolepis* (Euphorbiaceae), *Scolopia* (Flacourtiaceae), *Rhizophora* (Rhizophoraceae), *Lumnitzera* (Combretaceae), *Rhodomyrtus* (Myrtaceae), *Evolvulus* (Convovulaceae), *Lepidagathis* (Acanthaceae), *Cyrtococcum, Ichinanthus, Oryzopsis* (Graminae), *Epipremnum, Pristia* (Araceae), *Acanthephippium* (Orchidaceae).

In addition, *Tapeinidium* (Pteridaceae), *Pipterus* (Urticaceae) and *Rhyssopterys* (Malpighiaceae) belong to the group that is restricted to the eastern side of Kano's line. Among other genera, *Cibotium* (Dicksoniaceae), *Helminthostachys* (Ophioglossaceae), *Hernandia* (Hernandiaceae), *Thespesia* (Malvaceae) and *Tutcheria* (Theaceae) are distributed further northwards to Oki-no-erabu Is., while the northernmost limit of occurrence of *Dodonaea* (Sapindaceae) and *Flagellaria* (Flagellariaceae) is Tokunoshima Is.

The Tokara Straits, between Amami-Oshima Is. and the Yaku-Tane

FIG. 16. Typical tropical members of the Ryukyu flora and their distributional limits.

island group, form the second biogeographical demarcation line. This line corresponds to the so-called Watase line, which is based chiefly on palaeobiogeographical data for vertebrate animals, or to the so-called Aoki line, which is based chiefly on entomological studies and associated climatological data. Vascular plants show a clear similarity to the faunal distribution, with at least 41 genera having their northern limit on Amami-Oshima Is. The important members of this group are as follows: *Pileostegia* (Saxifragaceae), *Erythrina*, *Pongamia*, *Derris* (Leguminosae), *Murraya* (Rutaceae), *Croton*, *Macaranga* (Euphorbiaceae), *Maytenus* (Celastraceae), *Heritiera*, *Helicteres* (Sterculiaceae), *Schima* (Theaceae), *Barringtonia* (Lecythidaceae), *Bruguiera* (Rhizophoraceae), *Cerbera* (Apocynaceae), *Wendlandia* (Rubiaceae), *Aristida*, *Dactyloctenium*, *Urocloa* (Graminae), *Gahnia* (Cyperaceae), *Forrestia* (Commelinaceae), *Cymodocea*, *Diplanthera* (Potamogetonaceae) and *Thalassia* (Hydrocharitaceae).

In consideration of the presence of the many tropical genera listed above in the flora of the Ryukyu region, it appears natural to accept the generally held concept (for example, Mattick in *Eugler's Syllabus der Pflanzenfamilien*, 12th ed., 1964) that the Ryukyus, Formosa, south and southwest China, Indochina peninsula and eastern Himalayas, together constitute a Southeast Asian floristic region within the palaeotropical floristic kingdom. The representative endemic families of this region are the Rhoipteleaceae, Sargentodoxaceae and Pentaphyllacaceae. However, these families do not actually occur in the Japanese archipelago or Formosa. Moreover, any general theory on the affinities of the Ryukyu flora must take into account the distribution and differentiation of its integrated warm-temperate elements. Based on this approach, the Ryukyu floristic region is perhaps best regarded as a stereometrically differentiated zone intervening between the true temperate and tropical zones. That is to say, although the Ryukyu flora belongs essentially to the Himalayo-Sino-Japanese region and elements of this type occupy the mountainous areas, the escarpments and lowland areas are covered by tropical elements which have invaded chiefly from the south in recent geological time. Similar floristic patterns can be found, respectively, from the foothills of the Nepalese Himalayas, through the Shan area of Burma, upland Thailand, the mountains of the Si Kiang, Kwangsi and Kwantung regions, and Formosa.

In the Ryukyus, mountain slopes above 300 m a.s.l. are covered by dense forests of *Castanopsis cuspidata* var. *Sieboldii* and *Machilus Thunbergii*. These forests are very similar to those of Japan proper, having the species *Distylium racemosum*, *Elaeocarpus japonicus* and *Bladhia Sieboldi* in common, but differ slightly in that they also contain *Cyathea podophylla* and *Camellia lutchuensis*. In parts of these forests that have been subject to damage or human interference,

many species (other than those of *Machilus* and *Distylium*) are lost, and replaced by secondary components such as *Ficus Wightiana*, *Melia Azedarach*, *Elaeocarpus silvestris*, *Diospyros maritima*, *Fraxinus insularis*, *Arenga Eugleri*, etc. Characteristic lowland vegetational types include those of the upraised coral zone and estuarine mangrove forests. Representative species of the former are *Limonium Wrightii*, *Crossostephium chinense* and *Osteomeles anthyllidifolia*, while the latter occur in most typical form on Iriomote Is.

The fauna of the Ryukyus includes three endemic genera, represented by *Pentalagus furnessi* (Leporidae; Amami and Tokunoshima Islands), *Tokudaia oshimensis* (Muridae; Amami and Okinawa Islands) and *Mayailurus iriomotensis* (Felidae; discovered in 1965 on Iriomote Is.). However, among the plants there is only one truly endemic genus, *Satakentia* (Palmae), as mentioned above, which is also found on Iriomote Is.*

Among the many plants, the following appear to be well differentiated and firmly established as typical species of the Ryukyu archipelago, even though they may also be distributed in adjacent areas: *Pinus lutchuensis*, *Cycas revoluta*, *Cyclobalanopsis Miyagii*, *Rubus amamianus*, *Viola Iwagawai*, *V. Tashiroi*, *Alchornea liukiuensis*, *Bredia yaeyamaensis*, *Swertia Tashiroi*, *Salvia pygmaea*, *Ajuga elatior*, *Suzukia luchuensis*, *Viburnum suspensum*, *Aster Asa-Grayi*, *Pleioblastus linearis*, *Lilium longiflorum*, *Ophiopogon Jaburan*, etc.

On the other hand, another group can be distinguished whose member species occur mostly in limited areas and are, as relict forms, endemic. This group includes *Heterotropa Fudsinoi*, *Geotaenium gelasinum*, *Ilex dimorphophylla*, *Oxalis amamiana*, *Viola amamiana*, *Schizocodon rotundifolia*, *Vaccinium amamianum*, *Chikusichloa brachyanthera*, *Satakentia liukiuensis*. Also, perennial herbs of the Aristolochiaceae, especially of the *Asarum* group (now divided into *Asarum*, *Heterotropa* and *Geotaenium*), occur as about ten species, all of which are strictly endemic to different islands. This speciation is at present under study by the author, both from the geohistorical and pure biological viewpoint, with the expectation that the results will assist in clarifying fully the true origin and nature of the flora of the Ryukyu region.

Finally, it should be added that the Borodino Islands to the east have some species in common with the Bonin Islands. These include *Boniniella Ikenoi* and *Lycium griseolum* (see also section 2.5.8.).

* It should be noted that, formerly, the two genera *Tashiroea* and *Tetraplasia* were often referred to as endemic. However, they are now assigned to the genera *Bredia* (Melastomaceae) and *Damnacanthus* (Rubidaceae), respectively.

General References

1. *Ed.* R. W. Chaney, *Tertiary Floras of Japan*, pp. 262, The Collaborating Association to Commemorate the 80th Anniversary of the Geological Survey of Japan, Tokyo, 1963.
2. S. Fujita, *Establishment of the Japanese Archipelago* (Japanese), pp. 257, Tsukiji Shokan, Tokyo, 1973.
3. H. Hara, Contributions to the study of variations in the Japanese plants closely related to those of Europe or North America, part 1, *J. Fac. Sci. Univ. Tokyo Sect. III*, **6** (2), 29–96 (1952); part 2, *ibid.*, **6** (7), 343–91 (1956).
4. H. Hara, An outline of the phytogeography of Japan, pp. 96, *in* H. Hara and H. Kanai, *Distribution Maps of Flowering Plants in Japan*, facs. 2, Inoue Shoten, Tokyo, 1959.
5. H. Hara and H. Kanai, *Distribution Maps of Flowering Plants in Japan*, facs. 1, maps 1–100, 1958; facs. 2, maps 101–200, 1959.
6. S. Hatusima, *Flora of the Ryukyus* (Japanese), pp. 940, Okinawa Kyoiku Kenkyu-kai, Naha, 1971.
7. E. Horikoshi, Orogenic system of the Japanese islands, *Kagaku* (Japanese), **42** (12), 665–73 (1972).
8. T. Hosokawa, Hosokawa's line and its after-effects, *Acta Phytotax. Geobot.* (Japanese), **18**, 14–19 (1959).
9. K. Ichikawa, S. Fujita and M. Shimazu, *Nihon-Retto Chishitsukozo Hattatsushi* (Japanese), pp. 232, Tsukiji Shokan, Tokyo, 1970.
10. K. Ichikawa, T. Matsumoto and M. Iwasaki, Growth of the Japanese archipelago, *Kagaku* (Japanese), **42** (1), 181–91 (1972).
11. H. Kanai, Distribution patterns of Japanese plants, distributed on the Pacific side of central Japan, pp. 14, *in* H. Hara and H. Kanai, *Distribution Maps of Flowering Plants in Japan*, facs. 1, Inoue Shoten, Tokyo, 1958.
12. S. Kawano, Flora of Japan: its natural historical background, *Shokubutsu to Shizen* (Japanese), **1** (1), 22–24; (2), 9–12; (3), 10–14; (4), 16–17; (5), 11–13; (8), 8–10 (1967); **3** (1), 11–16 (1969).
13. S. Maebara, *Florula Austrohigoensis* (*Japanese*), 1931.
14. F. Maekawa, Notes on prehistorical naturalized plants, *Acta Phytotax. Geobot.* (Japanese,) **13**, 274–79 (1943).
15. F. Maekawa, Makinoesia and its bearing on oriental Asiatic flora, *J. Jap. Bot.* (Japanese), **24**, 9–96 (1949).
16. F. Maekawa, On some problems in Japanese plant geography, *Chiri* (Japanese), **6** (19), 1030–35 (1961).
17. F. Maekawa, Trans-palaeoequatorial distribution and bipolar distribution, *Sci. Rept. Tohoku Univ. Ser. 4* (*Biol.*), Kimura Commemorative Number, 335–40 (1963).

18. F. Maekawa, Floristic relation of the Andes to Eastern Asia with special reference to the trans-palaeoequatorial distribution, *J. Fac. Sci. Univ. Tokyo Sect. III*, **9** (5), 161–95 (1965).
19. F. Maekawa, Notes on trans-palaeoequatorial distribution, *Shizen* (Japanese), **22** (7), 55–64 (1967).
20. G. Masamune, *Shokubutsu Chirigaku Shinko* (Japanese), pp. 166, Hokuryukan, Tokyo, 1956.
21. H. Matsuo, A study of the so-called "Daijima-type" floras from Fukui, Ishikawa and Toyama Pref. Inner side of central Japan, *Ann. Sci. Kanazawa Univ.* (Japanese), **9**, 81–98 (1972).
22. S. Miki, On the change of flora in Eastern Asia since the Tertiary Period 1., *Jap. J. Bot.*, **11** (3), 237–303 (1941).
23. M. Minato and S. Ijiri, *Japanese Islands* (Japanese), 2nd ed., pp. 221, Iwanami Shoten, Tokyo, 1966.
24. *Ed.* A. Miyawaki, Vegetation of Japan compared with other regions of the world, *Encyclop. Sci. Technol.* (Japanese), vol. 3, pp. 535, Gakken, Tokyo, 1967.
25. T. Nasu, Aspects of Japanese plants and animals, *Biological Sci.* (Japanese), **24** (1), 1–10 (1972).
26. J. Ohwi, *Flora of Japan* (Japanese), pp. 1383, 1953; Pteridophyta, pp. 244, 1957; ed. 2, pp. 1560, Shibundo, Tokyo, 1965.
27. J. Ohwi, *Flora of Japan* (English edition), pp. 1067, Smithsonian Institute, Washington, 1965.
28. S. Oishi, The Mesozoic flora of Japan, *J. Fac Sci. Hokkaido Imp. Univ. Ser. IV*, **5** (2–4), 123–480 (1940).
29. M. Sato, Floristic and faunistic regions in biogeography, *Contrib. Biol. Lab. Coll. Lib. Arts Ibaraki Univ.* (Japanese), **7**, 7–27 (1969).
30. A. Sugimura, Boundary between plates near Japan proper, *Kagaku* (Japanese), **42** (1), 192–202 (1972).
31. T. Suzuki, The highest vegetation units of Japan and their areas, *Pedologist*, **10** (2), 1–7 (1966).
32. H. Takahashi, Fossa Magna element plants, *Res. Rept. Kanagawa Pref. Mus. Nat. Hist.* (Japanese), no. 2, pp. 90 (1971).
33. *Ed.* F. Takai, T. Matsumoto and R. Toriyama, *Geology of Japan*, pp. 279, University of Tokyo Press, Tokyo, 1963.
34. T. Tanai, *Tertiary Phytogeography of the Northern Hemisphere*, Commemorative Volume to Prof. Matsushima, 201–16, 1971.
35. T. Tanai, Tertiary history of vegetation in Japan, *Floristics and Paleofloristics of Asia and Eastern North America* (ed. A. Graham), p. 235–55, 1972.
36. M. Tatewaki, *Distribution of Plants* (Japanese), pp. 170, Kawate Shobo, Tokyo, 1948.
37. M. Tokuda, *Biogeography* (Japanese) pp. 199, Tsukiji Shokan, Tokyo, 1969.
38. *Ed.* T. Tuyama and S. Asami, *Nature in the Bonin Islands* (Japanese), pp. 271, Hirokawa Shoten, Tokyo, 1970.

39. J. A. Wolfe, An interpretation of Alaskan Tertiary floras, *Floristics and Paleofloristics of Asia and Eastern North America* (ed. A. Graham), p. 201–33, 1972.
40. T. Yamazaki, Plant distribution in Japanese islands, *Nat. Sci. and Mus.* (Japanese), **26**, 10–19 (1959).

3

Forest Vegetation Zones

Tsunahide SHIDEI*

1. General Survey
2. Principal Natural Forest Zones
    2.1. Subarctic/subalpine evergreen coniferous forest
    2.2. Cool-temperate broad-leaved deciduous forest
    2.3. Warm-temperate broad-leaved evergreen forest
    2.4. Subtropical broad-leaved evergreen forest
3. Special/Local Forest Types
    3.1. Coastal forest
    3.2. Dry-land forest
    3.3. Transitional forest
    3.4. Swamp forest and mangrove forest
4. Man-made Forests
    4.1. Artificial regeneration
    4.2. Natural regeneration

**Department of Agriculture, Kyoto University, Sakyo-ku, Kyoto-shi 606, Japan*

In order to give a general survey of the ecology and distribution of forests and forest types in Japan, it is convenient at the outset to distinguish three broad classes of forests. These may be summarized as follows: (1) those which are entirely natural communities (i.e. the typical, widely distributed zonal forests), (2) special natural communities which show strong effects from local conditions and/or which may be partially modified by human activity (special/local forests), and (3) tree plantations established purely for their timber production (man-made or artificial forests).

This chapter gives a systematic account of the major forest vegetation zones and types in Japan based on these three broad categories.

1. General Survey

Among natural forests in Japan, four principal zones may be distinguished according to the general physiognomy and climatic conditions: (1) subarctic/subalpine evergreen coniferous forest, (2) cool-temperate broad-leaved deciduous forest, (3) warm-temperate broad-leaved evergreen forest, and (4) subtropical broad-leaved evergreen forest. Another classification (that of Japanese plant sociologists, mainly of the Zürich-School) distinguishes Montpellier 3 types: (1) the *Vaccinium–Picea* class, (2) the *Fagus–Quercus* class, and (3) the *Camellia* class. The former two correspond in general to the physiognomic zones (1) and (2), respectively, while class (3) embraces all forests classified under the physiognomic zones (3) and (4). However, even though these warm-temperate (3) and subtropical (4) forests do have common elements (they consist mainly of evergreen oak species, including *Shiia*, with *Camellia* as the typical plant), certain genera and species such as *Cyathea*, *Alsophila*, *Pandanus*, *Ficus retusa*, etc., and the occasional mangrove forests at river mouths, occur only in the subtropical zone. For these reasons, the physiognomic classification into four types is preferred.

The special/local forests include a wide variety of types, plus "transitional" communities. For example, on both the Sea of Japan and Pacific seaboards (apart from Hokkaido), the costal strip, where it is sandy or covered by dunes, is occupied by pine trees (*Pinus Thunbergii* to the south and *P. densiflora* to the north). In Hokkaido, the coastal forest is made up of deciduous oak species. Also, although there are no truly arid areas in Japan, the comparatively dry areas (annual precipitation $<$1200 mm) often exhibit special forest types comprised of rather hardy tree species, e.g. the famous *Pinus densiflora* forests of the Seto Inland Sea and Iwate Plain, the deciduous oak forests of the Ishikari Plain, and the *Larix leptolepis* forests that occur in subalpine regions of the Nagano Basin.

At the junctions of the principal forest zones various special, transitional communities can be distinguished. For example, between typical warm-temperate and cool-temperate forests there may be a zone of broad-leaved deciduous forest containing *Quercus serrata* and *Q. acutissima* or *Zelkova serrata*, etc. Although these special forests can be assigned to the warm-temperate climatic zone, the winter temperature is sufficiently low to preclude the establishment of typical broad-leaved evergreen forest. Similarly, at the junction between subalpine evergreen coniferous and cool-temperate beech forests, a zone of hardy, broad-leaved deciduous forests may occur.

Since mangrove forest is not found very widely in Japan (there is no true tropical zone), it is also considered here as a special/local forest type. Three principal mangrove species occur in subtropical areas: *Bruguiera conjugata*, *Kandelia Candel* and *Rhizophora mucronata*. On Yakushima Is. the life-form is shrub-like; on Okinawa Is. the plants are about 2–5 m in height.

In the case of man-made forests, the principal species concerned are conifers, particularly *Cryptomeria japonica*, *Chamaecyparis obtusa*, *Pinus densiflora*, *Pinus Thunbergii* and *Larix leptolepis*. Normally such artificial forests have, or acquire, only one or two stories of undergrowth. This is in sharp contrast to most natural forests which, in the subarctic and warm-temperate zones for example, usually exhibit some 3–5 sublayers of smaller trees under the main canopy.

Finally, it is necessary also to mention the special undergrowth type featured by many Japanese forests. This is comprised of various "Sasa" (dwarf-bamboo) species, which are found in all forest zones and especially in cool-temperate beech forests. Generally speaking, tall *Sasa* species grow in the snowy districts on the Sea of Japan side of the Japanese mainland, where they build very dense, tall communities, so rendering natural regeneration within them very difficult under normal circumstances. The principal species are *Sasa kurilensis* and *S. paniculata*. On the Pacific Ocean side, the forests contain different dwarf-bamboo types, such as *Sasamorpha purpurascens*, *Sasa nipponica*, etc. The life-form in this case is different from that of the Sea of Japan side (which is typically tall and creeping, with shallow rhizomes), being erect, with deep rhizomes. *Sasa* species may show a very long life (reportedly with a duration of over 80 yr in some cases). When flowering time arrives, almost all individuals in the district bloom within a year or two of each other and then die. New young shoots sprout immediately from the remaining rhizomes and seeds, and, generally within a period of 2–3 yr, the new *Sasa* undergrowth so formed grows and spreads again into a dense community.

2. Principal Natural Forest Zones

2.1. Subarctic/subalpine evergreen coniferous forest

The principal tree species of the subarctic/subalpine evergreen coniferous forests differ between Honshu and Hokkaido, and also (in Honshu) between the Sea of Japan and Pacific Ocean sides of the island.

In Hokkaido, the three main species of tall trees of the subarctic zone are *Abies sachalinensis, Picea jezoensis* and *Picea Glehnii.* These trees are sometimes found intermixed, although *P. Glehnii* commonly occurs as pure stands on gravelly volcanic soil, on steep rocky slopes, and in swamps. The coniferous species are sometimes accompanied by *Betula Ermanii* and *Betula Maximowicziana,* which themselves form pure stands at the upper tree limit and in areas which have been fired. The variety *Abies sachalinensis* var. *Mariana* (Japanese name, *Aotodo*) occurs on high land in the Oshima Peninsula, southwest Hokkaido, and is to some extent mixed with *Fagus crenata.* It is therefore often assigned to the cool-temperate zone by forest ecologists, and in fact the subarctic/cool-temperate zonal boundary is usually drawn to the north of the Oshima Peninsula, beyond the natural area of occurrence of *A. sachalinensis* var. *Mariana.* Temperatures on the Oshima Peninsula are milder than those typical of the subarctic zone (the warmth index, i.e. the annual sum of positive differences between monthly mean temperatures and +5°C, is +66.7°C at Hakodate). The upper limit of subarctic coniferous forests in Hokkaido is *ca.* 1500 m a.s.l.; the lower limit is *ca.* 400 m a.s.l. in the south, dropping to sea-level in the north.

In Honshu, subarctic (subalpine) coniferous forests are found only at high elevations, mainly on the so-called Japanese Alps which run north-south through the centre of the island. In comparison with Hokkaido the number of coniferous tree species in this zone is high, with *Picea jezoensis* var. *hondoensis, Tsuga diversifolia, Abies Veitchii* and *A. Mariesii* being the main components. Also, in other special, limited regions, species such as *Picea Glehnii, P. Koyamai, P. Maximowiczii* and *Larix leptolepis* occur less commonly. For example, *Picea Glehnii* has recently beeni dentified in subalpine forest on Mt. Hayachine (northern Honshu), while *P. Koyamai* has a restricted area of occurrence on Mt. Yatsugatake (Nagano Pref.), and *P. Maximowiczii* is limited to the Chichibu mountain range, Mt. Yatsugatake and Mt. Kitadake (Chubu district, central Honshu). *Larix leptolepis* (the only larch species in Japan) is found on northern and southern slopes of the Japanese Alps and on Mt. Fuji, although

it is also cultivated at present in plantations over a very wide area of Hokkaido and Honshu due to its rapid growth as a young plant.

Tsuga diversifolia typically shows a preference for dry soils, and its distribution is thus limited mostly to steep or rocky mountain slopes and narrow ridges, where it may occur as pure stands. This type of subalpine forest is illustrated in Fig. 1.

FIG. 1. Natural *Tsuga diversifolia* forest of the subalpine zone (Mt. Shiga, Nagano Pref.).

FIG. 2. Subalpine *Abies Veitchii* forest close to the timber line (Mt. Ontake, Nagano Pref.).

Abies Veitchii forest (see Fig. 2) is widely distributed on the Pacific Ocean side of Honshu, but is very limited in northern areas and towards the Sea of Japan side, where the winter snow is heavy. The reasons for this phenomenon are not yet fully understood, although it appears that *A. Veitchii* is unable to grow up erect through thick snow deposits against the high incumbent pressure. Towards central inland areas of Honshu, where the snow deposits are thinner but the winter temperatures are still low, *A. Veitchii* occurs as mixed forests with *Abeis Mariesii*. The mountain ranges on the Sea of Japan side of the island are, however, occupied by *A. Mariesii* only, which here may occur as pure stands in the subalpine zone.

These subalpine coniferous trees are often mixed with broad-leaved deciduous species in the high tree layer. One such example is *Betula Ermanii*, which is a principal accompanying tree of subalpine coniferous forests (see Fig. 3). Moreover, at the boundary between the subalpine and alpine zones, *B. Ermanii* often forms pure horizontal stands which, in autumn, when the leaf-colour of deciduous trees changes rapidly, allow a rapid identification of the timber line on the mountain slopes.

In addition to the species mentioned above, there are also others which are known to occur in both the subalpine and cool-temperate zones of Honshu. These include *Thuja Standishii, Chamaecyparis obtusa* and *Pinus pentaphylla*.

The upper limit of the subarctic (subalpine) zone in central Honshu is *ca.* 2500 m a.s.l.; the lower limit may reach 1600 m a.s.l. but is in general higher. In the Kinki district and on Shikoku, subalpine forests occur only on mountains exceeding 1800 m in height, e.g. the Mt. Omine range in the Kinki district (max. altitude, 1915 m) and on Mt. Tsurugi (1955 m) and Mt. Ishizuchi (1981 m) in Shikoku. Here, natural *Abies Veitchii* forests occur but their total area of distribution is very limited.

FIG. 3. Subalpine coniferous forest mixed with *Betula Ermanii* (Mt. Hodaka, Nagano Pref.).

Almost all forests of the subalpine zone outlined above come under the jurisdiction of the Forestry Agency of Japan. Prior to the end of World War II they were for the most part left in their natural state; however, since the end of the war, in response to the rapidly growing demand for constructional timber and pulp, and also to the development of road constructional methods and logging machinery, a programme of clear-cutting and reforestation with trees such as *Larix leptolepis* has been promoted by the Bureau. This policy continues to the present, and Japan's natural subalpine coniferous forests are thus in danger of further widespread destruction. Also, the soil type characteristic of this zone is podsol, whose properties are in general rather unsuitable for the regeneration of forest by artificial planting (ref. Chapter 1, section 3), although brown forest soil is not uncommon in some areas. Certain plantations in the subalpine zone, or parts of them, have therefore had to be abandoned, and these are usually covered gradually by the dwarf-bamboo shrub, *Sasa*.

Sasa itself is however not a principal shrub of the subalpine zone, this niche being occupied in most cases by *Vaccinium Vitis-Idaea, V. axillare, Oxalis Acetosella, Cornus canadensis, Arctous alpinus* var. *dilatatum, Sorbus* spp., etc. In other words, the subalpine coniferous zone corresponds closely with the so-called *Vaccinium–Picea* class, as mentioned above.

In addition to the above-mentioned forests, there are some special willow communities on the banks of certain rivers, etc. in the subalpine zone. The principal species of these swampy forests are *Chosenia bracteosa, Salix sachalinensis, Toisusu Urbaniana, Populus Maximowiczii*, etc. In one particular case, where a very strong typhoon hit the Hokkaido and Tohoku districts (the so-called Toya-maru typhoon of 26–27 September 1954), large areas of subalpine coniferous forest in central Hokkaido were destroyed. After the wind-felled trees had been cleared, new *Salix* and *Populus* forests began to appear along riverbanks and hillsides, perhaps as a result of the moisture conditions of soil in these localities. In fact, it appears that as a result of the loss of the forest trees (and therefore the transpirations from their leaves), there was a tendency for more swampy forests to dominate in place of the former coniferous forests. This phenomenon is observed quite commonly in subalpine regions of Japan.

2.2. Cool-temperate broad-leaved deciduous forest

Typically, the vegetation of this zone is beech forest, admixed with large deciduous oak trees (*Quercus crispula*) and having a thick, tall *Sasa* undergrowth. There are two beech species in Japan, *Fagus crenata* and *Fagus japonica*, of which *F. japonica* occurs only as a forest tree in the lower (or warmer) parts of the cool-temperate zone. The two species are thus in general separate

as regards their area of occurrence. General features of *Fagus crenata* forest are illustrated in Fig. 4 and 5.

The composition of these forests varies between the Pacific Ocean and Sea of Japan sides of Honshu. For example, *Abies homolepis* is usually mixed

FIG. 4. *Fagus crenata* forest of the cool-temperate zone (northern Kyoto Pref.).

FIG. 5. Undergrowth of *Fagus crenata* forest showing the creeping shrub-type conifer *Cephalotaxus drupacea* var. *nana* (lower centre), which is characteristic of snowy districts on the Sea of Japan side of Honshu.

with the beech trees in forests on the Pacific Ocean side, and may sometimes grow as pure local stands in this region, particularly on steeper mountain slopes. Also, another typical feature on the Pacific Ocean side is the occurrence of pure *Abies firma* stands (of the warm-temperate zone) directly abutting against the typical beech zone. On the Sea of Japan side, on the other hand, the beech forests are always bounded by broad-leaved evergreen forests of the warm-temperate zone. These differences may be a result of the heavy winter snowfall on the Sea of Japan side of Honshu, in contrast to the comparatively dry winter on the Pacific seaboard. This conclusion is clearly supported by the fact that the above *Abies* species, like *A. Veitchii* in the subalpine zone, are unable to withstand high incumbent snow pressures, but are highly resistant to cool, dry climatic conditions.

The undergrowth of the beech forests in most localities consists largely of *Sasa,* although there is a difference of species between the two sides of Honshu. For example, on the Pacific Ocean side, there are several dwarf-bamboo species which have a deep rhizome distribution, such as *Sasa nipponica, Sasaella ramosa, Pleioblastus* sp., etc., whereas on the Sea of Japan side only shallow rhizome forms such as *Sasa paniculata* and *S. kurilensis* occur widely throughout the zone. This difference may perhaps be a result of the protective effect of the thick snow-cover against the severe winter conditions.

Since the *Sasa* undergrowth is typically very tall and thick, it presents a serious obstacle to both natural and artificial regeneration of beech trees in the cool-temperate zone. There are often many young beech seedlings in the *Sasa* layer but they are unable to grow above the height of the *Sasa* plants, probably due to the comparatively low light intensity in the shrub layer. This phenomenon is similar to that observed in subalpine coniferous forests with a thick *Sasa* growth; indeed, the *Sasa* layer of the cool-temperate forests is undoubtedly the greatest hindrance to forestry, and reforestation, in this region.

Apart from the broad-leaved deciduous forests, there are also a few (but excellent) stands of conifers in the cool-temperate zone. Pure forests include the *Thujopsis dolabrata* forests of the Tsugaru and Shimokita Peninsulas (Aomori Pref.), the *Cryptomeria japonica* forests of Akita Pref., of Yanase (Kochi Pref.) and of Shikoku, the *Chamaecyparis obtusa* forests of the Kiso district (southwestern Nagano Pref.), and the *Sciadopitys verticillata* forests of Mt. Koya (Wakayama Pref.). These forests were formerly the property of family clans, except the *Sciadopitys* forests which belonged to a Buddhist temple; all were thus protected against indiscriminate tree-felling, etc. It appears that the original, natural forests were in fact essentially mixed conifer–beech communities. These, under the clans' policy of protecting the conifers but allowing beech and other decidous trees to be taken by neighbouring farmers (such

trees were considered of lesser value than the conifers), were gradually converted to pure coniferous stands.

The number of different broad-leaved deciduous tree species mixed with *Fagus* in the sub-high tree layer of typical cool-temperate beech forests is comparatively large. The principal components are certain species of *Acer* (such as *A. mono, A. micranthum, A. japonicum,* etc.), *Acanthopanax sciadophylloides, Fraxinus lanuginosa, Tilia japonica, Kalopanax innovans, Sorbus alnifolia, Magnolia obovata,* etc. The number of tree species in the shrub layer is also large, including certain species of *Viburnum* and *Euonymus, Ilex macropoda, Rhus trichocarpa, Benzoin umbellatum, Hamamelis obtusata, Magnolia salicifolia,* etc. On the Sea of Japan side where there is deep snow in winter, certain dwarf broad-leaved evergreen shrubs and creeping-form trees are found. Of these, *Camellia rusticana* is one of the most well-known species occurring in beech forests in snowy districts. In the case of the grass layer, the common species are *Shortia uniflora, S. ilicifolia, Pyrola japonica, Mitchella undulata, Plagiogyria Matsumureana, Blechnum niponicum, Carex conica,* etc.

In certain valleys where the soil is very moist or wet, special forest types consisting of *Aesculus turbinata* (see Fig. 6) and/or *Cercidiphyllum japonicum* are found. Also, species such as *Betula grossa, Alnus Maximowiczii, Cornus Kousa* and *Euptelea polyandra* form small groups of forests on wet soil, where *Symplocos chinensis* is one of the dominant undergrowth species.

The cool-temperate beech forests are very widely distributed through the islands of Japan. The northern limit is, as mentioned above, at the northern neck of the Oshima Peninsula (along the so-called Kuromatsunai line, through Suttsu, Kuromatsunai and Oshamanbe). The beech trees in this area are mixed with *Abies sachalinensis* var. *Mayriana*, while to the northeast

FIG. 6. Mixed stand of *Aesculus turbinata* and *Cryptomeria japonica* in the cool-temperate zone (northern Kyoto Pref.).

there are many mixed broad-leaved deciduous forests without beech trees. The principal species of these latter forests, which occur mostly in lowland areas, are *Quercus crispula, Q. dentata, Ulmus Davidiana* var. *japonica, U. laciniata, Tilia Maximowicziana, Acer mono* and *Kalopanax innovans*. The undergrowth is very similar to that of the beech forests. Some phytosociologists prefer to include these broad-leaved deciduous forests in the cool-temperate zone, although in the opinion of the present author they might best be regarded as a transitional forest type between the subarctic and cool-temperate zones. The difference in warmth index across the northern demarcation line of beech forests is not so great; however, there is an appreciable (and critical) difference in cold indices. For example, the cold indices for Hakodate and Suttsu are −29.4 and −26.5°C, respectively, whereas that for Sapporo is as low as −35.2°C. Clearly, the cold winter temperatures to the north of the Kuromatsunai line prevent the establishment of beech forests in this region.

In northern Honshu, the lower limit of the cool-temperate zone is marked by beech forests at an elevation of *ca.* 500–600 m a.s.l. In central Honshu, this limit is at a somewhat higher elevation than in the north. Thus, almost all mountains in these districts support beech forests. The timber available from them is not particularly suitable as constructional material, and therefore the beech trees remained virtually uncut until the early 1900's, except for the limited amounts used for charcoal manufacture. However, the exploitation of hardwood trees, in preference, created a serious situation in Japanese forestry during the mid-1900's, due mainly to the fact that their total biomass (in natural forests) was appreciably less than that of beech trees (see Table 1). From the mid-1900's onwards, therefore, beech logs were gradually utilized on a wider scale for pulp, furniture, and as housing material, while during World War II beech wood became an important item for the manufacture of light aircraft. For these reasons, destructive clear-cutting

TABLE 1. Forest resources in Japan in the mid-1900's (unit, 1000 m^3)

		Coniferous trees	Broad-leaved trees	All trees
Natural forests	National forests	261,230	474,777	736,007
	Private forests	160,268	420,533	580,801
	All forests	421,498	895,310	1,316,808
Man-made forests	National forests	116,285	10,414	126,699
	Private forests	450,204	3,933	454,137
	All forests	566,489	14,347	580,836

of natural beech forests was begun, and the resultant open areas were replanted with *Cryptomeria* or larch trees. However, in an appreciable number of cases it proved impossible to establish man-made coniferous forests on these sites, and they were quickly reduced to unproductive *Sasa*-dominated land. This situation has continued to the present, and is accompanied by a considerable rise in the demand for, and price of, beech logs as pulp material. In southern parts of Japan, the lower limit of the beech forests is *ca.* 1000 m a.s.l. The best stands of beech trees are found on the Sea of Japan side of Honshu, particularly in Niigata Pref., Yamagata Pref., and Akita Pref. in the north.

As a whole the forests of the cool-temperate zone show lesser productivity than other forest zones. This is due primarily to the fact that the dominant species are deciduous trees, largely *Fagus*. Such trees do not carry out photosynthesis in winter, and their net production is thus low compared with evergreens, etc. (see Table 2).

TABLE 2. Net production of various forest types in Japan

Forest type	Mean net production (ton/ha.yr)	No. of materials sampled
Deciduous broad-leaved (subalpine and cool-temperate)	8.7 ± 3.0	64
Deciduous coniferous (larch, subalpine)	10.1 ± 4.4	44
Evergreen coniferous†1 (subalpine and cool-temperate)	13.5 ± 4.2	46
Pine†2 (cool- and warm-temperate)	14.8 ± 4.1	52
Cryptomeria (cool- and warm-temperate)	18.1 ± 5.6	92
Evergreen broad-leaved†3 (warm-temperate)	18.1 ± 4.9	21
Mean/total	13.9 ± 5.9	319

†1 Excluding pine trees and *Cryptomeria*. †2 Excluding the shrub-type pine, *Pinus pumila*.
†3 Excluding *Acacia* sp. (introduced from Australia and newly planted in Kyushu).

2.3. WARM-TEMPERATE BROAD-LEAVED EVERGREEN FOREST

The typical composition of forests of this zone is a mixed community of very many broad-leaved evergreen species, dominated by *Quercus* spp. and *Castanopsis* spp. in the upper canopy, and succeeded by a complex vertical profile containing several distinct tree layers. In general, two or three high tree layers can be identified in addition to the shrub layer. A typical forest of this zone is illustrated in Fig. 7.

Southwest Japan, i.e. the Chugoku district, Shikoku and Kyushu, was

FIG. 7. Broad-leaved evergreen forest of the warm-temperate zone (Kyushu).

formerly very widely covered by such forests. However, the warm-temperate zone in Japan has been extensively exploited since ancient times, following the introduction of primitive forms of agriculture. At first, shifting cultivation was undertaken on the slopes of hills in areas of evergreen forest. Then, with the introduction of rice cultivation from China via Korea, agricultural activity spread widely across the greater part of lowland Japan. As a result, the broad-leaved evergreen climatic climax forests of the warm-temperate zone were largely destroyed. At present, almost all lowland areas in this zone have been converted to rice paddies, farmland, etc., and the only remnants of the original forests are found in the precincts of Buddhist temples and Shinto shrines. The situation thus bears a close resemblance to lowland areas of the cool-temperate zone. Moreover, it is estimated that more than 80% of Japan's population now lives in the warm-temperate zone. The original hillside forests of agricultural areas have been progressively modified, not only through shifting cultivation but also through the utilization of their accumulated leaf litter as manure. These hillside regions are now occupied mostly by the Japanese red pine (*Pinus densiflora*) and by deciduous broad-leaved forests. A few areas of truly virgin forest are still present in the southern part of Kyushu (Osumi Peninsula) and on the small island of Yakushima in the Satsuma Islands, although even here the natural forests are being rapidly cut back in order to satisfy the ever-increasing demand for wood pulp.

The species composition of warm-temperate broad-leaved evergreen forests varies according to the local soil and climatic conditions. For example, where the soil is sandy and where there are strong sea breezes, Japanese black pine (*Pinus Thunbergii*) forests with an undergrowth of *Eurya emarginata*, *Pittosporum Tobira*, etc. occur. On the slopes of coastal hills, *Shiia* forests composed of *S. Sieboldii* (*Castanopsis cuspidata* var. *Sieboldii*) and *S. cuspidata* (*Castanopsis cuspidata*), mixed with *Machilus Thunbergii* and having an undergrowth of *Benzoin erythrocarpum, Litsea aciculata, Camellia japonica*, etc., are found. These forests occupy wet habitats, but are replaced by *Quercus phillyraeoides, Rapanea neriifolia*, etc. forests on ridges and in dry habitats. A second type of highland forest, consisting mainly of evergreen oak species, also occurs widely. The principal trees are *Quercus gilva, Q. salicina, Q. acuta, Q. glauca Q. myrsinaefolia* and *Q. sessilifolia*, which may be mixed with *Cinnamomum japonicum, Distylium racemosum, Symplocos prunifolia*, etc. In some localities, evergreen coniferous stands are present in the forests, the principal species being *Abies firma, Tsuga Sieboldii, Podocarpus macrophylla, P. Nagi, Torreya nucifera, Picea polita, Cryptomeria japonica, Sciadopitys verticillata* and *Chamaecyparis obtusa*. Of these, *Abies* (see Fig. 8) and *Tsuga* forests are the most common. The location of such coniferous stands is usually limited to stoney or steeply sloping mountainous habitats, and they generally comprise rather small groups of trees within the broad-leaved evergreen forests, except on some south-facing slopes where more extensive coniferous forests occur.

Of the species adapted to drier habitats, *Quercus phillyraeoides* has small, thick, leathery leaves that prevent excessive transpiration. It thus forms hard-leaved forests similar to the *Quercus ilex* stands around the Mediterranean Sea. The distribution of such hard-leaved forests in Japan is very limited, being restricted mainly to low coastal hills of the Seto Inland Sea and the

FIG. 8. *Abies* forest of the warm-temperate zone (Mt. Hiei, Kyoto).

southern, Pacific-facing coasts of the Chubu district, Shikoku and Kyushu. The coast of the Seto Inland Sea has one of the lowest annual rainfalls in Japan, plus a comparatively high evaporation rate, while the Pacific-facing coasts (although with high summer precipitation) become dry in winter, and show very high evaporation rates throughout the year due to the continuous onshore winds from the Pacific. These sea winds also cause drying of the forest soil, and thus serve to create conditions suitable for the appearance of hard-leaved evergreen oak species. The *Quercus phillyraeoides* forests are discussed again in section 3.1 below on coastal vegetation.

At low elevations on many hills in the warm-temperate zone, red lateritic soils are found (see Chapter 1, section 3). Forests in these areas have long served farmers as a source of material for the production of wood-ash, etc., which is in turn employed as mineral fertilizer. Trees, shrubs and leaves have been cut and gathered on a wide scale, burnt in the farmers' houses, and the ash then scattered on the land. As a result, the original forests have been lost or severely damaged, and many areas have been reduced simply to unproductive land. The typical tree is now *Pinus densiflora*, which occurs as rather poor stands on hills and plains in southern and southwestern Japan.

The warm-temperate zone also contains several famous areas of forestry with man-made *Chamaecyparis obtusa* and *Cryptomeria japonica* forests. These are discussed in detail in section 4.

2.4. Subtropical broad-leaved evergreen forest

The subtropical zone in Japan includes the long chains of small islands (the Nansei (Ryukyu) and Ogasawara Islands) which lie to the south of Japan proper between the latitudes 24° and 30°N. However, as a result of widespread land cultivation, and also due to heavy damage during World War II, most of the original, natural vegetation has been lost. Only Iriomote Is., which is now protected as one of Japan's National Parks, contains good examples of the natural subtropical forests of this region.

In the case of Ogasawara, the population was evacuated to Honshu just prior to World War II, and the islands remained virtually uninhabited (apart from some members of the US Armed Forces) until their recent return to Japan. The vegetation was thus left essentially in an undisturbed state for about 20 yr, and it was expected that the original subtropical flora would reestablish itself. However, one exotic tree species of the Leguminosae (*Leucaena leucocephala*), which was seeded as a camouflage plant for Japan's fortresses, became dominant, and the recovery of the natural vegetation was greatly inhibited. At present, the primary forest types are represented in only a few, limited areas.

The natural forests of the subtropical zone can be classified into two broad

groups. The first constitutes the coastal vegetation, and is composed of *Cycas revoluta, Pinus lutchuensis, Ficus Wightiana, F. retusa, Pandanus boninensis, Livistona subglobosa*, etc. on Okinawa Is. Also, at the mouths of rivers, where the sea- and river-water merge, swamp forests containing mangrove species are found. The principal species are *Brugiera conjugata, Kandelia Candel* and *R. mucronata*. The second forest type is mountain forest, which bears a close resemblance to that of the warm-temperate broad-leaved evergreen zone. The principal species are *Shiia Sieboldii, Quercus phillyraeoides, Q. salicina, Machilus Thunbergii, Elaeocarpus japonicus*, etc., sometimes mixed with *Podocarpus macrophylla*, the wooden fern *Cyathea boninsimensis*, and the climbing plant *Pandanus*, etc. (see Fig. 9). This flora is well represented on Iriomote Is.

There are no high mountains in the Ryukyus (the maximum elevation is only about 500 m a.s.l), and the climate is thus oceanic in most respects. The coastal forests have largely been destroyed, and are now replaced by agricultural and residential land, although occasional bushy areas with *Cycas, Pinus, Pandanus*, etc. do remain. In populated areas of Okinawa Is., *Garcinia subelliptica* (Japanese name, *Fukugi*) is planted along the coast as a windbreak against the many typhoons which hit the island destroying crops and houses. *G. subelliptica* was introduced from China and now forms part of the characteristic landscape of the Okinawan coast. Recently, *Casuarina equisetifolia* has also been planted as a windbreak.

The original mountain forests in many parts of the Ryukyus have been destroyed more or less completely, and converted to pineapple plantations or artificial pine forests. However, where agricultural activity has been abandoned for some reason, the vegetation has reverted first to *Miscanthus* grassland and then gradually to *Pinus lutchuensis* forests. Areas for artificial pine plantation are usually prepared by firing, followed by the planting of pine

FIG. 9. Broad-leaved evergreen forest of the subtropical zone showing the wooden fern *Cyathea* (Ishigaki Is.).

seedlings. However, the firing technique is particularly dangerous in regions of naturally regenerated forest, and may often spread outside its intended limits. The natural and artificial pine forests of Okinawa Is. are illustrated in Fig. 10 and 11.

Although *Pinus lutchuensis* and *Casuarina equisetifolia* are at present the most important species of tree plantations, *Podocarpus macrophylla* and *Cryptomeria japonica* were previously the principal trees of man-made forests. These latter

FIG. 10. Natural subtropical pine forest (Okinawa Is.).

FIG. 11. Subtropical forest on Okinawa Is. showing pine plantations with *Miscanthus* undergrowth in the foreground.

species, in contrast to the wind-resistant *P. lutchuensis* and *C. equisetifolia*, were grown because of their ability to withstand termite damage. *Podocarpus macrophylla* plantations are still present on Ishigaki Is. (of the Yaeyama archipelago), although their total extent is very limited. *Cryptomeria japonica* plantations are present in some parts of central and northern Okinawa Is. In the case of *C. japonica*, seedlings were propagated from cuttings and then planted directly into moderately moist forest soil in narrow valleys having natural protection against high winds and typhoons. It is thought that the method of propagation of *C. japonica* by cutting was probably introduced from Kyushu, since this species is not a native tree of Okinawa Is.

3. Special/Local Forest Types

3.1. Coastal forest

In Japan, several special forest types are found in coastal regions where there are strong prevalent winds (see also Chapter 5). The most important types occur on sandy seashores, at the mouths of large rivers, on the shores of large lakes, and on the windward sides of hill-slopes facing the open sea, etc. Characteristically, there is a high evapo-transpiration rate from these forest stands, and the species which comprise them display a strong resistance to desiccation.

In southern Japan, *Pinus Thunbergii* forests with an evergreen shrub layer occur on both the Pacific and Sea of Japan coasts (see Fig. 12), while in central districts, such as the eastern Kanto coast and Sea of Japan side of the

FIG. 12. Coastal *Pinus Thunbergii* forest that has been treated by selective cutting (Kagawa Pref.).

FIG. 13. *Quercus phillyraeoides* coastal forest (Kochi Pref.).

Hokuriku district, the coastal forests are mixed stands of *P. Thunbergii* and *P. densiflora*. In northern Honshu, pure stands of *P. densiflora* with a broad-leaved deciduous undergrowth occur, while in Hokkaido the coastal pine species are generally replaced by deciduous oaks such as *Quercus crispula* and *Q. dentata*. One exception is the natural occurrence of *Pinus densiflora* on the slopes of Mt. Komagatake. Also, in many parts of Japan (including Hokkaido), *P. densiflora* and other coastal forests have been replaced by artificial stands of *P. Thunbergii*, since the latter displays especially strong resistance to severe, salty winds and desiccation, and is able to flourish in poor sandy soils.

Another special coastal-hill forest type is *Quercus phillyraeoides* forest (see Fig. 13), described briefly above in section 2.3. The indivudual trees are not tall (up to *ca.* 10 m in height). The branching of the upper canopy is very complex and interwoven, providing a comparatively smooth (or streamlined) surface towards the wind. *Rapanaea neriifolia* commonly occurs in the sub-high tree layer, while *Pittosporum Tobira* and other evergreen tree species are found in the shrub layer. The charcoal produced from the timber of *Q. phillyraeoides* (*Bincho-zumi* in Japanese) is of very high quality and used domestically as fuel.

3.2. Dry-land forest

Although Japan in general has a high annual precipitation, certain areas are comparatively dry, and thus show several special forest types. For example, *Pinus densiflora* forests are very common in the dry regions (including small islands) around the Seto Inland Sea. These districts contain abundant exposures of heavily weathered granite, and the resultant poor soil has helped

to ensure a wide distribution of *P. densiflora* forests, a fact which is no doubt related to the ancient and modern development of civilizations in the region. Also, in northern Japan, in the upper parts of the dry and mild Kitakami river basin, there are several famous areas of natural *P. densiflora* forest. In particular, the trees at the foot of Mt. Iwate are noted for their fine, straight trunks, and as a whole the forests provided excellent timbers for constructional purposes.

In dry districts of Nagano Pref. and Yamanashi Pref., there are numerous natural stands and plantations of *Larix leptolepis*, that is in addition to the common *Pinus densiflora* forests. *L. leptolepis* is able to withstand the dryness and coolness of winter in these areas, and occurs principally in the subalpine zone of the Japanese Alps, on Mt. Asama, Mt. Yatsugatake and Mt. Fuji. As mentioned above, this larch species has been widely used for reforestation and artificial plantations in Japan since the war, and in fact, following the introduction of *L. leptolepis* seeds into Europe and America in the early 1900's, it has also enjoyed considerable popularity there. It is said that hybrid trees of the Japanese larch and European larch show excellent growth and resistance to disease.

Around the dry, low-lying Ishikari Plain in Hokkaido, there were formerly wide areas of special, natural deciduous oak forests, consisting mainly of *Quercus crispula*. However, since the wood of *Q. crispula* makes excellent timber, most of the oak trees were felled before or during World War II. In the early days, a large part of the timber was exported to Europe (particularly Britain) for use as coffin boards, etc. Although *Q. crispula* is, as mentioned above, widely distributed throughout the beech forests of the cool-temperate zone, the quality of the wood in Honshu is inferior to that of Hokkaido, no doubt due in part to the differences in climatic and environmental conditions.

3.3. Transitional forest

Transitional forests occur at the boundaries between principal forest zones. For example, at the junction of the cool-temperate and warm-temperate zones, natural broad-leaved deciduous forest can be found in several localities. This special forest type consists of tree species such as *Zelkova serrata, Quercus acutissima, Q. variabilis, Q. serrata, Castanea crenata, Aphanathe aspera, Celtis sinensis, Styrax japonica*, etc. According to Kira, such forests are distributed in regions where the summer temperature is sufficiently high for the growth of broad-leaved evergreen trees, but the winter temperature is so low as to preclude their actual establishment. *Fagus crenata*, the type species of cool-temperate forest, is entirely absent from these transitional forests, and they are therefore generally assigned to the warm-temperate zone. *Fagus japonica*, the second Japanese beech species, occurs as transitional forests towards the

southern (warmer) limit of the cool-temperate zone. The trees provide excellent wood for use as fuel, and therefore in intermediate climatic districts, *F. japonica* was formerly grown in artificial forests as coppice stands. Of the other species listed above, *Quercus acutissima* has been planted for the purpose of wood-charcoal manufacture, while natural *Zelkova serrata* stands still remain as part of the famous Musashino Forest of the Kanto district. Some other *Quercus* stands also remain in various parts of the Kanto Plain.

Other transitional forests are found in the cool-temperate zone, at or near its margin with the subarctic (or subalpine) zone, and in the subalpine zone at or near its margin with the alpine zone. For example, broad-leaved deciduous forests occur in certain localities in the upper (cooler) parts of the cool-temperate zone. The principal tree species are *Acer mono, Ulmus Davidiana, U. laciniata, Quercus dentata, Q. crispula, Tilia japonica*, etc. This transitional forest type is particularly common in Hokkaido, occurring between the cool-temperate *Fagus* forest and subarctic *Abies–Picea* forest regions.

In the subalpine zone, birch forests (*Betula Ermanii*) occur along the margin of typical evergreen coniferous forests, and their upper parts are usually defined as the timber line. Above this, certain mixed deciduous communities of creeping trees such as *Alnus Maximowiczii, Sorbus sambucifolia, Betula Ermanii*, etc. are found. These represent a transitional zone bordering on the higher, alpine zone.

In general, transitional forests of the types described above occur at points along the margins of major forest zones where there is a wide difference in summer and winter temperatures, i.e. the climate is more "continental" in character. The principal factor controlling the distribution of transitional communities is nevertheless the winter temperature.

3.4. Swamp forest and mangrove forest

Throughout the temperate zone there are occasional tree communities, composed of species such as *Alnus japonica* and *Salix gracilistyla* or *S. sachalinensis*, that thrive in areas of wet soil, normally along riverbanks, etc. However, due to the fact that these trees lack the aerial roots of true swamp species, the communities cannot strictly be regarded as swamp forest.

The soil of typical swamp-lands is rich in organic matter and low in oxygen, the latter being consumed in the decomposition of accumulated organic debris. Swamp trees, such as the mangrove, thus typically have aerial roots.

Mangrove forests are found in subtropical and tropical areas at the mouths of rivers where sea- and fresh-water converge (see Fig. 14). In Japan, they are limited to the Nansei (Ryukyu) archipelago, which lies in the subtropical zone. The northern limit of mangrove species is on Yakushima Is., just to the south of Kyushu and at the southern limit of typical warm-temperate forests.

FIG. 14. Riverside mangrove forest of the subtropical zone (Okinawa Is.).

On Yakushima Is., the small plant height gives rise to a shrubby community, which is not strictly mangrove forest. On the nearby island of Tanegashima, the single mangrove species, *Kandelia Candel*, also ocurrs. The plant height of these subtropical communities increases southwards, as does the number of mangrove species. For example, on the islands south from Okinawa Is., at least three species form swamp forests at the mouths of rivers. Nevertheless, even here the tree height is only about 1/3 to 1/2 that of tropical mangrove forests. The principal genera on the Yaeyama Islands are *Brugiera, Rhizophora, Kandelia, Lumnitzera*, etc.

The different mangrove genera and species show a clear pattern of distribution along the river banks, both away from the river mouths and away from the shore. In general, *Kandelia Candel, Rhizophora mucronata* and *Bruguiera conjugata* forests occur successively upstream, while species of *Rhizophora, Kandelia* and *Bruguiera* occur successively inland away from the submerged parts of the shore. The dominant factor deciding this pattern of distribution is the sea- to fresh-water ratio. The swamp forests also contribute significantly to erosional control along the shores and mouths of the rivers.

In Southeast Asia, mangrove trees are employed for charcoal manufacture, and tannic acid for leather-making is extracted from the bark. In Japan, the small size of the mangrove shrubs and trees renders them unsuitable for utilization in these ways, although in restricted areas they are used as fence material. Some of the mangrove forests on Okinawa Is. are now protected; however, due to road construction programmes, the projected industrialization, and the pollution that is likely to result, it appears that the already limited examples of this rather rare forest type will suffer seriously in the future.

4. Man-made Forests

Man-made forests in Japan are here divided into two types according to their regeneration method, i.e. artificial planting and seeding or natural regeneration. The former method is utilized for creating high forests, such as "sugi cedar" forests (*Cryptomeria japonica*), "hinoki cypress" forests (*Chamaecyparis obtusa*), "matsu" forests (*Pinus densiflora* and *P. Thunbergii*) and "karamatsu" forests (*Larix leptolepis*). Sugi and hinoki plantations are the most common in Japan, and the young plants used for creating new forests are generally nursery-grown seedlings, although in some cases tree cuttings are employed. For example, in Kyushu there is widespread use of cuttings in the case of *Cryptomeria japonica*, since this species easily roots from cuttings. The natural regeneration method is employed in most cases for creating coppice forests, where naturally germinated sprouts from cut stumps are cultivated. Such coppice forests are widely distributed in temperate parts of Japan where broad-leaved trees are dominant. The wood is often used for fuel or charcoal manufacture, although recently there has been some decline in coppice forests due to the increasing use of electric power and kerosene for heating and cooking. Occasionally, the natural regeneration method is also employed for creating high forests. For example, *Pinus densiflora* and *P. Thunbergii* forests are often regenerated from natural seedlings.

The present extent of man-made forests in Japan, including artificially planted and naturally regenerated forests, is summarized in Table 3 (*cf.* previous data in Table 2). It can be seen that at present man-made forests constitute about 35% of the total forest area, or about 33% of the total stem volume. The ratio of coniferous to broad-leaved forests is 1.03:1 (as stem volume) for all forests in Japan, compared with the very high ratio of 46.4:1 for man-made forests alone. Clearly, plantations of coniferous trees form the major proportion of man-made forests in Japan.

National forests constitute about 32% of the total forest area, or 42% of the total stem volume. The wood volume per unit area in these forests is thus on average appreciably higher than that in the private forests. Man-made forests form about 24% of the total area of national forests but only 15% of their total stem volume, the reason being that the man-made plantations of national forests are still young, i.e. when compared with those of private forests.

The total area occupied by the islands of Japan is about 37×10^6 ha. Forests cover a very high percentage (68.1%) of this, a figure which ranks among the highest in the world. Table 4 illustrates the extent of present forest areas in various countries with major forestry operations. When considered in

TABLE 3. Area and standing volumes of trees in present-day Japanese forests (1971)† (units, 1000 ha and 1000 m³, respectively)

		Area	Standing volume			
			Coniferous trees	Broad-leaved trees	All trees	Bamboo
Man-made forests	National forests	1951	116,654	9 859	126,513	0
	Private forests	6912	534,446	4 169	538,615	148
	All forests	8863	651,100	14,028	665,128	148
All forests	National forests	8028	379,192	476,304	855,496	—
	Private forests	17,195	674,183	549,393	1,223,576	—
	All forests	25,223	1,053,375	1,025,697	2,079,072	—

† Data from *Japanese Governmental Forestry Statistics* (Japanese), 1972.

TABLE 4. Present status of forest-land in the world (unit of area, 10^6 ha)

Country	Total land area	Forested land area	Forest cover (%)
Korea	9.76	6.69	68.6
Japan	37.0	25.2	68.1
Malaysia	36.3	24.0	66.1
Finland	33.7	21.7	64.4
Thailand	51.4	28.1	54.7
Sweden	45.0	23.4	52.0
Canada	998	443	44.4
U.S.S.R.	2240	910	40.6
Austria	8.38	3.35	40.0
U.S.A.	936	307	32.8
West Germany	24.7	7.15	28.9
Norway	32.4	8.99	27.7
New Zealand	26.9	7.37	27.4
East Germany	10.8	2.95	27.3
Australia	770	209	26.9
Poland	31.2	7.68	24.6
Switzerland	4.13	1.01	24.5
France	55.2	11.6	21.0
All world	13,528	4126	30.5

terms of the population, however, Japan's forests represent an extremely small average area (about 0.2 ha) per capita, a figure which is among the lowest in the world. Also, with the increasing population and heavy demand for extra constructional timber and wood pulp, Japan is now forced to im-

TABLE 5. Area distribution of principal private and national forests in Japan (unit of area, 1000 ha)

District	Private forests		National forests		Total area
	Area	% of total	Area	% of total	
Kanto	1621	66.5	816	33.5	2437
Chubu	3419	75.7	1095	24.3	4514
Kinki	2147	96.1	88	3.9	2235
Chugoku	2178	94.0	138	6.0	2316
Shikoku	1190	87.4	183	12.6	1373
Kyushu	2146	81.7	480	18.3	2626
Okinawa	94	70.1	40	29.9	134
Hokkaido	2278	42.4	3100	57.6	5378
Tohoku	1927	53.7	1661	46.3	3588
Total	17,000	69.1	7601	30.9	24,601

port more than 55% of its logs. The principal sources are Southeast Asia, the U.S.S.R., Canada and the United States.

As shown above, almost 70% of Japan's forests have private ownership. These forests are concentrated mostly in southern areas (see Table 5) which together support over 80% of the population. The majority of the privately owned forests are the property of farmers, and as such have long suffered considerable damage through the use of wood charcoal and leaf-litter as mineral fertilizers. This is especially true on hills, etc. close to the intensely cultivated valleys and plains. However, in cases where shifting cultivation was replaced by forestry a long time ago, famous forests still remain. These include the sugi forest reserves of Obi (Miyazaki Pref.) and Hita (Kumamoto Pref.) in Kyushu, Chizu (Tottori Pref.), Yoshino (Nara Pref.) and Kitayama (Kyoto Pref.) in western Honshu, Tenryu (Shizuoka Pref.) in central Honshu, and Nishikawa (Saitama Pref.) and Sambu (Chiba Pref.) in the Kanto district (see Fig. 15).

Apart from the various coniferous trees mentioned above, other species are grown in artificial plantations. These include the broad-leaved "konara" (*Quercus serrata*) and "kunugi (*Q. acutissima*), which are cultivated by the natural regeneration method. Also, in the warm-temperate zone, the "shii" (*Castanopsis cuspidata*) and certain evergreen oak species form man-made coppice forests. Table 6 summarized the present status of man-made forests in Japan. The two dominant trees, sugi and hinoki, are characterized respectively by their excellence and ease of growth and their provision of good quality constructional timbers. Both are readily obtainable as seedlings from nurseries, and both can be planted and grown in open fields with a good rate of survival.

FORESTS:

1. Aomori (*Thujopsis*)
2. Akita (Sugi)
3. Nambu (Pine)
4. Kiso (Hinoki)
5. Yanase (Sugi)
6. Yakushima Is. (*Yaku-sugi*)

FORESTRY RESERVES:

① Aizu (Broad-leaved deciduous)
② Sambu (Sugi)
③ Nishikawa (Sugi)
④ Nagano (Larch)
⑤ Tonami (Sugi)
⑥ Noto (*Thujopsis*)
⑦ Tenryu (Sugi)
⑧ Owase (Hinoki)
⑨ Yoshino (Sugi)
⑩ Tane (Sugi)
⑪ Kitayama (Sugi)
⑫ Chizu (Sugi)
⑬ Daisen (Pine)
⑭ Oki (Sugi)
⑮ Kito (Sugi)
⑯ Kuma (Sugi)
⑰ Hita (Sugi)
⑱ Kikuchi (Sugi)
⑲ Obi (Sugi)

FIG. 15. Important forests and forestry reserves in Japan.

TABLE 6. Extent of man-made forests in Japan by species and regeneration method†

Regeneration method	Coniferous trees	Broad-leaved trees	Area (ha)	(%)
Artificial	Sugi		127,967	24.9
	Hinoki		101,212	19.7
	Matsu		38,004	7.4
	Karamatsu		40,563	7.9
	Others		44,369	8.6
		All species	2250	0.4
			354,365	68.9
Natural	All species		159,749	31.1
			514,114	100.0

† Data for major forestry areas, 1970.

4.1. ARTIFICIAL REGENERATION

4.1.1. Sugi (Cryptomeria japonica) and hinoki (Chamaecyparis obtusa) forests

As mentioned above, the sugi is very widely distributed in Japan. It occurs from the border of Akita Pref. and Aomori Pref. in northeast Honshu to the island of Yakushima south of Kyushu, so covering most of Japan's principal islands except Hokkaido, the dry areas of the Seto Inland Sea, and southern parts of Kyushu. The principal natural distribution zone is on hill and mountain slopes facing the Sea of Japan side of the islands, although it is found in several areas facing the Pacific. In the cool-temperate zone, the sugi is mixed with beech trees, especially on the Sea of Japan side, while in the warm-temperate zone it is usually mixed with evergreen oak species. In such forests, the sugi occurs as comparatively small groups of trees that propagate as seedlings or vegetatively.

The Akita national forest consists of very fine stands of naturally regenerated sugi more than 250 yr old (see Fig. 16). These forests were protected by the Akita feudal clan more than 100 yr ago. Also, in the Kochi national forest, there are some very old, fine sugi forests which were formerly under the protection of the Tosa clan. The areas of sugi forest on Yakushima Is. contain large trees several thousands of years old (called *Yaku-sugi*).

Naturally distributed sugi trees can be divided into two types according to their life-form and mode of propagation. On the Sea of Japan side, where deep snow accumulates during the winter, the so-called *Ura-sugi* is dominant. It has dark green leaves, which project from the twig axis at a low angle and curl inwards. The form of the tree crown is narrow and sharp, and in many cases the colour of the heartwood is reddish. Such *Ura-sugi* trees are easily able to propagate vegetatively, by germinating vegetative roots when their lower branches are bent under heavy snow cover and pressed into the uppermost layer of the soil.

FIG. 16. Old, mature sugi plantation (Akita Pref.).

The Pacific Ocean side of Japan is occupied primarily by the less abundant *Omote-sugi*. This has light green leaves, which project from the twig axis at a high angle and have a relatively straight form. The tree crown is wide and oblique, and the heartwood is comparatively dark in colour. Vegetative propagation in the *Omote-sugi* is always very difficult.

These two types of sugi are not found exclusively on the Sea of Japan and Pacific Ocean sides, respectively, but rather, they represent two "ecotypes" whose relative distribution is governed by climatic and other conditions, and in particular by the amount of snow cover. Fig. 17 illustrates the relationship between snow depth, annual precipitation and proportion of *Ura-sugi* in sugi forests of Japan.

The number of plants per unit area used in artificial plantations varies very much according to the locality. For example, in the Yoshino forestry reserve of Nara Pref., about 10,000–12,000 seedlings are planted per hectare, whereas in the Obi forestry reserve of Miyazaki Pref., only 1000–1500 cuttings are planted per hectare. The origin of these differences is not fully clarified, but appears to be related to differences in the changeover from shifting cultivation to forestry, which occurred in most of the famous localities. In general, the density of planting was very low in the early stages of changeover. The first trees were usually planted about 3–4 yr after commencing cultivation, when the nutrients in the soil were at low levels due to intensive farming, but the open space on these young plantations continued to be used as farmland for a variable period. In the case of Yoshino, the initial planting density was also low, but soon it became the rule to plant at a very high

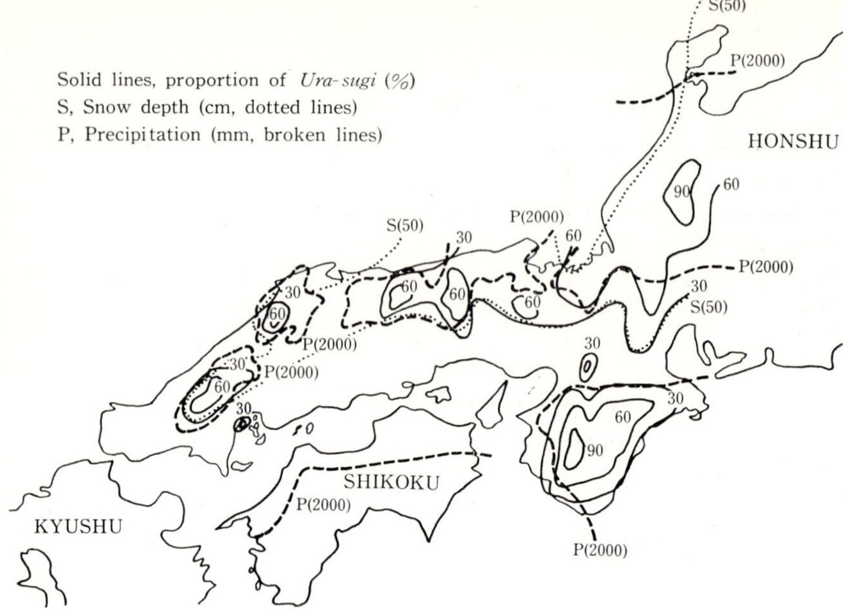

FIG. 17. Relationship between mean maximum snow depth, annual precipitation and proportion of *Ura-sugi* forest in southeast Japan.

density, in order to satisfy the economic demands of the forestry management.

The natural distribution of the hinoki in Japan is also very wide, although somewhat more restricted than that of the sugi. It occurs more towards the Pacific Ocean side of Honshu, with a northern limit to the north of the mountain ranges of the Kanto district and a southern limit of Yakushima Is. The most famous natural hinoki forests are those of Kiso-dani (Nagano Pref.), which were formerly controlled by the Owari clan of Nagoya and maintained as almost pure forests, forming their most important property.

The hinoki tree is easily subject to snow damage, which accounts for its limited extent as natural forests and artificial plantations on the Sea of Japan side of Honshu. The principal kind of damage due to snow is called "*Roshi-byo*", which is the leaking of sap from damaged joints of branches. When the deposited snow is sufficient to cover the tree crown, the branches tend to be pulled down as the snow settles. The force of this settling is often sufficient to break or damage the junctions between branches or between branches and the main stem, causing *Roshi-byo* and perhaps even tree death.

There are many hinoki plantations on the Pacific side of Honshu, among which the most famous is that of Owase (Mie Pref.), which ordinarily produces small logs for use as pillars. In some cases, sugi trees are planted on the lower parts of the mountain slopes, and the hinoki on the upper slopes,

while *Pinus densiflora* is grown on ridges where the moisture content of the soil is low. The reason for this pattern of planting is that the sugi prefers the moist soil of valleys, whereas the hinoki is able to grow on comparatively dry soil. In the central parts of some mountain slopes, the two trees are grown together. In general, the speed of growth of the sugi is greater than that of the hinoki, but the latter produces higher quality logs.

Some famous, specific forestry reserves in Japan will next be described.

(a) *The Kitayama forestry reserve.* The Kitayama forests are situated in the proximity and to the north of Kyoto, along the branch river (Kiyotaki) of the Katsura River. From the viewpoint of forestry, this area has several unique features. Two kinds of polished logs are produced, i.e. polished pillar logs and small logs for use as rafters. The forests yielding these two types of logs are treated in different ways. Polished pillar logs are produced by the clear-cutting of plantations of uniform age or by sprouting from living stumps, whereas the small rafter logs are produced only by the sprouting method. Polished logs are also manufactured in the Yoshino forestry reserve, but the quality of the Kitayama logs is higher and the forestry more intensive there.

The two kinds of polished logs are employed in the construction of special Japanese houses by the so-called *Sukiya-zukuri* technique, as exemplified by the small log huts used for the tea ceremony, and as alcove posts (i.e. the ornamental tokonoma pillars of Japanese houses). These polished logs are very expensive in comparison with other constructional timbers, and the special forestry techniques employed for their manufacture were developed in order to satisfy the special demands of the old capital of Kyoto.

The centre of the Kitayama forestry region lies between the villages of Nakagawa and Ono-go. This area is surrounded by steep mountains, the lower slopes of which are used as special sugi plantations (see Fig. 18). The

FIG. 18. Artificial stand of *Kitayama-sugi* (Kyoto Pref.).

upper slopes are covered with coppice trees and *Pinus densiflora* forests. The total area of the sugi plantations is much less than that of other famous private forest areas, being only about 800 ha. Of this, approximately 60% is used for pillar log production and the remainder for rafter production. The planting density of trees in the uniform-aged plantations used for pillar log production is very high (>10,000 trees/ha). Pruning begins at an early stage (within a few years of planting) and is carried out intensively every two years until the felling period. The purpose of this pruning is to produce straight, untapered, knot-free stems, and also serves to limit trunk diameter growth. In general, a period of over 30 yr is required for the sugi trees to reach 10 cm in diameter at breast-height. Such dense, uniform forests thus do not require any periodic thinning, since the pruning process tends to keep the size of the tree crown almost constant.

For the production of the smaller rafter logs, a special form of the sugi tree (known as *Dai-sugi*) is prepared. Each *Dai-sugi* consists of many stems, which grow from a single cut stump. The planting density of such trees is thus much lower than in the case of the clear-cutting system, being only about 4000–5000 trees/ha. At the first pruning (in the preparation of the *Dai-sugi*), several of the branches in the lower crown are left intact, but the tops of the remaining branches are cut and the stumps left to sprout. The sprouts so produced are thus all of the same age. The pruning process is then repeated periodically to allow the growth of a series of sprouting stems. These are later cut from the *Dai-sugi* by the so-called selective cutting method, whereby stems of marketable size are taken in turn as they grow up. Sprouting from the original lower branches may continue for a very long time, the oldest living *Dai-sugi* trees of Kitayama being estimated at more than 200 yr old. They sometimes have more than 50 sprouting stems each. This type of sugi is now often used as an ornamental tree for Japanese traditional-style gardens, i.e. quite apart from its use in producing polished logs.

The yields and production process of polished logs also have several unique features. In the case of the *Dai-sugi*, to produce a rafter log that is 3 cm in upper diameter and 3–4 m long, a period of 15–25 yr is required, while to produce a pillar log that is 10 cm in upper diameter and 3–4 m long, a period of 30–50 yr is required. This compares with a final cutting age of about 30–40 yr in the case of the clear-cutting system. The cutting seasons are spring and autumn, when debarking can be carried out with ease. In summer, heavy rainfall facilitates the growth of fungi on the surface of debarked logs, producing stained spots, and so the cutting is stopped.

After cutting and debarking, the logs are seasoned within the forest for several days. The crown foliage is left intact during this period, so as to accelerate reduction of the water content of the logs without causing cracking

of the wood. The crown foliage is next clipped, and the logs then stored in special warehouses for the long, slow seasoning period. When this is complete, the surface of the seasoned logs is hand-polished using wet, well-weathered granite sand.

Apart from their use in the traditional houses mentioned above, the polished pillar logs and rafter logs of Kitayama now have a wide market throughout Japan, for both the building of Japanese-style wooden houses and the construction of Western-style houses. The production areas are being enlarged annually, and some of the wood is also exported to customers overseas.

(b) *Yoshino forestry reserve.* The Yoshino district is situated in the southern part of Nara Pref., southwest Honshu, close to the old towns of Osaka, Kyoto and Nara. The area is mountainous, has high precipitation throughout the year, moderate temperatures, and a comparatively deep, fertile soil. It is thus well suited for the establishment of sugi and hinoki plantations.

The highest peaks are Mt. Odai (1695 m) and Mt. Sanjogadake (1720 m) on the southern side of the Yoshino district. The sugi and hinoki plantations are situated at 300–1300 m a.s.l. and have a total area of *ca.* 200,000 ha, of which 60% is coniferous forest and the remainder coppice forest. On the summits of mountains above 1550 m a.s.l., natural subalpine coniferous forests are found, while mountain beech–deciduous oak mixed forests also occur above 1200 m. The sugi and hinoki plantations thus lie in the warm-temperate zone at lower elevations, and in the cool-temperate zone at higher elevations. The total annual precipitation varies greatly according to elevation, with a mean of 2000–3000 mm and a maximum (on Mt. Odai) of about 8000 mm.

In ancient times the area was inhabited by hill-side farmers who carried out shifting cultivation on accessible slopes. However, the close proximity of Osaka, Kyoto and Nara gradually led to the development of forestry in the region, to satisfy the increasing demand for constructional and other timber. Also, lacquer-ware crafts were developed in the Yoshino district, so requiring a local supply of fine wood. The forestland was cut back and timbers supplied to the large towns by timber merchants. The turning point occurred about 250–300 yr ago, when farming began to take a secondary place and merchants in the towns began furnishing funds to the forest-dwellers to plant useful trees on their land. This rapidly led to the enlargement of the artificial plantations, and a special relationship was built up between the forest owners, who became the forest rangers or managers, and their financiers in the towns. The occupation of most of the local villagers has shifted to forest labourer.

The characteristic treatment method of plantations in the Yoshino dis-

trict consists of dense planting, repeated but weak thinning, and long-term rotation, so allowing the production of large, fine, uniform timbers. These logs were formerly transported to the market place as river rafts, but the sole method now is by logging truck. The ratio of sugi to hinoki forests is about 7:3, although in some places mixed plantations are also grown. Both the sugi and hinoki are raised from seeds gathered from local forests, and the seedlings are generally planted at a density of over 10,000 per hectare. Thinning is begun early, and repeated weakly and often. Since the timber markets are relatively close to the forests, even the smallest logs (known as coin-logs; top diameter, 2.5 cm) are saleable. Medium-sized timbers resulting from the thinning process are used as posts, etc. The large, high quality timbers were formerly employed in the production of *sake* (Japanese rice wine)-barrels, although now the principal use is for ceiling boards and decorative, sawn pillars. A limited part of the Yoshino forestry reserve is occupied by dense plantations producing polished logs similar to those of Kitayama. However, due to the high cost of labour in maintaining areas under intense silviculture, the tree density on the plantations is gradually being reduced elsewhere to about 6000 per hectare.

(c) *Other old and famous forestry reserves.* Other old and famous sugi and hinoki plantations in Japan, together with their essential characteristics, may be summarized as follows (see also Fig. 15).

(1) *Sambu* (Chiba Pref.), noted for its special race of sugi cuttings.
(2) *Nishikawa* (Saitama Pref.), characterized by very intensive sugi cultivation with a comparatively short rotation period.
(3) *Tonami* (Toyama Pref.), famous for its special race of sugi, the so-called *Boka-sugi*, which grows prolifically, roots easily, and produces slender timbers suitable for use as telegraph poles.
(4) *Tenryu* (Shizuoka Pref.), consisting of steep, rocky hillside sugi plantations that help control erosion and landslips, and yield high quality timbers.
(5) *Owase* (Mie Pref.), comprised of hinoki plantations, and repeatedly producing small pillar logs. The hinoki forests have a very thick canopy, and in general lack undergrowth species due to the low light intensity within the plantation. The forest floor is thus easily eroded by rainwater penetrating the leaf canopy. The area represents a typical example of low-productivity hinoki forest.
(6) *Tane* (Shiga Pref.), noted for its selective cutting system with plantation of sugi seedlings.
(7) *Chizu* (Tottori Pref.), famous for the use of sugi cuttings from naturally sprouted and rooted branches produced beneath snow cover, and for its intensive pruning system.

FIG. 19. Artificial plantation of sugi: clone complex after cutting (Kyushu).

(8) *Hita* (Kumamoto Pref.), characterized by its use of special sugi clone complexes.

(9) *Obi* (Miyazaki Pref.), famous also for its use of sugi clone complexes, and for the production of the *benko* timbers employed in the construction of Japanese-style wooden ships. The planting density is only about 1000 trees/ha, and the tree-trunks thus have widely-spaced annual rings, yielding a very soft kind of timber that is able to withstand and absorb severe shock.

One of the clone complexes of the Kyushu sugi plantations is illustrated in Fig. 19.

4.1.2 Japanese larch (Larix leptolepis) plantations

As described in section 2.1, the natural distribution of the Japanese larch is limited to subalpine slopes of central Japan. The principal areas are the northern and southern Japanese Alps (Shinshu district, especially Nagano Pref.) and higher altitude slopes on Mt. Fuji (Yamanashi Pref.). The climate in these areas is comparatively dry and continental in type, and the growing site of the larch is typically loamy soil, steep stony hillsides, or areas of new (immatured) soil. Since the Japanese larch is a pioneer tree species, it thrives best in open spaces with plentiful sunlight and rather dry soil.

Formerly, the larch was planted artificially only in the Shinshu district. However, following the extensive cutting back of forests during World War II, plantations were begun and rapidly developed in many parts of Japan, such as Hokkaido, Ou, and other central districts. As mentioned above, the principal reason for selecting the larch as a plantation species was its rapid growth at the young stage. When the war ended, rapid reforestation was

given high priority, and so both private and government policy favoured the planting of quick-growth tree species such as the larch.

From that time on, the larch became one of Japan's principal forestry species, although recent attacks by shoot-blight disease have damaged many young plantations in Hokkaido and Ou. The total annual planting area is also being reduced gradually now, since the market for larch timbers from artificial plantations is less than that for conifers. The quality of the wood from natural larch forests allowed its wide use as sawn timber; however, the smaller logs produced by plantations are inferior in quality and less suitable for use as either sawn timber or constructional material. They are unsuitable as wood pulp, and satisfy only a restricted market for pit-props, wooden constructional piles, etc.

Although at first many difficulties were encountered in bringing up larch seedlings as new forestry plants, e.g. their high sensitivity to soil-borne diseases, manure, etc., these problems were essentially overcome by the early 1900's. For this reason, as mentioned above, the Japanese larch was introduced into Europe and America as a new plantation species.

4.1.3. Bamboo plantations

There are three main bamboo species in Japan, which were introduced originally from the Chinese continent via Okinawa Is. and Kyushu. These are *Phyllostachys bambusoides, P. nigra* and *P. pubescens*. The cut bamboo stems have many uses, such as for fencing, basketware, ladders, fishing-rods, fans, various ornamental wares, etc. Also, the young shoots of *P. pubescens* are edible and cultivated intensively for human consumption in many districts, mostly in southern Japan, such as the western hills of Kyoto. *P. pubescens* grows best on red soils, and is thus found as groves in regions where such soil abounds. *P. bambusoides* and *P. nigra* are cultivated in wet regions close to rivers or in plains. The total distribution area of *Phyllostachys* groves covers most of Japan apart from Hokkaido and the northern part of the Ou district. However, due to the extensive destruction of bamboo groves during and since World War II, and their replacement by agricultural fields and residential areas, their occurrence is now far more limited. Fig. 20 shows part of a *P. bambusoides* grove in Hyogo Pref.

It is said that the bamboo in Japan blooms every 80–100 yr, although the normal flowering interval is still not known exactly. About 10 yr ago, *Phyllostachys bambusoides* began flowering in southern Japan and still continues at present. Once a particular area of bamboo has flowered, almost all of the individual stems in the stand die, except for the tops of some new rhizomes. From the remaining rhizomes and scattered seeds, new plants or stems develop. Newly reforested bamboo is very short at the early stage, but within 10 yr the stand reaches about the same height as the older stands. Manuring

FIG. 20. Man-made bamboo grove of *Phyllostachys bambusoides* in a snowy district of Hyogo Pref.

is the best treatment for such recovering groves. The soil of bamboo groves in Japan is manured every year, in the winter, and covered with a thick layer of rice straw. In particular, the treatment of groves of *P. pubescens* is very intensive. They are periodically covered with fresh red soil throughout, to a depth of several centimeters, since they are otherwise unable to continue production of good, edible shoots.

Bamboo groves are generally treated by the selective cutting method. Stems more than 3–4 yr old are often cut individually from the groves, while individual stems can usually grow for 7–8 yr before death. The bamboo forests are also used for flood control or erosion control along rivers, since the subterranean rhizome network can withstand such calamities well.

4.2. Natural regeneration

Natural regeneration methods are divided into two types, i.e. natural regeneration by the natural distribution of seeds from parent trees, and natural regeneration by vegetative sprouting or rooting from stumps or creeping branches. To produce high tree forests, natural regeneration from seeds is the normal method, while the regeneration of coppice forests is generally by vegetative propagation from stumps.

The total extent of naturally regenerated high forests in Japan is comparatively small, since the commonly employed forestry method is clear-cutting and artificial planting of seedlings. However, as mentioned above, both *Pinus densiflora* and *P. Thunbergii* forests are in some areas regenerated through the natural distribution of seeds from the parent trees. Both of these pine species require ample sunlight, and both have winged seeds that are readily scattered naturally into clear-cut parts of the forests. Furthermore,

these species tend to grow in poor forest soil with scant undergrowth, circumstances which facilitate the growth of new seedlings.

The pine forests are distributed widely in the coastal belt, on river banks, and on hillsides in the vicinity of agricultural and residential areas. Their wood and leaf-litter are often employed as fuel, and the ash as fertilizer for rice paddies, etc. Treatment is generally by limited-area clear-cutting or strip/belt cutting. In Shikoku, there are some special regeneration areas of pine forest which are treated by the selective cutting method and regenerated from seeds. Until recently, the pine logs and branches were used in the Japanese tile-manufacturing industry, as a colouring source, and also, as fuel, in primitive open pans producing salt. In these cases, the treatment method was selective cutting and intensive pruning of branches, to ensure sufficient light intensity throughout the forests and to facilitate natural regeneration. However, new industrial and chemical methods are now replacing the older techniques for tile and salt manufacture, leading to a gradual shift in the treatment method of the pine forests and a change in the uses of the timber they produce.

In some districts on the Sea of Japan side of Honshu, high sugi forests are regenerated by natural vegetative sprouting or rooting from stumps or creeping branches. The method of natural regeneration from sprouting stumps is termed "*Ritsujo*" regeneration in Japanese. It is a common type of regeneration among broad-leaved coppice forests, but a very special situation in the case of high coniferous forests. The sugi displays many local races in Japan, and certain of these have a tendency to propagate by sprouting from stumps. One such example is the famous *Kuma-sugi* of Akita, Niigata, Nagano, Toyama, Ishikawa and Fukui Pref., i.e. of snowy districts on the Sea of Japan side of Honshu.

The sugi is also able to propagate naturally by rooting from creeping lower branches (so-called "*Fukujo*" regeneration in Japanese). As mentioned above, on the Sea of Japan side of Honshu most young forest trees are buried by the deep winter snows, and their branches are pressed down into the upper soil layer. Such branches may take on a creeping form and readily propagate roots from their lower parts. If they find adequate open space for growth, they themselves then develop into new independent plants. After selective cutting of sugi forests with this regenerative capability, natural regeneration of new trees can occur with little difficulty. In fact, in certain localities on the Sea of Japan side, sugi forests are still treated by the selective cutting method and allowed to regenerate by the *Fukujo* process. Sugi trees with this characteristic can also form small groups of new plants (clones) from a single parent tree, and the forests which ultimately result are often complex stands of numerous overlapping clones. The tree form of stands regenerated

by the *Fukujo* process has certain special features, such as a narrow, slender crown with numerous branches, and retains the original bent lower stem even when the tree has become large and mature.

About 20 yr ago, there were abundant hillside coppice forests throughout Japan. Their wood was employed as fuel, although this role is now filled mostly by electricity and oil. Moreover, timbers of such coppice forests are suitable for the production of edible mushrooms such as the "shii-take" (*Lentinus edodes*), "hira-take" (*Pleurotus ostreatus*), "nameko" (*Pholiota nameko*), etc. In fact, such artificial cultivation of edible mushrooms has increased very rapidly recently, offering a continued use for wide areas of coppice forest. The coppice stands are generally treated by clear-cutting and regenerated naturally from sprouting stumps, although in a few districts there is selective cutting and regeneration from vegetative sprouts. The main tree species are deciduous and evergreen oaks, mixed with *Castanopsis* sp., *Carpinus* sp., *Fagus* sp., etc. The most useful oak species are *Quercus acutissima, Q. serrata* and *Q. crispula.* The Forestry Agency of Japan also hopes to improve the coppice stands of coniferous high trees, since their timber is in great demand for use in construction projects and in the pulp and paper industries.

In the last 10 yr, about 390,000 ha of forestland has been newly regenerated per annum by artificial means, of which about 300,000 ha represents the replacement of coppice stands of high trees. These figures correspond to about 77% of the total area of the artificial plantations. The natural regeneration areas represent about 240,000 ha.

References

1. S. Hatusima, *Flora of Okinawa* (Japanese), Okinawa Ass. Biol. Education, 1971.
2. Y. Hayashi, The natural distribution of important trees indigenous to Japan, *Bull. Gov. Forest Exptl. Sta.* (Japanese; English summ.), 1951–54.
3. Japanese Forestry Technical Association, *Forestry Lexicon* (Japanese), Maruzen, Tokyo, 1972.
4. T. Makino, *Makino's New Illustrated Flora of Japan* (Japanese), Hokuryukan, 1961.
5. *Ed.* A. Miyawaki, Vegetation of Japan compared with other regions of the world, *Encyclop. Sci. Technol.* (Japanese), vol. 3, pp. 535, Gakken, Tokyo, 1967.
6. M. Nakajima *et al.*, *Practical Handbook of Woody Plants* (Japanese), Tokyo, 1961.
7. M. Numata, *Illustrated Plant Ecology* (Japanese), Tokyo, 1969.
8. J. Ohwi, *Flora of Japan* (Japanese), Shibundo, Tokyo, 1965.
9. A. Okazaki, *Forestry in Japan*, Oregon Univ., 1964.
10. T. Suzuki, *Preliminary Studies on Japanese Forest Zones* (Japanese), 1961.
11. K. Yoda, *Forest Ecology* (Japanese), Tsukiji Shokan, Tokyo, 1971.

4

Grassland Vegetation

Makoto NUMATA*

1. Grassland Types and Classification
2. Grassland Types and Their Relation to Macroclimate
3. The Course of Grassland Successions in Japan
    3.1. General
    3.2. Relation to edaphic-biotic factors
    3.3. The ecological gradient and "degree of succession"
4. Detailed Descriptions of Grassland Types in Japan
    4.1. *Miscanthus sinensis* type grassland
    4.2. *Zoysia japonica* type grassland
    4.3. *Pleioblastus distichus* var. *nezasa* type grassland
    4.4. *Sasa* type grassland
    4.5. *Pteridium* type grassland
    4.6. Deteriorated stages with dominant *Erigeron* spp.
    4.7. Other minor grassland types

**Department of Biology, Faculty of Science, Chiba University, 1-33 Yayoi-cho, Chiba-shi 280, Japan*

Most of the native grasslands of Japan are, in the sense of Tansley, seminatural grasslands. They are not genuinely virgin, untouched natural communities but communities established under the disruptive influence of human activity and domestic animals, i.e. they are subject to mowing, firing, grazing, trampling, etc. The grassland vegetation is in general not the climatic climax but rather seral stages of plant successions controlled by edaphic, biotic and other factors.[1] Nevertheless, the alpine grasslands, and those of windswept mountaintops, can be regarded as truly natural communities in the same way as the climatic climax forests of Japan.

A survey of grassland vegetation throughout the country led to the establishment of three principal zones (A, B and C; see Table 1). These correspond

TABLE 1. Grassland vegetation zones in Japan based on a survey of all areas under mowing (meadow) or grazing (pasture)

Grassland zone	Meadow	Pasture
A	*Sasa* type, *Calamagrostis* type (A_m)	*Poa* type, *Festuca* type (A_p)
B	*Miscanthus* type (B_m)	*Zoysia* type (B_p)
C	*Miscanthus* type, *Pleioblastus* type (C_m)	Inland *Pleioblastus* type, coastal *Zoysia* type (C_p)

approximately to (i) the subarctic, (ii) the cool- plus warm-temperate (Sea of Japan and coast of Kyushu), and (iii) the warm-temperate (Pacific side) plus subtropical climatic zones, respectively (see Fig. 1). The boundary between A and B coincides with the northernmost limit of the *Fagus crenata* climax zone (see Fig. 2), although other broad-leaved deciduous forests are distributed further northeast into Hokkaido. However, the boundary between B and C deviates rather from that between the deciduous and evergreen broad-leaved forest zones (see also Fig. 2).

This chapter gives a general account of grassland types and successions in Japan and discusses their relationship to climatic and other conditions, reviewing the content of two recent articles[2,3] by the present author.

1. Grassland Types and Classification

Grassland in the broad sense includes grasslands (in the narrow sense), herblands, forblands, savanna, etc.[7] Such broad-sense grassland can be classified into a number of divisions and subdivisions[3] based on different viewpoints, as follows:

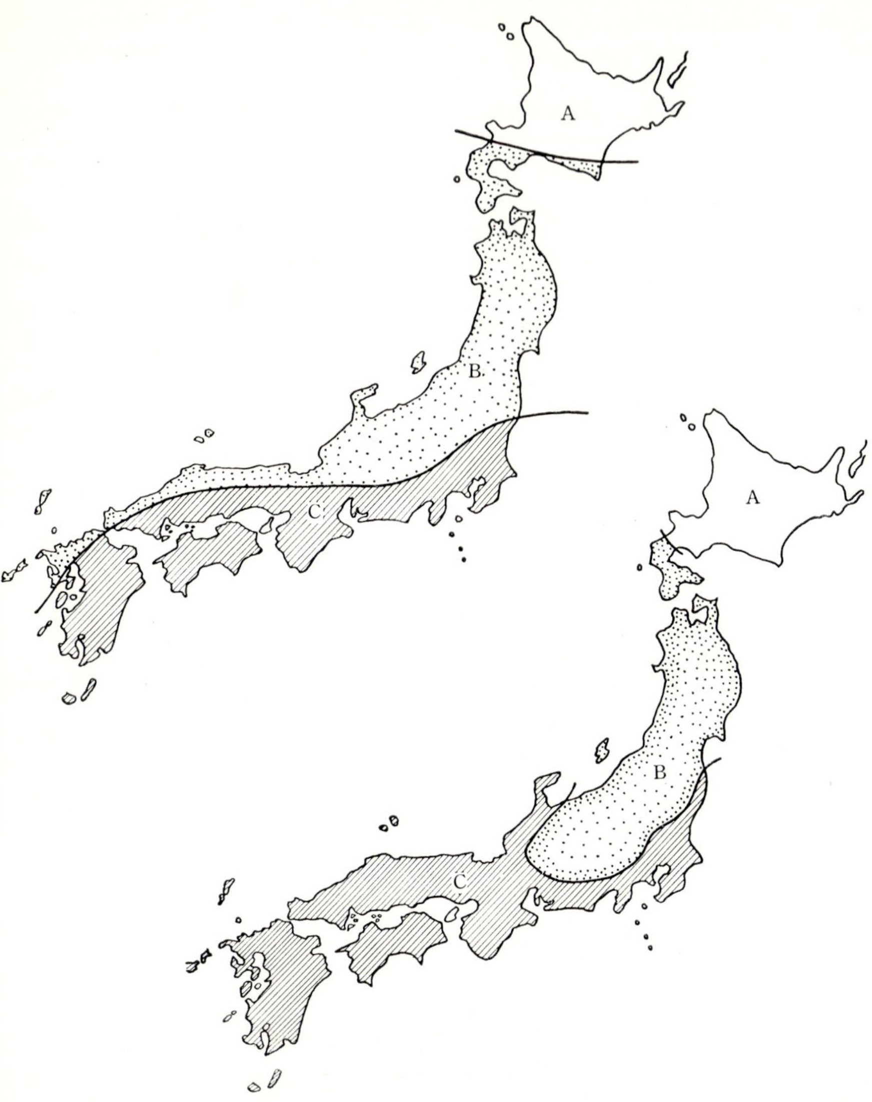

FIG. 1. Grassland vegetation zones in Japan (see Table 1 for key). The zone C extends down through the Ryukyu island chain.

FIG. 2. Climax forest zones in Japan (after data in ref. 4–6). A, Needle-leaved coniferous forest (subarctic); B, broad-leaved deciduous forest (cool-temperate); C, broad-leaved evergreen forest (warm-temperate to subtropical). Zone C extends down through the Ryukyu island chain.

(1) *Grassland in the botanical sense*

1.1. Local names: prairie, plain, steppe, pampa, campo, llano, savanna, veld, etc.

1.2. Based on moisture amount in the environment: mesophytic, hygrophytic, aquatic, xerophytic grassland, etc.

1.3. Based on salinity: fresh-water, salt-water, brackish-water (estuary), coastal grassland, etc.

1.4. Based on the substratum type: loamy soil, sandy soil, gravelly soil, peat grassland, etc. (also rocky desert, coastal desert, etc.).

1.5. Based on plant successions: pioneer, seral, climax grassland.

1.6. Based on general physiognomy and floristic composition: grassland in the narrow sense, forbland, herbland, Staudenfluren, bambooland, dwarf-bambooland, sedgeland, rushland, vineland, savanna, etc.

1.7. Based on dominant plants or their life-form: *Miscanthus sinensis*, *Zoysia japonica*, tussock, evergreen, summer-green, winter-green grassland, etc.

1.8. Based on plant life duration: annual, perennial grassland.

1.9. Based on grass height: tall, medium, short grassland.

1.10. Based on elevation and geomorphology: lowland, mountain, highland, alpine, hilltop, leeward, windward grassland; snow-patch carpets, alpine mats, etc.

(2) *Grassland in the agricultural sense*

2.1. As an administrative term: artificial, natural (wild), improved grassland, etc.

2.2. Based on lack/existence of biotic alteration: natural, semi-natural, man-made (sown) grassland.

2.3. Based on selected biotic factors: meadow (mown grassland), pasture (grazed grassland), abandoned (old) fields, grassland cultivated for thatch, etc.

2.4. Based on degree of management: extensive (range), improved, intensive grassland, etc.

2.5. Based on soil fertility: fertile, sterile, barren grassland, etc.

2. Grassland Types and Their Relation to Macroclimate

As an example of the relationship between climatic type (particularly precipitation) and grassland type, the data of Walter[8] can be cited for "savanna". In such climax grassland the annual precipitation averages 200–800 mm, with a dry season of duration 7–11 months. The annual production of dry matter (1000–6000 kg/ha on average) shows a clear linear relationship to annual precipitation (over a range of 100–600 mm).[9] Apart from the

climax grassland, all other grasslands present are of pioneer or seral stages, and the grassy or herbaceous taxa are sensitive to both moisture and temperature. Should the annual rainfall be more than 800 mm, the duration of the dry season less than 7 mouths, and the annual humidity index[10] greater than 0.7, the climax grassland does not develop and an apparent grassland gradually merges into forest with the invasion of arborescent forms. The grassland types in both the forest climax and grassland climax regions are restricted and typified by a number of dominant species.

In a similar manner, several principal grassland types can be distinguished in Japan in relation to the macroclimatic zones. These are shown in Table 2,[3] together with the dominant genera and species.

TABLE 2. Principal grassland types in Japan and their dominant genera and species

Climatic zone	Early stage forbland	Orthoseral grassland	Plagioseral grassland
Subarctic (A)†	*Agrostis alba* *Erigeron canadensis*	*Calamagrostis Langsdorffi* *Miscanthus sinensis* *Sasa* spp.	*Poa pratensis* *Festuca ovina* var. *vulgaris*
Cool-temperate (B)	*Persicaria vulgaris* *Erigeron* spp.	*Miscanthus sinensis* *Sasa* spp.	*Zoysia japonica*
Warm-temperate (C)	*Ambrosia artemisiaefolia* *Erigeron* spp.	*Miscanthus sinensis* *Imperata cylindrica* var. *Koenigii* *Pleioblastus distichus* var. *nezasa*	*Pleioblastus distichus* var. *nezasa* *Zoysia japonica*
Sub-tropical (C)	*Digitaria adscendens* *Paspalum dilatatum*	*Miscanthus floridulus* *Pleioblastus* spp.	*Cynodon dactylon* *Zoysia tenuifolia*

† (A, B, C) show the corresponding three vegetation zones indicated in Table 1.

3. The Course of Grassland Successions in Japan

3.1. GENERAL

Based on studies of the course of grassland successions in Japan, Oseko[11] originally suggested the main course to be as follows: the waste stage ⇆ *Zoysia* stage ⇆ *Miscanthus* stage ⇆ forest stage with branches of *Sasa*, *Lespedeza*, cryptogam (fern), brushwood, coppice, and pinewood stages. Later, a similar course of seral stages was recognized by Yoshida[12,13] and by Shimada.[14] However, Numata[1] has shown that this scheme applies only to cool-temperate regions of northeast Japan.

TABLE 3. Relationship between climatic zones and the course of secondary successions in Japan

Stage	SUBARCTIC ZONE	
	Orthoseral grassland	Plagioseral grassland
I	*Digitaria violascens* Link. *Persicaria vulgaris* Webb. et Moq. *P. perfoliatum* L.	*Poa annua* L. *Cerastium caespitosum* Gilb. var. *ianthes* Hara *Persicaria vulgaris* Webb. et Moq. *Rumex Acetosella* L.
II	*Erigeron* spp. *Agrostis palustris* Hudson *Artemisia montana* Pamp. *Rudbeckia laciniata* L. *Epilobium angustifolium* L.	*Poa pratensis* L. *Phleum pratense* L. *Trifolium repens* L. *Hydrocotyle ramiflora* Maxim. *Artemisia margaritacea* Benth. et Hooker f.
III	*Sasa nipponica* Makino et Shib. *S. Veitchii* Rehd. *S. kurilensis* Makino et Shib. *Miscanthus sinensis* Anders. *Calamagrostis Langsdorffii* Thrin. *Thalictrum Thunbergii* DC.	*Sasa nipponica* Makino et Shib. *Miscanthus sinensis* Anders. *Pteridium aquilinum* Kuhn
IV	*Sorbus commixta* Hedl. *Vaccinium Smallii* A. Gray *Sambucus Sieboldiana* Bl. var. *Miquelli* Hara *Aralia elata* Seem. *Rhus orientalis* Schneid.	*Lespedeza bicolor* Turcz. *Rosa rugosa* Thunb. *Hydrangea paniculata* Sieb. *Cerastrus orbiculatus* Thunb. var. *strigillosus* Makino *Rubus phoenicolasius* Maxim.
V	*Betula platyphylla* Sukat. var. *japonica* Hara *Alnus tinctoria* Sarg. var. *velutina* Hara *Quercus mongolica* Fisch.	
VI	*Abies sachalinensis* Fr. Schm. *Picea jezoensis* Carr. *Taxus cuspidata* Sieb. et Zucc.	

Stage	COOL-TEMPERATE ZONE	
	Orthoseral grassland	Plagioseral grassland
I	*Digitaria adscendens* Henr. *D. violascens* L. *Persicaria vulgaris* Webb. et Moq. *P. nodosum* L.	*Poa annua* L. *Digitaria violascens* L. *Rumex obtusifolius* L. *Carex nubigera* D. Don subsp. *albata* T. Koyama *Agrimonia Eupatria* L. var. *pilosa* Makino
II	*Erigeron annuus* Pers. *E. canadensis* L. *Agrostis palustris* Hudson *Artemisia vulgaris* L. var. *indica* Maxim.	*Zoysia japonica* Steud. *Carex nervata* Fr. et Sav. *Ranunculus japonicus* Thunb. *Hydrocotyle ramiflora* Maxim. *Luzula campestris* DC. var. *capitata* Miq. *Geranium Thunbergii* Sieb. et Zucc.
III	*Miscanthus sinensis* Anders. *Arundinella hirta* Tanaka *Spodiopogon sibiricus* Trin. *Pteridium aquilinum* Kuhn	*Viola obtusa* Makino *Miscanthus sinensis* Anders. *Pteridium aquilinum* Kuhn. *Sasaella ramosa* Makino

TABLE 3—*Continued*

Stage	COOL-TEMPERATE ZONE	
	Orthoseral grassland	Plagioseral grassland
IV	*Stephanandra incisa* Zabel *Symplocos chinensis* Druce var. *leucocarpa* Ohwi *Viburnum dilatatum* Thunb. *Rhus javanica* L. *Corylus Sieboldiana* Blume	*Lespedeza bicolor* Turcz. var. *japonica* Nakai *Salix Bakko* Kimura *Weigela hortensis* C.A. Mey *Rhododendron japonicum* Suring. *Salix vulpina* Anders.
V	*Betula platyphylla* Sukat. var. *japonica* Hara *Pinus densiflora* Sieb. et Zucc. *Pinus Thunbergii* Parl.	
VI	*Fagus crenata* Blume *Acer Miyabei* Maxim. *Quercus crispula* Blume	

Stage	WARM-TEMPERATE ZONE	
	Orthoseral grassland	Plagioseral grassland
I	*Ambrosia artemisiaefolia* L. *Digitaria adscendens* Henr. *Persicaria Blumei* Gross. *Setaria viridis* Beauv.	*Poa annua* L. *Digitaria adscendens* Henr. *Microstegium vimineum* A. Camus *Plantago asiatica* L.
II	*Erigeron annuus* Pers. *E. canadensis* L. *E. sumatrensis* Retz. *Oenothera parviflora* L. *Bromus unioloides* H.B. et Kunth *Chenopodium album* L. *Artemisia vulgaris* L. var. *indica* Maxim.	*Pleioblastus distichus* Muroi et H. Okam. var. *nezasa* Muroi *Zoysia japonica* Steud. *Artemisia vulgaris* L. var. *indica* Maxim. *Hydrocotyle maritima* Honda
III	*Imperata cylindrica* Beauv. var. *Koenigii* Dur. et Sch. *Miscanthus sinensis* Anders. *Pleioblastus Chino* Makino *P. distichus* Muroi et H. Okam. var. *nezasa* Muroi *Gleichenia dichotoma* Hook.	*Pleioblastus distichus* Muroi et H. Okam. var. *nezasa* Muroi *Miscanthus sinensis* Anders. *Pteridium aquilinum* Kuhn
IV	*Viburnum erosum* Thunb. *Rhododendron dilatatum* Miq. *Rh. Kaempferi* Planch. *Deutzia crenata* Sieb. et Zucc. *Clethra barbinervis* Sieb. et Zucc.	*Lespedeza bicolor* Turcz. var. *japonica* Nakai *L. cyrtobotrya* Miq. *Rosa multiflora* Thunb. *Lyonia Neziki* Nakai et Hara *Pieris japonica* D. Don *Smilax China* L. *Wikstroemia Gampi* Maxim.
V	*Pinus densiflora* Sieb. et Zucc. *P. Thunbergii* Parl. *Quercus serrata* Thunb. *Q. dentata* Thunb.	
VI	*Machilus Thunbergii* Sieb. et Zucc. *Castanopsis cuspidata* Schottky *C. cuspidata* var. *Sieboldi* Nakai *Pasania edulis* Oerst. Evergreen oak (*Cyclobalanopsis acuta* Oerst., *C. gilva* Oerst., etc.)	

In the case of grazing land (pasture), the *Zoysia* stage is represented principally in cool-temperate regions, although it does also occur in coastal warm-temperate regions of southwest Japan. It is generally replaced by the *Poa pratensis* stage in subarctic regions and by the *Pleioblastus* stage in warm-temperate regions. *Zoysia japonica* dominates cool-temperate pastures in the *Fagus crenata* zone (Tohoku, and corresponding altitudinal zones); its total distribution covers most of Japan from southern Hokkaido to Kyushu. However, pastures with a physiognomy resembling the *Zoysia* type in southern districts of Japan are generally dominated by *Pleioblastus*, the growth form of which is very similar to that of *Zoysia*. Naturally growing *Pleioblastus* has an average height of 2–3 m, but forms growing under grazing are only about 10 cm or so in height. Further, Itôw[15,16] has reported the occurrence of *Miscanthus* and *Zoysia* type grasslands in northwest Kyushu which lack *Pleioblastus distichus* var. *nezasa*. In central areas of the island (at an elevation of $>$700 m a.s.l. and $>$20 km from the coast), *Pleioblastus* and *Pteridium* type grasslands occur, but in the coastal zone (at an elevation of $<$300 m a.s.l. and $<$5 km from the coast) grasslands without *Pleioblastus* are found.

In the case of mown grassland (meadow), *Miscanthus sinensis* is distributed even more extensively than *Zoysia japonica* throughout Japan. However, in subarctic regions of northeast Hokkaido, where the competing power of *M. sinensis* against *Sasa* is weak and where the tempo of progression of secondary successions is slow, wide-ranging homogeneous meadow is not established. Here, *Sasa* is the dominant genus and *M. sinensis* is suppressed.

The above relationships are clear from the general schemes for principal grassland zones and types shown in Tables 1 and 2 and Fig. 1. In addition, it is possible to define with more precision the course of secondary successions in relation to the principal climatic zones, as shown in Table 3. Here, five stages are distinguished: I, the deteriorated or 1st (year) pioneer stage; II, short-grass type pasture (in the plagiosere) or 2nd (year) pioneer stage; III, tall-grass type meadow or perennial grass stage; IV, shrubby stage; V, pioneer tree stage; VI, climax stage. In each case, the dominant and principal other species are indicated for both orthoseral and plagioseral grassland types.

3.2. Relation to edaphic-biotic factors

As mentioned above, edaphic and biotic conditions have a strong influence on grassland vegetation (within the general framework of climatic conditions). For example, the grassland type at any locality varies with the rate of decomposition of organic matter within the soil at the locality. Thus, tall-grass type meadow dominated by *Miscanthus sinensis*, etc. exhibits a thicker humus layer and higher contents of moisture, humus, total nitrogen, avail-

able nitrogen and exchangeable bases, when compared to the short-grass type pasture dominated by *Zoysia japonica*.[17] Also, *Pleioblastus* type meadows and pasture often have lower contents of humus, nitrogen, phosphate and calcium than the *M. sinensis*-dominated meadows.

Land under active grazing is usually trampled sufficiently to yield compact soil, which is typically covered by the *Zoysia* type vegetation and often has a thin surface-soil layer (dependent on the rate of erosion). For these reasons, it in general exhibits a comparatively low content of moisture, humus, calcium, magnesium and exchangeable bases, but does have a high content of total nitrogen, available nitrogen (especially as nitrate) and available phosphate. These features are particularly well developed in the upper soil layers.

In the case of sown pastureland, the pioneer stage is represented by mixed stands of orchard grass, ladino clover, Italian ryegrass, etc., and throughout the country (apart from Hokkaido) such pasture is invaded by numerous weeds. It thus exists as a deteriorated semi-natural grassland following the natural succession. The growth of northern-type forage plants (orchard grass, red clover, etc.) declines during the summer months due to the high temperatures and occasional drought periods.[18,19] Also, decline of leguminous plants in comparison with grasses is common in Japan during the early stages of establishment of sown pastureland. This is not based simply on the compensative role of legumes to grasses,[20] but on a variety of other factors, such as shortening of the life-span of northern-type forage plants under warm climatic conditions, deficiency of phosphate in the volcanic ash soils (*ando*) distributed widely through Japan, etc. Nevertheless, under other conditions, the reverse situation (where the growth of grasses is inhibited instead of legumes) is known. This may be due to high moisture and/or phosphate content in the soil, insect damage (e.g. by *Prodenia litura*), etc. It is thus clear that the many relationships between the course of secondary grassland successions and edaphic-biotic factors in Japan require both detailed research in local areas and comparative research on a broad scale.[1]

3.3. The ecological gradient and "degree of succession"

As shown in the tables above, the basic distribution of vegetation types from north to south in Japan in *Sasa nipponica–S. Veitchii* or *Calamagrostis* (type A_m)—*Miscanthus sinensis* (type B_m)—*M. sinensis* or *Pleioblastus distichus* var. *nezasa* (type C_m) in the case of meadows, and *Poa pratensis* or *Festuca* (type A_p)—*Zoysia japonica* (type B_p)—inland *Pleioblastus distichus* var. *nezasa* or coastal *Zoysia* (type C_p) in the case of pastures, corresponding to the climatic regions.[1,21-23] However, each such grassland type exhibits a range of parallel ecological gradients. For example, grasslands belonging to the same *Zoysia*

type are diversified according to their precise standing in the local plant succession. Also, the simple classification into *Zoysia*, *Miscanthus* types, etc. ignores the fact that there are transitional communities between the main types. Thus, the location of each grassland or grassland type in the main succession should be defined in some quantitative way. For this purpose, the author has proposed the parameter "DS", the so-called "degree of succession". However, even though Pichi-Sermolli[24] had suggested an index for the attainment or establishment of maturity in plant communities based on the frequency percent of all species, the author's DS is not based on the frequency percent but instead on a number of other characteristics of the plant community, as follows:

$$DS = \{(\Sigma cld)/n\} v$$

where c is the climax adaptation number in the range 1 to 5;* l, the life span of the constituents; d, the relative dominance (SDR); n, the number of species; and v, the ground cover (in the range 0 to 1, where 100% cover ≡ 1).[25] l (in years) is taken as 1 (Th), 10 (Ch, H and G), 50 (N) and 100 (M and MM) according to the life-form, based on data for the longevity of plants. SDR is a measure of the relative importance of the constituent species of a community, expressed by summing and averaging ratios of phytosociological characteristics, such as the two-factor SDR of $(C'+H')/2$ (%) where C' and H' are the cover ratio (%) and height ratio (%), respectively.[26]

4. Detailed Descriptions of Grassland Types in Japan

In relation to the general account given above of grassland types and successions in Japan, and of their relationship to climatic, biotic and other factors, detailed descriptions are now presented for the principal grassland types mentioned, including floristic composition lists, SDR and DS data.

4.1. Miscanthus sinensis type grassland

Miscanthus sinensis type grassland (Fig. 3) is the most representative of tall-grass type meadows in Japan, where tall (high) grass is that defined by Hanson[27] (a class of grasses, 6–8 ft or more in height, distinct from medium and short grasses). Apart from the typical *M. sinensis* meadows distributed widely throughout the subtropical to subarctic zones, there are also meadows dominated by *Miscanthus floridulus* (in subtropical to tropical regions), those domi-

* In the present article, the term c is neglected since its value for grassland species is close to unity.

FIG. 3. Representative tall-grass type grassland dominated by *Miscanthus sinensis* (Kawatabi, northeast Honshu).

nated by *M. sinensis* var. *condensatus* (particularly in coastal areas of the warm-temperate zone), and, although very much more limited in extent, those dominated by *M. sacchariflorus*, *M. tinctorius* or *M. oligostachys*. Forms and varieties of *Miscanthus* also extend to islands in the southern Pacific Ocean.

An example of the floristic composition of typical *Miscanthus sinensis* type grassland is given in Table 4. In general it is a mesic semi-natural grassland, partly subject to mowing, and the remainder mainly seral stages of secondary successions. Only in rare instances is it used for grazing, due to the fact that the growing point of *M. sinensis* is located rather high above the ground and its regenerative capacity is comparatively small. Therefore, mowing during the growing season also has a strong deleterious effect on regeneration. (The normal cutting season is September or October, after flowering.) The life-form of *M. sinensis* is ChD_1R_3t, with large bunches. Parts of the aerial shoots overwinter (remaining green), although the upper part of the plant has already withered and died. Moreover, windy localities near the summits of mountains or hills are often covered by tall-grass type (or dwarf-bamboo type) grassland regardless of the forest climax zone.

The ground cover ratio (v) for *Miscanthus sinensis* type grassland is almost 1.0, although certain degenerate phases show values appreciably less than 1.0. The total number of species (n) is normally about 25, rather less in degraded phases, and occasionally much greater in the case of certain very heterogeneous communities. The summed dominance ratio (SDR) of *M. sinensis* usually approximates to about 100, but is lower when there are competing co-dominants such as the shrubby species *Lespedeza bicolor*. The relative SDR of *M. sinensis* (SDR′ (%), the ratio to the sum of SDR values for all

TABLE 4. Floristic composition of *Miscanthus sinensis* type grassland (example from JIBP Area, Kawatabi, Miyagi Pref.)

Species	SDR
Miscanthus sinensis Anders.	100
Lespedeza bicolor Turcz. var. *japonica* Nakai	60
Pteridium aquilinum Kuhn	43
Hydrangea paniculata Sieb.	37
Aster scaber Thunb.	32
Eupatorium japonicum Thunb.	26
Astilbe odontophylla Miq.	25
Lysimachia clethroides Duby	23
Iris ensata Thunb. f. *spontanea* Makino	21
Hypericum erectum Thunb.	20
Inula salicina L. var. *asiatica* Kitam.	20
Artemisia japonica Thunb.	18
Potentilla Freyniana Bornm.	14
Carex lanceolata Boott	13
Petasites japonicus Maxim.	13
Hosta undulata Bailey var. *erromena* F. Maekawa	13
Desmodium racemosum DC.	13
Lastrea Thelypteris Bory	10
Lespedeza pilosa Sieb. et Zucc.	10
Agrimonia pilosa Ledeb.	10
Ixeris dentata Makino	10
Carex Maximowiczii Miq.	9
Viola mandshurica W. Becker	5
Polygonatum sachalinense Fr. Schm.	5
Reynoutria japonica Houtt.	4
Cirsium japonicum DC.	3

plants in the stand) is generally about 15 or so. However, when the number of competing species is large, this value may be considerably less. Based on such data, and on values for the life span of the constituents (l), a distribution curve for the degree of succession (DS) in *M. sinensis* type grasslands was plotted (see curve 4 of Fig. 4, which also includes comparable plots for the other five principal grassland types discussed in this section). The DS distribution for *Miscanthus* type grasslands approximates to a normal curve, with the mode at *ca.* DS=500 and a total range of DS values from *ca.* 100 to 1000, where the lower figures (100–200) are for early stages and the higher figures (900–1000) for advanced stages of *Miscanthus* type grasslands that include many tree saplings.

4.2. ZOYSIA JAPONICA TYPE GRASSLAND

Zoysia japonica type grassland (Fig. 5) is the most representative of short-grass type pastures in Japan; and the *Zoysia* grasslands as a whole are the most representative pasture type in Japan (especially of cool-temperate regions). Species of *Zoysia* occur widely throughout the country and also extend

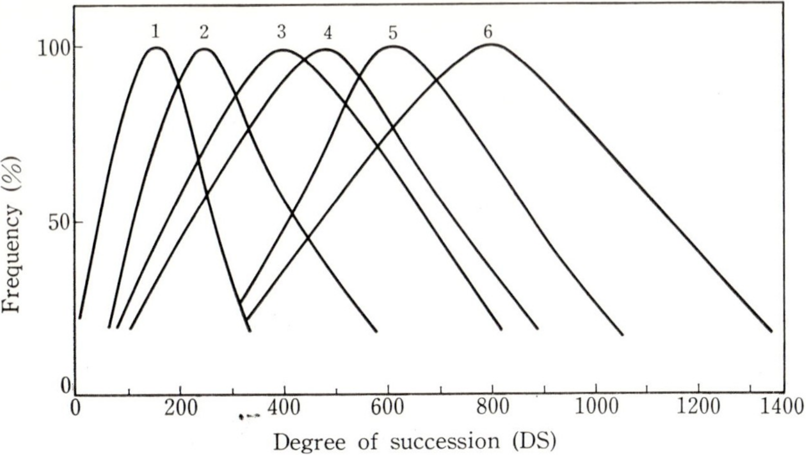

FIG. 4. Distribution curves of DS for grassland vegetation types. 1, *Erigeron* type; 2, *Zoysia japonica* type; 3, *Pteridium aquilinum* type; 4, *Miscanthus sinensis* type; 5, *Pleioblastus distichus* var. *nezasa* type; 6, *Sasa* type.

FIG. 5. Representative short-grass type grassland dominated by *Zoysia japonica* (Hakkoda, northeast Honshu).

to islands in the southern Pacific Ocean. However, *Zoysia japonica* is comparatively limited in extent as a dominant plant, occurring mostly in areas under the influence of biotic pressures from grazing and trampling (as discussed in section 3.1 above). Also, even though the *Zoysia* grassland can be said to be roughly equivalent to the cool-temperate climatic zone, it may (together with associated communities) be more correctly assigned to a circum-cool-temperate zone.[27] One possible zonal classification is given in Table 5, which also shows the climatically corresponding forest types.

A recent comparison of northern *Zoysia* pasture (Hokkaido) with southern *Zoysia* pasture (Kyushu) by the present author,[28] also indicated the Zoysion japonicae to be essentially cool-temperate. The *Zoysia* pastures in the central mountains of Kyushu belong to the *Fagus* zone of southern Japan. The *Zoysia* pastures on coastal lowlands,[15,16] even though their occurrence may at first sight appear to be anomalous, can probably be understood as having developed under the influence of the Sea of Japan (esp. San'in)-type climate (which is essentially "cool-temperate") or as a phenomenon of settlement in coastal regions at the margin of the main distribution area of the species.[29] *Dichondra repens* (Convolvulaceae), a consitituent of *Zoysia* pastures in Kyushu, can be equated with the northern *Trifolium* and southern *Desmodium* as seral equivalents. Also, the Zoysion of temperate Asia may correspond ecologically to the Sporobolion, Paspalion, Cynodonion, etc. of tropical Asia. Most *Zoysia* pasture in northwest Kyushu is of the polydominant precursor type,[30] including the genera *Zoysia*, *Arundinella*, *Miscanthus*, *Pleioblastus*, *Imperata*, etc. (all of which are potential dominants of Japanese grasslands), and a monodominant biotic disclimax of the *Zoysia* type or *Miscanthus* type, etc. is established under the influence of appropriate biotic pressures.

The northern limit of *Zoysia* type pasture has been confirmed as southern Hokkaido, including Oshima Peninsula and Cape Erimo. Other parts of the island are covered with *Poa pratensis* type pasture (as mentioned above), and in mountainous pastures of central Honshu corresponding to those of Hokkaido, *Festuca ovina* var. *vulgaris* type grassland occurs.

It should be noted that the distribution of *Zoysia japonica* as a species is dif-

TABLE 5. Scheme showing correspondence between grassland and forest types in relation to climate

Grassland	Forest	Climate
Geranio–Zoysietum	Abietum Mariesii	Subarctic
Violo–Zoysietum	Saseto–Fagetum	Cool-temperate
Erigero–Zoysietum	Abietum Mariesii and Saseto–Fagetum	Subarctic and cool-temperate
Arundinello–Zoysietum	Fagion crenatae and Tsugion Sieboldii	Cool- and warm-temperate

ferent from that of *Zoysia japonica*-dominated grasslands (*cf.* the difference between the vertical distribution of *Cynodon dactylon* in east Nepal and the distribution of *Cynodon dactylon*-dominated pastures[31]). Also, *Zoysia japonica* does not in general appear in the ordinary course (orthosere) of secondary successions, but flourishes in plagioseral grasslands under the influence of biotic pressures from grazing, trampling, etc.[2] As described in section 3.1, Oseko's suggested course[11] for grassland successions in Japan (waste stage ⇆ *Zoysia* stage ⇆ *Miscanthus* stage, where the leftward direction indicates increasing pressure from biotic factors) is a suitable scheme for plagioseral grasslands in the cool-temperate zone, although it is generally replaced by the course, annual stage ⇆ *Pleioblastus* stage ⇆ *Miscanthus* stage, in the warm-temperate zone.[1,32]

An example of the floristic composition of typical *Zoysia japonica* type grassland is given in Table 6. The life-form of *Z. japonica* is ChD_4R_1p, with

TABLE 6. Floristic composition of *Zoysia japonica* type grassland (example from Mt. Asama, Gunma Pref.)

Species	SDR
Zoysia japonica Steud.	100
Kummerowia striata Schindl.	50
Metanarthecium luteo-viride Maxim.	47
Rhododendron japonicum Suringer	42
Trifolium repens L.	38
Miscanthus sinensis Anders.	37
Spiraea japonica L. fil.	34
Viola mandshurica W. Becker	29
Sanguisorba officinalis L.	28
Arundinella hirta Tanaka	28
Agrostis clavata Trinius	27
Potentilla Freyniana Bornm.	26
Berberis Thunbergii DC.	24
Drosera rotundifolia L.	23
Carex nubigera D. Don subsp. *albata* T. Koyama	20
Geranium Thunbergii Sieb. et Zucc.	18
Lactuca dentata Makino	16
Salix integra Thunb.	16
Fimbristylis tristachya R. Br. var. *subbispicata* T. Koyama	16
Cerastium glomeratum Thuill.	15
Malus Sieboldii Rehd.	13
Rosa multiflora Thunb.	12
Gentiana triflora Pall. var. *japonica* Hara	11
Cirsium Tanakae Matsum.	11
Plantago camtschatica Chamisso	10
Patrinia villosa Juss.	10
Spiranthes sinensis Ames	8
Carex nervata Franch. et Sav.	8

stolons or rhizomes, and the species can be regarded as a northern vicariant corresponding to the tropical *Cynodon dactylon.*

The ground cover ratio (v) for *Zoysia japonica* type grassland is almost 1.0, although certain degenerate (strongly grazed or trampled) phases show values less than 1.0. The total number of species (n) is extremely variable (from *ca.* 6 to 63, with a sample mean of 24), although homogeneous, good-quality *Z. japonica* pasture does display a moderate and more constant number (Table 6). The summed dominance ratio (SDR) of *Z. japonica* also varies much more widely than that for *Miscanthus sinensis,* ranging between 50 and 100 in stands where *Z. japonica* is dominant. The value of SDR′ is about 15 or so, and so comparable to that for *M. sinensis.* The DS distribution for *Zoysia japonica* type grasslands has the mode at *ca.* DS=250 and a total range of DS values from *ca.* 100 to 500 (curve 2, Fig. 4).

4.3. PLEIOBLASTUS DISTICHUS VAR. NEZASA TYPE GRASSLAND*

The dwarf-bamboo type grasslands of Japan can be divided into two main classes, viz. the *Pleioblastus* type pasture or meadow of warm-temperate regions and the *Sasa* type meadow of subarctic and cool-temperate regions, where the former can be equated with the *Castanopsis cuspidata* var. *Sieboldii,* evergreen oak (*Cyclobalanopsis*) zone and the latter with the subarctic *Picea jezoensis–Abies sachalinensis* and cool-temperate *Fagus crenata* zones. Besides the two principal genera of dwarf-bamboos, an intermediate type (*Sasaella*) is also found. It is closely akin to the genus *Arundinaria,* retaining much of the form of ancestral dwarf-bamboos.

The author has observed *Arundinaria* in the Nepal Himalayas, where it occurs as a shrubby layer beneath the forest canopy, in the absence of typical dwarf-bamboo type pastureland. The species of *Arundinaria* range over a wide area of lowland and highland in Nepal and neighbouring countries, and the leaves are taken as forage. Similar species appear to be distributed widely as ancestral forms in both North and South America, and in Africa. The author has previously reported on the *Arundinaria* grasslands of central Brazil.[33]

In Japan, *Sasaella* occurs in a very narrow band at the point of contact between *Sasa* and *Pleioblastus* type communities (e.g. in the vicinity of Mt. Hakone, Kanagawa Pref.), and so cannot be regarded as sufficiently widespread to constitute a separate, third, dwarf-bamboo zone.

An example of the floristic composition of typical *Pleioblastus distichus* var. *nezasa* type grassland (see Fig. 6) is given in Table 7. Among other species of *Pleioblastus, P. Chino* is distributed mainly between the Kanto (northern

* Synonyms: *Pleioblastus Yoshidake* Nakai, *P. variegatus* var. *viridis* f. *glabra* Makino; some authors also use the generic name *Arundinaria* instead of *Pleioblastus.*

FIG. 6. Dwarf-bamboo type grassland dominated by *Pleioblastus distichus* var. *nezasa* (Mt. Kuju, Kyushu).

TABLE 7. Floristic composition of *Pleioblastus distichus* var. *nezasa* type grassland (example from Mt. Kuju, Oita Pref.)

Species	SDR
Pleioblastus distichus Muroi et H. Okam. var. *nezasa* Muroi	70
Miscanthus sinensis Anders.	59
Salix subopposita Miq.	28
Pteridium aquilinum Kuhn	27
Lespedeza cyrtobotrya Miq.	25
Cymbopogon Goeringii Honda	24
Angelica longeradiata Kitag.	22
Salix Saidaeana Seemen	18
Artemisia vulgaris L. var. *indica* Maxim.	17
Haloragis micrantha R. Br.	14
Lysimachia clethroides Duby	14
Carex nervata Franch. et Sav.	12
Agrostis clavata Trinius	10
Sanguisorba officinalis L.	10
Pinus densiflora Sieb. et Zucc.	7
Polygala japonica Houtt.	4
Viola pumilio W. Becker	2

warm-temperate) and Tohoku (cool-temperate) regions. *Zoysia* type pastures replace *P. Chino* or *Sasaella ramosa* grassland in cool-temperate areas, according to the intensity of grazing. However, in warm-temperate lowlands the same *P. Chino* type grassland degenerates into annual type grassland dominated by *Persicaria Blumei*, *Digitaria adscendens*, *Microstegium vimineum*, etc.

The ground cover ratio (v) for *Pleioblastus distichus* var. *nezasa* type pasture or meadow is usually 1.0. Even under strong grazing pressure (inc. over-

grazing), where the average plant height is appreciably reduced, the ground cover is not so adversely affected (in contrast to *P. Chino* where, as mentioned, there is reversion to the annual type pioneer stage). The total number of species (*n*) for *P. distichus* var. *nezasa* type grassland is *ca.* 20. The summed dominance ratio (SDR) of *P. distichus* var. *nezasa* is comparatively high (of the order of 70), except under conditions of severe overgrazing. The value of SDR′ for *Pleioblastus* spp. is in general about 15 or so, and is thus comparable to those for *Miscanthus sinensis* and *Zoysia japonica*. The DS distribution for *P. distichus* var. *nezasa* type grasslands has the mode at *ca.* DS=600 and a total range of DS values from *ca.* 400 to 1000 (curve 5, Fig. 4).

4.4. Sasa type grassland

Sasa type grassland (Fig. 7) is typically represented by northern forest-pastures, pastures and meadows, particularly those in Hokkaido. Indeed, the dominant dwarf-bamboo species of Hokkaido grassland is generally *Sasa Veitchii* or *Sasa nipponica*, the former being the most widespread. Such *S. Veitchii* grassland, which may establish itself in place of burnt or felled forest-land, remains in a comparatively constant state when properly mown, but degenerates if used as grazing land. In cases where the *Sasa* withers or denudation follows overgrazing, an artificial pasture composed of forage grasses and legumes (introduced from northern/central Europe) can easily be established on a base of cattle dung or by artificial seeding. However, this forage type pasture is kept permanently only in Hokkaido and is retained in proper balance to the grazing requirement. Deteriorated *Sasa* type pasture that has been overgrazed may also progress to *Betula platyphylla* var. *japonica*, *Quercus crispula* or *Q. mongolica* forests, or alternatively, to *Pteridium aquilinum* type

FIG. 7. Dwarf-bamboo type grassland dominated by *Sasa senanensis* (Sarobetsu, northern Hokkaido).

grassland. The latter is quickly eatablished in areas selectively grazed by horses or cattle, since *P. aquilinum* is unpalatable to these animals.

In 1950, Yoshida[12] reviewed Oseko's suggested course[11] for grassland successions (see section 3.1), and proposed the *Sasa* stage as a representative pasture, including a dwarf-bamboo type grassland under the *Pleioblastus* stage. However, as shown above, these two grassland types should be considered as distinct. Yoshida[13] also recognized the successional courses, *Sasa* stage ⇆ *Miscanthus* stage (under proper use) and *Miscanthus* stage ⇆ *Sasa* stage ⇆ *Zoysia* stage, or *Miscanthus* stage ⇆ *Zoysia* stage (under excessive use). However, the competition between *Miscanthus* and *Sasa* is affected by climatic as well as biotic factors, and the competing power of *Miscanthus sinensis* is in general sufficiently weakened in northeastern Hokkaido to prevent the establishment of a homogeneous *M. sinensis*-dominated grassland. Thus, the coactive relation between *Miscanthus* and *Sasa* in the subarctic zone is different from that in the cool-temperate zone, and Yoshida's scheme cannot be given general applicability.

An example of the floristic composition of typical *Sasa* (*S. Veitchii*) type grassland from Hokkaido is given in Table 8.

The ground cover ratio (v) for *Sasa* type pasture is about 1.0 under weak grazing, but drops to *ca.* 0.3 or 0.4 under strong grazing. The total number of species (n) is extremely variable (from *ca.* 5 to 28, with a sample mean of 13), but is in general rather lower than that of other grassland types due to the strong dominance of *Sasa*. The summed dominance ratio (SDR) of *Sasa* is *ca.* 80 to 100 (unless lowered through grazing or other causes), and the value of SDR′ is within the wide range of 10 to 40. In typical grassland, however, SDR=100 and SDR′=30, values which are somewhat high when compared

TABLE 8. Floristic composition of *Sasa* type grassland (example from Shari-machi, Hokkaido)

Species	SDR
Sasa Veitchii Rehd.	88
Miscanthus sinensis Anders.	73
Pteridium aquilinum Kuhn	72
Petasites japonicus Maxim. var. *giganteus* Hort.	62
Cacalia hastata L. var. *glabra* Ledeb.	51
Phleum pratense L.	51
Artemisia montana Pamp.	45
Acer Miyabei Maxim.	39
Salix Bakko Kimura	35
Anaphalis margaritacea Benth. et Hook. fil.	35
Hypericum erectum Thunb.	34
Eupatorium sachalinense Makino	32
Taraxacum officinale Weber	4
Viola verecunda A. Gray	3

with the other grassland types mentioned above. The DS distribution for *Sasa* type grasslands has the mode at *ca.* DS=800 (when data for *Sasa* type undergrowth of forests are excluded) and a normal range of DS values from *ca.* 200 to >1000 (curve 6, Fig. 4), where DS=200 to 500 represents grasslands under very strong grazing conditions and values over 1000 (up to 3000) mainly correspond to the *Sasa* type undergrowth of forests.

4.5. Pteridium type grassland

The species *Pteridium aquilinum* is very widely distributed among the grasslands of Japan; however, those communities actually dominated by *P. aquilinum* are, through grazing or burning, of a deteriorated type. The genus is particularly common in short-grass type pastures dominated by *Pleioblastus distichus* var. *nezasa* or *Zoysia japonica*, and these have (on a purely physiognomic basis) sometimes been referred to as *Pteridium* type grasslands even though *P. aquilinum* is not the dominant plant when expressed in terms of SDR. True *Pteridium* type grasslands have a transitional character, and Oseko[11)] supposed them to constitute a seral stage corresponding to the soil fertility of the *Lespedeza* and *Sasa* stages between the waste (probably annual) and forest stages.

An example of the floristic composition of typical *Pteridium* type grassland (see Fig. 8) is given in Table 9. Fresh plants of *P. aquilinum* (in particular) are poisonous, due to the possession of enzyme with B_1-deactivating activity, although in parts of the Tohoku region (Aomori Pref.) this bracken is mown and the aerial shoots, mixed with rice-bran, are used as forage.

The ground cover ratio (v) for *Pteridium* type grassland is almost 1.0, and the total number of species (n) is variable (from *ca.* 9 to 41) but normally about 20 or so. The summed dominance ratio (SDR) of *Pteridium aquilinum*

FIG. 8. Bracken type grassland dominated by *Pteridium aquilinum*.

TABLE 9. Floristic composition of *Pteridium* type grassland (example from Kawatabi, Miyagi Pref.)

Species	SDR
Pteridium aquilinum Kuhn	100
Hydrocotyle ramiflora Maxim.	46
Zoysia japonica Steud.	43
Lactuca dentata Makino	22
Miscanthus sinensis Anders.	22
Dryopteris Thelypteris A. Gray	21
Rubus parvifolius L.	16
Trifolium repens L.	16
Lycopus Maackianus Makino	15
Patrinia scabiosaefolia Link	15
Lespedeza cuneata G. Don	15
Carex lanceolata Boott	15
Haloragis micrantha R. Br.	11
Potentilla Freyniana Bornm.	11
Prunella vulgaris L.	8

is generally within the range 60 to 100, and the value of SDR′ about 20–30 (but with a total range of 10 to 60). The DS distribution for *Pteridium* type grasslands has the mode at *ca.* DS=400 (i.e. between the *Zoysia* and *Miscanthus* types) and, excluding special cases, a total range of DS values from *ca.* 200 to 800 (i.e. a somewhat narrower range than that in *Miscanthus* type meadows; see curve 3, Fig. 4).

4.6. DETERIORATED STAGES WITH DOMINANT ERIGERON SPP., ETC.

The following species are common as dominants of pioneer stages of secondary successions in Japan: *Erigeron annuus*, *E. canadensis*, *Ambrosia artemisiaefolia*, *Digitaria adscendens*, *Persicaria Blumei*, *P. vulgaris*, *Setaria viridis*, etc. These pioneer stages are not dealt with in general here, but rather only the initial stages after tree-felling, denudation of turf, tethering and trampling, overgrazing, etc. (i.e. the "waste type" of Oseko[11]) are considered.

An example of the floristic composition of *Erigeron*-dominated grassland (see Fig. 9) is given in Table 10.

The ground cover ratio (v) for such pioneer stages is generally less than 1.0, and the total number of species (n) is variable (from *ca.* 7 to 34) but normally about 15–20. The summed dominance ratio (SDR) of the dominant plant is generally within the range 60 to 100, and the value of SDR′ about 11 to 37 (where larger SDR′ values apply to communities with smaller n, and smaller SDR′ values to those with larger n). In the latter case (of low n), there is an overwhelming predominance of dominants such as *Erigeron canadensis*, *Digitaria adscendens*, *Rumex obtusifolius*, *Agrostis palustris*, etc. Some of the dominants in the deteriorated stages are common to pioneer species of

TABLE 10. Floristic composition of *Erigeron* type grassland (example from Ota-shi, Shimane Pref.)

Species	SDR
Erigeron sumatrensis Retz.	88
Artemisia vulgaris L. var. *indica* Maxim.	65
Erigeron annuus Pers.	55
Erigeron canadensis L.	45
Dactylis glomerata L.	35
Rhus javanica L.	34
Agrostis palustris Hudson	33
Miscanthus sinensis Anders.	27
Rubus crataegifolius Bunge	26
Setaria viridis Beauv.	25
Siegesbeckia pubescens Makino	10
Trifolium repens L.	9
Cassia mimosoides L. var. *nomame* Makino	8
Hypericum erectum Thunb.	6

FIG. 9. Grassland dominated by *Erigeron annuus* (central Honshu). The foreground shows a fringing belt of *Pleioblastus Chino*.

old-field successions (e.g. *Erigeron canadensis*, *E. annuus*, *E. sumatrensis*, *Digitaria adscendens*, *Persicaria Blumei*, *P. nodosum*, etc.), whereas others, including trampled plants, are different (e.g. *Poa annua*, *Plantago asiatica*, *Paspalum Thunbergii*, *Rumex obtusifolius*, *Trifolium repens*, *Glysine Soja*, *Agrostis palustris*,

etc.). The DS distribution for *Erigeron* type grasslands has the mode at *ca.* DS=150 and a normal range of DS values from *ca.* 70 to 250 (curve 1, Fig. 4). These figures are rather high when compared with those for pioneer stages of old-field successions, due to the comparatively large number of perennial species usually present in the deteriorated stages. For example, the mean DS values for the *Persicaria vulgaris* and *Erigeron* stages of a normal old-field succession are only 64.4 and 127.5, respectively.[34]

4.7. Other minor grassland types

Besides the six principal grassland types described above, there is also a variety of other minor grassland types. Typical dominants are as follows:

Grassy type: *Arundinella hirta*, *Imperata cylindrica* var. *Koenigii*, *Ischaemum anthroides*, *Calamagrostis hakonensis*, *C. Epigeios*, *Hakonechloa macra*, etc.

Sedge type: *Carex leucochloa*, etc.

Forb type: *Thalictrum Thunbergii*, *Artemisia vulgaris* var. *indica*, *A. vulgaris* var. *vulgatissima*, *Lotus corniculatus* var. *japonicus*, *Comanthosphace sublanceolata* f. *hakonensis*.

Fern type: *Osmunda cinnamomea*.

Shrubby type: *Lespedeza bicolor*, *L. bicolor* var. *japonica*, etc.

It is expected that the precise ecological position and status of these minor types can be given in terms of the general principles used to define the six principal types.

References

1. M. Numata, Ecology of grasslands in Japan, *J. Coll. Arts Sci., Chiba Univ.*, **3** (3), 327–42 (1961).
2. M. Numata, Progressive and retrogressive gradients of grassland vegetation measured by degree of succession, *Vegetatio*, **19**, 96–127 (1969).
3. M. Numata, Quaternary history of grasslands, *Bull. Biogeogr. Soc. Japan*, **26**, 21–27; **27**, 1–8 (1971).
4. S. Honda, *On the Forest Zones of Japan* (Japanese), Ikeda Shoten, Tokyo, 1900.
5. A. Miyawaki and S. Itow, Phytosociological approach to the conservation of nature and natural resources in Japan (paper presented at the Divisional Meeting of Conservation), *11th Pacific Sci. Congr.*, Tokyo, 1966.
6. T. Suzuki, The highest vegetation units of Japan and their areas, *Pedologist* (Japanese; English summ.), **10** (2), 1–7 (1966).
7. W. T. Penfound, A physiognomic classification of vegetation in conterminous United States, *Bot. Rev.*, **33**, 289–326 (1967).

8. H. Walter, *Die Vegetation der Erde in ökophysiologischer Betrachtung, Bd. 1, Die tropischen und subtropischen Zonen,* VEB Gustav Fischer Verlag, Jena, 1961.
9. H. Walter and O. H. Volk, *Grundlagen der Weidewirtschaft in SW-Afrika*, Stuttgart-Ludwigsburg, 1954.
10. J. Papadakis, *Climatic Tables for the World*, Buenos Aires, 1961.
11. M. Oseko, *Studies on Grasslands in Japan* (Japanese), Korinkai, Tokyo, 1937.
12. S. Yoshida, A study on the grassland types and the plant succession of Bokuya in Japan, *Bull. Inst. Agr. Res. Tohoku Univ.* (Japanese; English summ.), **2**, 347–67 (1950).
13. S. Yoshida, Ecological studies on Bokuya in Japan, *ibid.*, **7**, 191–207 (1956).
14. Y. Shimada, An analytical study on the successional change of the *Miscanthus* type vegetation in native grassland owing to cattle and sheep grazing, *Rept. Stud. Upland Farm. Kawatabi Farm, Tohoku Univ., 1957–60*, 57–63, 1961.
15. S. Itow, Preliminary notes on grassland types and their distribution in Kyushu, Japan, *Bull. Fac. Lib. Arts, Nagasaki Univ.*, **9**, 25–31 (1968).
16. S. Itow, Centello–Zoysietum japonicae, a grazed grassland community in Kyushu, Japan, *Jap. J. Ecol.*, **20**, 53–59 (1970).
17. Agricultural Research Council of the Ministry of Agriculture and Forestry, *Methods of Surveying and Measuring the Semi-natural Grassland Vegetation* (Japanese), p. 66, 1959.
18. S. Kawanabe, The climatic adaptation of northern type perennial forage crops, *Anim. Husband.* (Japanese), **10**, 1011–16 (1956).
19. S. Kawanabe, Summer depression and cultural limit of perennial forage crops, *ibid.*, **11**, 132–38 (1957).
20. D. E. McCloud and G. O. Mott, Influence of association upon the forage yield of legume-grass mixtures, *Agron. J.*, **45** (2), 61–65 (1953).
21. T. Suganuma, Phytosociological studies on the semi-natural grassland used for grazing in Japan, I. Classification for grazing land, *Jap. J. Bot.*, **19** (2), 255–76 (1966).
22. T. Suganuma, Phytosociological studies on the semi-natural grassland used for gazing in Japan, II. Ecological strata in pasture vegetation, *Bot. Mag. Tokyo*, **80** (946), 145–60 (1967).
23. *Ed.* A. Miyawaki, Vegetation of Japan compared with other regions of the world, *Encyclop. Sci. Technol.* (Japanese), vol. 3, pp. 535, Gakken, Tokyo, 1967.
24. R. Pichi-Sermolli, An index for establishing the degree of maturity in plant communities, *J. Ecol.*, **36**, 85–90 (1948).
25. M. Numata, Judging grassland vegetation by degree of succession and index of grassland conditions, *Kagaku* (Japanese), **32** (12), 658–59 (1962).
26. M. Numata, Some remarks on the method of measuring vegetation, *Bull. Marine Lab., Chiba Univ.*, **8**, 71–78 (1966).
27. H. C. Hanson, *Dictionary of Ecology*, Philosophical Library, New York, 1962.
28. M. Numata, *Zoysia japonica* pastures in the northernmost and southernmost borders (Japanese), Typescript, 1–16, 1969.
29. M. Numata, Ecological background and conservation of the Japanese islands, *Micronesica*, **5** (2), 295–302 (1969).

30. M. Numata, Grassland vegetation in the vicinity of Choshi, *Bull. Marine Lab., Chiba Univ.* (Japanese; English summ.), no. 4, 39–40 (1962).
31. M. Numata, Grassland vegetation in eastern Nepal, *Ecol. Study and Mount. of Mt. Numbur in East Nepal* (ed. M. Numata) (Japanese; English summ.), p. 74–94, Chiba University, 1965.
32. M. Numata, Ecological studies of dwarf-bamboo type grassland in Japan, *Bull. Grassland Ecol. Res. Group* (Japanese), **6**, 4–16 (1965).
33. M. Numata, Bamboo in Brazil, *Bamboo* (Japanese), **6**, 55–58 (1967).
34. I. Hayashi, Studies on the plant succession in Sugadaira, central Japan, 1. *Res. Rept. Sugadaira Biol. Exptl. Sta.*, **1**, 1–18 (1967).

5

Maritime Vegetation

Kazuo ISHIZUKA*

1. Macroclimate and the Distribution of Maritime Vegetation
2. Vegetation of Salt Marshes
    - 2.1. Salt marshes in eastern Hokkaido
    - 2.2. Southern salt marshes
    - 2.3. Mangrove forests
3. Vegetation of Sandy Beaches and Coastal Sand Dunes
    - 3.1. Foreshore communities and embryonic dunes
    - 3.2. Mobile dune communities
    - 3.3. Fixed dune communities
        - *3.3.1. Hokkaido*
        - *3.3.2. Northern Honshu*
        - *3.3.3. Southern parts of Japan proper*
        - *3.3.4. Subtropical regions*
    - 3.4. Vegetation of dune slacks
4. Vegetation of Coastal Cliffs and Scree

**Faculty of Liberal Arts, Yamagata University, 1-4-12 Koshirakawa-cho, Yamagata-shi 990, Japan*

Japan consists of four main islands plus more than 3600 smaller islands, including the Ryukyus and Bonin Islands, with a total combined coastline of *ca.* 26,500 km, along which many types of maritime vegetation flourish. The vegetational types are influenced more or less profoundly by their close proximity and exposure to the sea. Salt marshes, sandy beaches and coastal sand dunes together fringe the coastal plains, whereas sea cliffs with scree, and shingle beaches, constitute the main habitats of rocky coastal vegetation. The salt marshes are occupied by a distinct group of halophytes whose distribution is dependent on tidal effects and the soil salinity. Conditions on sandy beaches on the other hand are controlled by wave action and by the instability of the substratum; here, salt-tolerant psammophytes or halophytes are found. On sand dunes, the erosion and accumulation of wind-blown sand, and the deposition of salt from evaporated spray, play an important role in creating habitats suitable for the growth of psammophytes. On sea cliffs and scree, the effects of the sea are confined to the salt spray action and desiccation which accompany strong onshore winds.

The maritime provinces of Japan, especially those of coastal plains in the warm-temperate zone, have long been densely populated, i.e. since the introduction of rice cultivation about 2000 yr ago. As a consequence of this, the maritime vegetation has continuously been subject to human interference. Especially since the end of World War II, Japan's accelerated industrialization and urbanization have had a destructive effect on the ecosystem of coastal areas, and serious problems related to nature conservation have now arisen in many areas.

1. Macroclimate and the Distribution of Maritime Vegetation

The horizontal vegetational zones of Japan are usually defined and named according to the major climax forests. As described in Chapter 3, they are, from north to south, as follows: (1) subarctic evergreen coniferous forest, in northeast Hokkaido; (2) cool-temperate broad-leaved deciduous forest, in southwest Hokkaido and northern Honshu; (3) warm-temperate broad-leaved evergreen forest, in southern parts of Japan proper; and (4) subtropical broad-leaved evergreen forest, in the Ryukyus and Bonin Islands. On the other hand, it is pointed out by Numata[1] in his discussion of altitudinal vegetational zones that the number of actual altitudinal vegetational zones in terms of secondary forests or semi-natural grasslands is less than the number defined simply according to climax forests, and their borders are ambiguous.

In the horizontal distribution of maritime vegetation in Japan, it is also

true to say that there is a smaller number of *vegetational* zones, though in terms of *specific* zones it is still possible to recognize all the zones as defined by climax forests. As described below, the subarctic and cool-temperate zones are essentially identical as regards the vegetational types found in salt marshes, mobile coastal dunes, and dune thickets and forests, except those that encroach on the climatic climax. However, subarctic salt marshes and mobile dune vegetation can be defined by the occurrence of certain, specific boreal or arctic species.

The common environment of the sea has the effect of making the distribution of various plant species rather ubiquitous throughout different climatic zones. This ubiquity in turn causes an ambiguity in the borderlines between the vegetational zones. The greatest ambiguities are apparent in the vegetation of sandy beaches and mobile dunes. The borderlines are somewhat better defined in the case of salt marsh vegetation and dune forests. Nevertheless, in the salt marshes of Japan, certain halophytes are known that are distributed throughout the coasts of the northern hemisphere. These are *Suaeda maritima*, *Spergularia marina*, *S. rubra*, *Salicornia europaea*, *Triglochin maritimum* and *Carex subspathacea*. Moreover, *Glaux maritima* var. *obtusifolia* and *Honkenya peploides* var. *major* are other coastal species found throughout the northern hemisphere. On mobile sand dunes, *Elymus mollis* is distributed as far as N. America, while its close relative, *E. arenarius*, is commonly found on sandy coasts of northern Europe.

2. Vegetation of Salt Marshes

In tidal estuaries and lagoons into which sand and mud are carried by the tide and laid on flat shores, i.e. where wave action is at a minimum, the characteristic vegetation of salt marshes is developed. In Japan, such habitats have, for the most part, been transformed into paddy fields or littoral industrial areas by reclamation. However, at present the natural conditions of salt marshes can still be observed over extensive areas of eastern Hokkaido.

The distribution and growth of halophytes along the coasts of Hokkaido, Honshu, Shikoku and Kyushu reveal a rather clear distinction into two groups, northern and southern. The northern group is distributed mainly in Hokkaido and northern Honshu, and consists of *Salicornia europaea*, *Spergularia marina*, *Triglochin maritimum*, *Glaux maritima* var. *obtusifolia*, *Potentilla egedei* var. *groenlandica*, *Puccinellia kurilensis*, *Carex subspathacea* and *C. ramenskii*. Among these species, the last three are confined to subarctic salt marshes in eastern Hokkaido. The southern group consists of *Limonium tetragonum*, *Suaeda maritima*, *S. asparagoides*, *S. japonica*, *Artemisia Fukudo*, *Zoysia sinica* var. *nipponica*, etc. and is found mainly on the coasts of central and western Honshu,

Shikoku and Kyushu. The muddy shores of subtropical lagoons and estuaries in the Ryukyus are occupied by mangrove forests.

2.1. Salt marshes in eastern Hokkaido[2,3]

The salt marsh vegetation of eastern Hokkaido is best developed on the Sea of Okhotsk coast (Abashiri Prov.), along the shores of Lakes Komuketo, Saroma and Notoro, and on the coasts of the Notsuke Straits (Nemuro Prov.), on the compound sand spits of Notsukezaki (Fig. 1) and Lakes Furen and Onneto. Salt marsh vegetation is also found in Akkeshi Bay (Kushiro Prov.), which faces the Pacific Ocean. In most cases, the effects of horse grazing in summer are more or less conspicuous. On the whole, there is a distinct difference in zonal arrangement of vegetation between the Okhotsk and Notsuke Straits–Pacific Ocean sides of the island.

In the salt marshes of the Notsuke Straits and Pacific Ocean coasts, the pioneer community occurring in the lowest parts is dominated by *Triglochin maritimum*. It often tends to form low tussocks, although most of these tussocks are probably the result of destruction of more continuous turf through horse trampling. Under natural conditions, *Carex subspathacea* turf (Fig. 2) develops on the landward side of the *Triglochin* community, though it is invaded by *Potentilla egedei* var. *groenlandica*, *Puccinellia kurilensis*, *Glaux maritima* var. *obtusifolia*, *Juncus gracillimus*, etc. under grazing. *Puccinellia* also invades the bare and sporadic mud flats (called *salt pans*) formed naturally among the

FIG. 1. Aerial view of the compound recurved spit of Notsukezaki (eastern Hokkaido). The island of Kunashiri (of the Kuril group) can be seen in the background. (Photograph by courtesy of the Asahi Shimbun, Tokyo)

FIG. 2. *Carex subspathacea* community on the Shunkunitai salt marsh (eastern Hokkaido).

Carex or *Triglochin* communities. With increasing distance from the sea, where fresh water is supplied underground from the dune forests lying behind, the *Carex* turf adjoins freshwater moor with *Phragmites communis* and *Calamagrostis neglecta* var. *aculeolata*. A transitional zone with predominant *Calamagrostis Epigeios* is also generally found on the landward side of the *Carex* turf. Sandy flats under grazing are dominated by *Juncus gracillimus*, in association with *Glaux maritima* var. *obtusifolia*, *Carex pumila*, etc., and generally abut onto fixed sand dune communities.

Salicornia europaea forms a wide community in the lowermost zone of the salt marshes along the Sea of Okhotsk coast, and is associated with *Spergularia marina* on its landward side. The community adjoins a *Puccinellia kurilensis* community on wetter sandy soil, or a *Glaux maritima* var. *obtusifolia* community in drier habitats. Under grazing pressure, *Juncus gracillimus* forms a narrow belt in the upper parts. A transitional community with predominant *Calamagrostis Epigeios* also occurs here between the salt marshes and sand dunes. On the shores of the lagoon Komuketo, a community dominated by *Carex ramenskii* has been reported. It occupies habitats similar to that of the *C. subspathacea* community of the Notsuke Straits–Pacific Ocean region.

Phragmites communis, which is the common dominant species of freshwater marshes and low moors throughout Japan, can tolerate considerable amounts of salt water. Communities of it develop on muddy tidal marshes throughout Hokkaido and Japan proper, although this species is more frequent in habitats affected by fresh water, where it is admixed with or has an undergrowth of halophytes.

2.2. Southern salt marshes[4)]

In southern parts of Japan proper, salt marsh vegetation is observable only as small fragments of natural stands, or as semi-natural communities such as those of salt farms or abandoned reclaimed land. It is developed on the coasts of the Pacific Ocean, Seto Inland Sea and East China Sea, but is seldom present on the coasts of the Sea of Japan due to the low amplitude of sea-level changes occurring between ebb and flow tides.

There is much evidence to suggest that the wide tidal marshes in larger bays have been gradually reclaimed for rice cultivation during historical time. For example, the progress of land reclamation near the mouth of the Kiso River is inferred to total about 6 km in the last 350 yr.[5)] The coast of Ariake Bay in Kyushu is also famous for its active reclamation in the past. Before World War II, salt marsh vegetation was frequently found on the coasts of the Seto Inland Sea, especially on salt farms. It was also present on the shores of the bays of Sendai, Tokyo and Ise, and in the smaller bays of the Shima Peninsula and northern Kyushu. In recent years, reclamation of

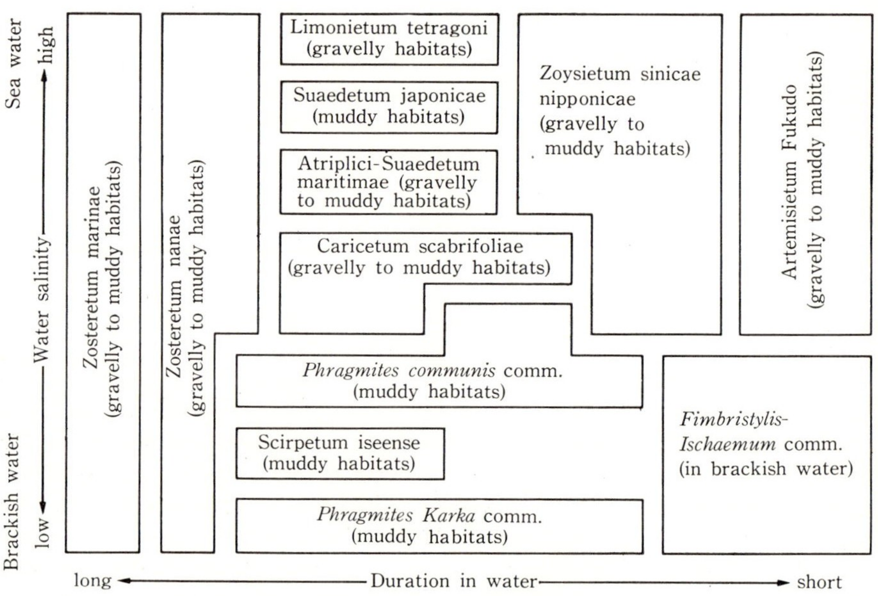

FIG. 3. Schematic diagram illustrating relationships between salt marsh communities and habitat conditions in central and western Japan (after Miyawaki and Ohba, 1969).

tidal marshes has so accelerated through the construction of littoral industrial sites, that the salt marsh vegetation remaining intact at present has become very precious from the viewpoint of conserving natural ecosystems.

Various types of southern salt marsh vegetation, except those of the Sendai Bay area, were studied by Miyawaki and Ohba[4]. In Fig. 3, the distribution of salt marsh communities is shown in relation to habitat conditions, i.e. in relation to water salinity, the duration of submergence in saline water, and the soil texture. It should also be noted that in Matsushima Bay (of the Sendai Bay area), Triglochinetum maritimae is widely developed on muddy soils at the lowermost part of the salt marshes.

2.3. Mangrove forests[6]

In subtropical regions of the Ryukyus, mangrove forests cover muddy tidal marshes of lagoons and estuaries. The best development is found on Iriomote Is (see Fig. 4). Here, six species of mangrove are known: *Kandelia Candel*, *Bruguiera conjugata*, *Sonneratia alba*, *Rhizophora mucronata*, *Avicennia officinalis* and *Lumnitzera racemosa*. They are arranged successively from *Avicennia officinalis* and *Sonneratia alba* at the water-front, through *Rhizophora mucronata* and *Kandelia Candel*, to *Bruguiera conjugata* at the inner portion, which adjoins a land community of *Pandanus tectorius* and *Barringtonia racemosa*. There are differences in dominant species of these forests, not only between Iriomote Is. and the main island of Okinawa, but also between the western and eastern coasts.

Of the above mangrove species, *Bruguiera conjugata* and *Kandelia Candel* are distributed as far as Amami-Oshima Is., and the latter further north of the

FIG. 4. Open mangrove forest of *Rhizophora mucronata* on a tidal marsh around the estuary of the River Nakama (Iriomote Is.). (Photograph by courtesy of K. Odaki)

islands of Yakushima[7] and Tanegashima. At Kiire, in southernmost Kyushu, a stunted stand of *K. Candel* is also known.[8]

(1) Eel-grass communities[3,4]
- *Ass.*: Zosteretum marinae: throughout Japan proper and Hokkaido.
- *Ass.*: Zosteretum nanae: throughout Japan proper and Hokkaido, in sea water shallower than the former.

(2) Salt marsh vegetation in eastern Hokkaido[2,3]
- *Ass.*: Salicornietum europaea (=Salicornietum brachystachyae)
- *Ass.*: Triglochinetum maritimi.
- *Ass.*: Salicornieto–Spergularietum marinae.
- *Ass.*: Glaucetum marinae var. obtusifoliae.
- *Ass.*: Puccinellietum kurilensis.
- *Ass.*: Caricetum subspathaceae (=Puccinellio kurilensis–Caricetum subspathaceae).
- *Ass.*: Caricetum ramenskii.
- *Ass.*: Juncetum gracillimi.

(3) Vegetation of southern salt marshes[4]
- *Ass.*: Atriplici–Suaedetum maritimae: southward from Sendai Bay.
- *Ass.*: Suaedetum japonicae: northern Kyushu.
- *Ass.*: Limonietum tetragoni: southward from the Sendai Bay area.
- *Ass.*: Zoysietum sinicae nipponicae: southward from Sendai Bay.
- *Ass.*: Artemisietum Fukudo: westward from the Kii Peninsula.
- *Ass.*: Triglochinetum maritimi: the Sendai Bay area.
- *Ass.*: Caricetum scabrifoliae: throughout Japan proper.
- *Ass.*: Scirpetum iseensis: Ise Bay region, central Honshu.

3. Vegetation of Sandy Beaches and Coastal Sand Dunes

Throughout the long coastline of Japan, sandy beaches fringe the coastal plains that face the open sea. In Honshu, the artificial planting of *Pinus Thunbergii* has been practiced widely on coastal dunes since feudal times for protecting inland agricultural and urban areas. Whitish sandy beaches fringed by dark-green pine forests have thus long formed the characteristic scenery of sandy coasts in Japan (see Fig. 5). However, the recent utilization of sandy beaches for tourist activities, the clearing and cultivation of areas of coastal pine forests, together with the construction of new harbours and littoral industrial areas, have brought about large-scale destruction of the natural ecosystem of sandy beaches and the coastal dune vegetation.

FIG. 5. The sand spit of Ohsu (Fukushima Pref.). Wide sandy beaches with low dunes occupy the recently formed sand spit (near foreground), whereas *Pinus Thunbergii* forests are established on the older parts of the spit in the distance, as well as on the islets of the tidal lagoon (Lake Marsukawa-ura) on the right.

Low, narrow sand dunes are found on almost all sandy beaches, though extensive dune areas are restricted to particular sites where long-lasting supplies of sand are ensured by large local rivers draining into the sea. In Hokkaido, the coastal dune areas of the subarctic coniferous forest zone are represented by the dunes of northwest Sarobetsu, and those of the sand bars and spits fringing the Sea of Okhotsk coast and the coasts of the Notsuke Straits. Extensive dune areas are also found in the deciduous forest zone at the mouth of the Ishikari River. In Honshu, they are relatively more extensive on coasts of the Sea of Japan side. Besides the famous dune area of the San'in-kaigan National Park (Tottori Pref.), there are many other wide dunes on the fringes of the coastal plains. The relatively few dune areas on the Pacific coast include those on the eastern shores of the Shimokita Peninsula, northern Honshu, which are noted for their relatively wide extent and well-preserved condition.

3.1. Foreshore communities and embryonic dunes[9–12]

A wide variety of plant and animal remains are washed ashore and piled in small heaps on sandy beaches, especially at high tide. This drift material, like growths of plants, acts as an obstacle to windblown sand, and is gradually buried beneath low piles of sand. The sand of such drift-lumps often contains many more seeds than the sand of bare flats, and moreover, the buried, decaying organic debris serves to supply nutrients to and increase

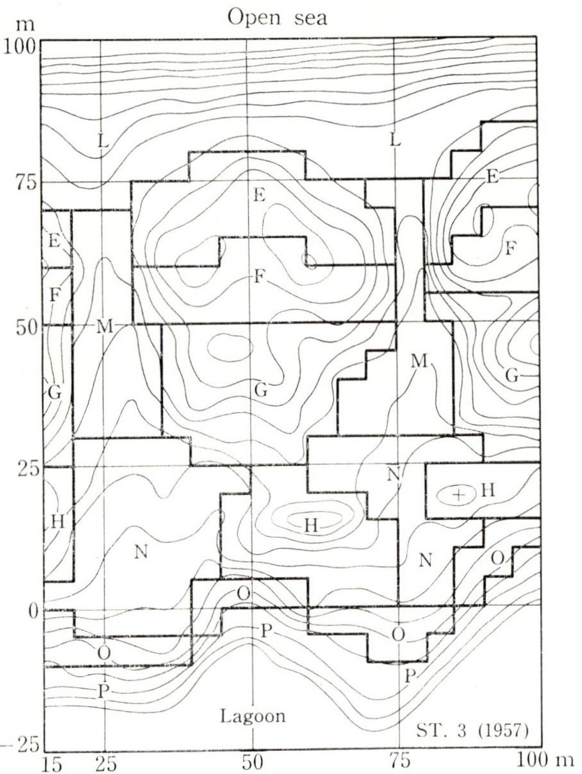

FIG. 6. Map of part of the sand bar at Gamo (Miyagi Pref.). Contours are at intervals of 20 cm, and topographic units are indicated by capital letters.

the water-retaining capacity of the soil. Therefore, seedlings commonly grow on the tops of such drift-lumps. Although the seeds of inland plants generally form the major proportion of the buried seed population, the seedlings are almost solely strand plants.

In northern Honshu and Hokkaido, annual halophytic strand plants such as *Salsola Komarovi*, *Atriplex Gmelini* and *A. subcordata* often grow abundantly on tidal drift, forming a conspicuous but ephemeral community on the foreshore. *Mertensia asiatica*, *Artemisia Stelleriana* and *Honkenya peploides* var. *major* also inhabit foreshore or dune-face sand, though the last plant is more frequently found on shingle beaches.

Slow-growing seedlings of perennial dune plants such as *Elymus mollis*, *Carex Kobomugi*, *Calystegia Soldanelloides*, *Ixeris repens*, *Glehnia littoralis*, *Ischaemum*

anthephoroides, etc. also occur on tidal drift. Moreover, unless there is disturbance by exceptionally high tides or waves, the growth of these plants favours the accumulation of wind-borne sand and the formation of tiny incipient or embryonic dunes within a few years. Where flat beach ridges are newly formed, the embryonic dunes grow and fuse into low dunes within a decade or so. These are then normally beyond the reach of invading waves. Nevertheless, the dunes are often interrupted by wave channels through which tidal floods pass. These channels usually take the form of shallow valleys (see Fig. 6).

3.2. Mobile dune communities[9,10,13)]

Many of the principal components of mobile dune vegetation are distributed rather widely throughout Japan proper and Hokkaido. The common species are *Elymus mollis*, *Carex Kobomugi*, *Zoysia macrostachya*, *Lathyrus maritimus*, *Ischaemum anthephoroides*, *Ixeris repens* and *Glehnia littoralis*. *Zoysia*, *Ixeris* and *Glehnia* are also found on the subtropical beaches of the Ryukyus.

The mobile dunes of the cool-temperate and subarctic forest zones are characterized by the occurrence of *Linaria japonica*, the occurrence of *Carex macrocephala* in northeastern Hokkaido, and of comparatively vigorous growths of *Elymus mollis* as a dune-building plant. *Elymus mollis*, *Carex Kobomugi*, *Ixeris repens*, etc. are found in the unstable zone with loose sand on the seaward slopes and crests of the dunes. *Elymus* and *Carex* accumulate sand most actively due to the upward growth of their rhizome systems (see Fig. 7). A change in dominance, from *Elymus* to *Carex*, is often observed

FIG. 7. Mobile dune communities of the sand spit of Gamo (Miyagi Pref.). *Carex Kobomugi* occupies the surface of the nearest dune; in the middle distance, vigorous growths of *Elymus mollis* have accumulated sand to form lower, incipient dunes.

along the coasts of northern Honshu during the course of development of mobile dunes. This change is related to a decrease in the influence or frequency of tidal floods. *Calystegia*, *Ischaemum*, *Lathyrus* and *Linaria* are confined rather to the semi-stable zone of the more protected backslopes of the dunes.

In the warm-temperate forest zone of central and southwest Honshu, Shikoku and Kyushu, the dune vegetation is characterized by growths of *Vitex rotundifolia*, *Hibiscus Hamabo* and *Wedelia prostrata*. The active dune builders in the unstable zone are *Carex Kobomugi* and *Wedelia prostrata*. Although *Elymus mollis* is found in western Honshu and Kyushu, its growth is restricted to nitrophilous habitats, where the original dune vegetation has been destroyed by human influence. In the semi-stable to stable zone on the landward side of the dunes, *Vitex rotundifolia*, *Hibiscus Hamabo*, *Rosa Wichuraiana*, etc. often form low thickets.

The dune vegetation of the subtropical forest zone in the Ryukyus and Amami Islands is characterized by the occurrence of *Ipomoea pes-caprae*, *Spinifex littoreus*, *Cassytha filiformis*, *Zoysia sinica*, *Thuarea involuta*, etc.

3.3. Fixed dune communities

Old dunes, protected by mobile dunes or wide sandy beaches on the seaward side, gradually become stabilized and covered with a continuous carpet of vegetation. In Japan, the vegetation of such fixed dunes is, in general, able to develop to the arboreal stage, forming thickets or forests. Artificial planting of *Pinus Thunbergii* plays an important role in the consolidation of the dunes of Honshu, Shikoku and Kyushu. The so-called "grey dunes", a type of treeless, fixed dunes consolidated chiefly by mosses and lichens, are not represented in Japan.

The natural dune forests and thickets of Japan proper have since ancient times been subject to strong human interference. Local documents often contain evidence of ancient dune forests that were completely cut down for fuel and then afforested by customary silvicultural practice, under the strict control of the feudal lords, with *Pinus Thunbergii*. The natural conditions of typical dune forests are now observed only in Hokkaido and certain parts of the Shimokita and Tsugaru Peninsulas in northernmost Honshu.

3.3.1. Hokkaido[14,15]

Rosa rugosa thicket is the most widespread of the scrub types found on fixed dunes in Hokkaido. It is often associated with *Malus baccata* var. *mandshurica*. Under natural conditions, this *Rosa* thicket occurs as a narrow belt between the sandy beaches and dune forests, although the zone is generally considerably expanded by cutting back the forests. It is usually utilized for grazing, and semi-natural maritime grassland often develops there. In

FIG. 8. *Quercus dentata* thicket (right) and maritime grassland on the Saroma sand spit (northeast Hokkaido).

summer, the thicket and grassland are also ornamented with other wild flowers such as *Lilium maculatum* var. *dauricum*, *Hemerocallis yezoensis*, *H. Middendorffii*, *Thermopsis lupinoides*, *Convallaria Keiskei*, etc.

Quercus dentata forest is commonly found on fixed dunes along the coast of Hokkaido, though it is replaced by *Q. mongolica* var. *grosseserrata* on the coasts of the Notsuke Straits. Under natural conditions, pure stands of *Quercus dentata* are developed along the beaches at a distance of 100–150 m from the shore. The height of the trees in the forward belt is 1.0–1.5 m, and they are often characterized by a shrubby form or by their occurrence as small trees with crooked, gnarled stems. The tree-crowns in such places give the appearance of having been shaved off level with a hair-clipper (see Fig. 8). The height of the trees increases inland, where they are sparsely mixed with other trees such as *Acer mono*, *Kalopanax pictus*, *Salix Bakko* and *Populus tremula* var. *Davidiana*. Here, *Quercus dentata* is often replaced by *Quercus mongolica* var. *grosseserrata*.

Dune forests of *Picea Glehnii* are rarely found in eastern Hokkaido. However, one particular example is that on the Shunkuntai sand bar of Lake Furen (see Fig. 9). Here the *Picea* forest occurs on ill-drained flats between the dunes and dune slacks (see Fig. 10). Well-drained habitats on the tops of the dunes are occupied by the climatic climax forest of *Abies sachalinensis*. On the compound spit of Notsukezaki,[15] which lies to the north of Lake Furen, *Picea Glehnii* and *Abies sachalinensis* are found admixed sparsely in a dune forest of *Quercus mongolica* var. *grosseserrata*, *Betula Ermanii*, etc. In this forest, *Picea jezoensis* occurs only very rarely. On the fixed coastal dunes of the Sarobetsu Plain, northwest Hokkaido, pure forests of *Abies sachalinensis* are widely developed behind foredunes covered by *Quercus dentata*.

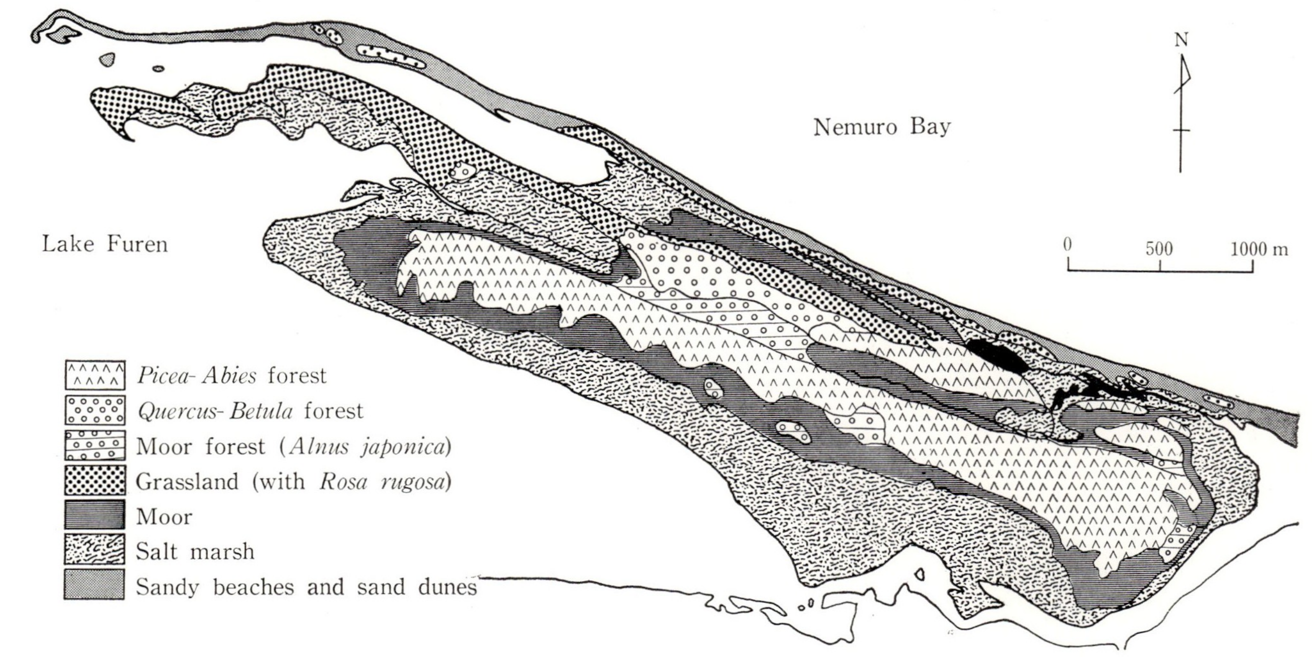

FIG. 9. Vegetation map of the Shunkunitai sand bar of Lake Furen (eastern Hokkaido).

FIG. 10. *Picea Glehnii* forest on the fixed sand dune of Shunkunitai (eastern Hokkaido).

3.3.2. *Northern Honshu*

In the extensive dune area of Sarugamori on the eastern coast of the Shimokita Peninsula,[16] the older dunes are covered either by *Pinus densiflora* forests or by low forests of *Tilia japonica* and *Quercus mongolica* var. *grosseserrata*. The succession of ligneous vegetation on the fixed dunes is thought to originate from *Juniperus conferta* thicket, which then changes to *Pinus densiflora* forest with an undergrowth of *Juniperus*. The *Pinus–Juniperus* forest is next invaded by several broad-leaved deciduous trees, and modified to *Pinus densiflora–Quercus mongolica* var. *grosseserrata* forest. It should be noted that *Pinus Thunbergii* is found only as young plantations on newer dunes. The *Tilia–Quercus* forest is thought to be the topographic climax on the seaward side of the older dunes. It includes small stands of *Thujopsis dolabrata* var. *Hondai* and scattered trees of *Fagus crenata*, indicating that the climatic climax forest of *Fagus* and *Thujopsis* was probably developed on the oldest, inland parts of the dunes. This is further supported by pollen records and by the occurrence of standing tree remains of *Thujopsis* under mobile dunes.

Another extensive dune area, known as Byobuyama, is also developed near the northern extremity of Honshu, on the western coast of the Tsugaru Peninsula.[17] Here, *Quercus dentata* forests extensively cover the seaward slopes of several dune ridges. Old plantations of *Pinus Thunbergii* are also found. On the Pacific coast of Iwate Pref., small sandy beaches are present in the

numerous bays of the Rikuchu-kaigan National Park. This coastline is also noteworthy as one lacking indigenous growths of *Pinus Thunbergii.*

Dune forests of *Pinus Thunbergii* are found widely on sandy coasts of the southern parts of northern Honshu, viz. in Akita Pref. and Yamagata Pref. on the Sea of Japan side, and in Miyagi Pref. and Fukushima Pref. on the Pacific side.[18] Although *P. Thunbergii* is often freely reproducing in older stands, there is some evidence that it was planted in feudal times to control the damaging effects of salt spray and high tides. In northern Honshu, mature stands of *Pinus Thunbergii* are generally invaded by deciduous trees, mainly one of several oak species such as *Quercus serrata*, *Q. dentata* or *Q. mongolica* var. *grosseserrata*, and sometimes by *Pinus densiflora* on the Pacific coast.

3.3.3. Southern parts of Japan proper[18]

In the warm-temperate zone of Honshu, Shikoku and Kyushu, fixed dunes are almost solely occupied by plantations of *Pinus Thunbergii.* Although they were created artificially by afforestation, it is generally accepted that the native distribution of this pine lies approximately over the same area. The pine is frequently observed to invade naturally into dunes already consolidated by shrubs such as *Vitex rotundifolia*, *Juniperus conferta*, etc.

The successional sequence of *Pinus Thunbergii* forest usually originates from young plantations of *Pinus–Imperata cylindrica.* Undergrowths composed of *Vitex rotundifolia* or *Juniperus conferta* may indicate a stage somewhat more advanced than *Imperata.* Older stands have an undergrowth of broad-leaved evergreen shrubs or subtrees such as *Pittosporum Tobira*, *Euonymus japonica*, *Eurya japonica*, *Fatsia japonica*, *Camellia japonica*, *Myrsine Seguinii*, etc. The most advanced stage is represented by *Pinus Thunbergii–Machilus Thunbergii* forest, in which *Machilus* is admixed abundantly in the tree and subtree layers, occasionally with *Cyclobalanopsis glauca*, *Neolitsea sericea*, *Quercus serrata*, etc. Such forest is observed more frequently on the dunes of southern parts of Shikoku and Kyushu.

Where the undergrowth has been frequently disturbed by clearing or raking up of fallen leaves, the forest floor has predominant *Miscanthus sinensis* or open growths of various mosses belonging to the genera *Hypnum*, *Dicranum*, *Rhacomitrium*, *Pogonatum*, *Polytrichum*, etc. On the landward side of the dune forests, especially those of the Pacific Ocean side, mature stands of *Pinus Thunbergii* are frequently observed to be invaded gradually by *Pinus densiflora.*

3.3.4. Subtropical regions[13]

On the sandy coasts of the Ryukyus and Bonin Islands, the thickets of fixed sand dunes and shingle beaches are represented by *Vitex rotundifolia*,

Messerschmidia argentea and *Scaevola taccada*. In the Ryukyus, they adjoin forests of *Pandanus tectorius* on the landward side. Dune forests of *Hernandia peltata* and *Terminalia cattapa* are found in the Bonin Islands.

3.4. Vegetation of dune slacks

When sandy beaches are successively accreted through the formation of new beach ridges, parallel rows of dune ridges are often formed, stretching parallel to the shoreline. The damp or wet hollows between these dune ridges, where the ground-water reaches or approaches close to the sand surface, are known as *slacks*. Dune marshes, low moor or pond vegetation are characteristic of these dune slacks, although salt marsh vegetation may develop where the surface of the sand is low enough to be covered by spring high tides. Topography of this type is frequently met with throughout Japan, but most of the natural dune slacks have been used for cultivating rice. The natural condition of the vegetation is observable only on the sandy coasts of northern and western Hokkaido.

Two rows of dune slacks develop among the three parallel dune ridges on the Sunkunitai sand bar of Lake Furen (see Fig. 9). Low moors dominated by *Phragmites communis*, *Carex* spp. and *Calamagrostis neglecta* var. *aculeolata* are found widely. The most advanced stage of the moor succession is represented by an abundant occurrence of *Oxycoccus quadripetalus*, *Eriophorum vaginatum* and three species of *Sphagnum* (*S. Girgensohnii*, *S. squarrosum* and *S. fimbriatum*) under the *Phragmites*. Fen forest of *Alnus japonica* occurs sporadically, while *Calla palustris*, *Cicuta virosa*, *Lysimachia thyrsiflora*, etc. are found constantly in the moor. Several rows of dune slacks are also present in the wide dune area of Sarobetsu, northwest Hokkaido, and abound in freshwater ponds. They are characterized by abundant growths of *Nuphar pumilum*.

(1) Foreshore communities[11]
Ass.: Salsola–Atriplicetum subcordatae: northern Honshu and Hokkaido.

(2) Mobile dune communities[19]
Ass.: *Elymus mollis–Carex macrocephala* association: northern and eastern Hokkaido.
Ass.: *Elymus mollis–Carex Kobomugi* association: northern Honshu and Hokkaido.
Ass.: *Wedelia prostrata–Carex Kobomugi* association: unstable zones in central and western Honshu, Shikoku and Kyushu.
Ass.: *Zoysia macrostachya–Fimbristylus sericea* association: semi-stable zones in southern parts of Japan proper.
Ass.: *Ixeris repens–Zoysia sinica* var. *sinica* association: unstable zones in the Ryukyus.

Ass.: *Ipomoea pes-caprae–Spinifex littoreus* association: semi-stable zones in the Ryukyus.

(3) Thickets on fixed dunes[19]

Ass.: *Rosa rugosa* association: Hokkaido and northern Honshu.

Ass.: *Vitex rotundifolia* association: southern parts of Japan proper and the Ryukyus.

Ass.: *Thuarea involuta–Cassytha filiformis* association: a grassy type on fixed dunes in the Ryukyus.

(4) Dune forests[18,19]

Ass.: *Angelica anomala–Quercus dentata* association: Hokkaido.

Ass.: Piceetum Glehnii: Hokkaido.

Ass.: *Pinus densiflora–Quercus serrata* association: northern and central Honshu.

Ass.: *Pinus Thunbergii–Quercus mongolica* var. *grosseserrata* association: Sea of Japan side of northern Honshu.

Ass.: *Pinus Thunbergii–Machilus Thunbergii* association: central and western Honshu, Shikoku and Kyushu.

4. Vegetation of Coastal Cliffs and Scree

Cliffs are formed as a result of wave erosion by the open sea at points where hills and mountains reach the coast. At the foot of such cliffs, talus or scree often develops through the accumulation of detached debris. Sea cliffs, scree and shingle beaches constitute the habitats of rocky coastal vegetation, although in Japan the vegetation of coastal cliffs and scree has not yet received a great deal of attention from ecologists or phytosociologists.

The sea cliffs of northern and western Hokkaido are known for their abundance of alpine or boreal species at low altitudes. For example, from the cliffs around the Shiretoko Peninsula, eastern Hokkaido, the following species have been reported: *Juniperus chinensis* var. *Sargentii*, *Dianthus superbus*, *Stellaria ruscifolia*, *Cochlearia oblongifolia*, *Draba borealis*, *Sedum Ishidae*, *Trifolium lupinaster*, *Primula modesta* var. *Matsumurae*, etc. At the tops of the cliffs, *Acer mono* var. *velutinum* forests are commonly found. The seaside scree vegetation is here represented by tall-herb communities composed of *Reynoutria sachalinensis*, *Filipendula kamtschatica*, *Artemisia montana*, *Urtica platyphylla*, *Cacalia hastata* var. *orientalis*, *Petasites japonicus* var. *giganteus*, *Calamagrostis Langsdorffii*, etc.

The succession of the seaside scree vegetation around Mt. Hakodate, southern Hokkaido, originates from an open community of *Lotus corniculatus* var. *japonicus* and *Lathyrus maritimus*. A tall-herb community of *Reynoutria*

FIG. 11. *Juniperus chinensis* var. *Sargentii* on scree, at Shitsukari (Shimokita Peninsula).

sachalinensis, *Artemisia montana*, etc., and *Miscanthus sinensis* sward develop on more stabilized slopes. Most stable habitats are occupied either by dwarf-bamboo thickets of *Sasa senanensis* and another species of *Sasa* sect. Crassinodi, or by low forests of *Quercus dentata* and *Morus bombycis*.

On the seaside cliffs of the Shimokita Peninsula, *Chrysanthemum yezoense*, *Sedum verticillatum*, *Lilium macuratum* var. *dauricum*, *Allium Schoenoprasum* var. *schoenoprasum*, *Lysimachia mauritiana*, *Cyrtomium falcatum*, etc. inhabit the rock crevices. Low thickets of *Juniperus chinensis* var. *Sargentii* are formed in more stable habitats, in which *Empetrum nigrum* var. *japonicum* is often found (Fig. 11). Forests composed of *Acer mono* var. *velutinum*, *Quercus mongolica* var. *grosseserrata* and *Tilia japonica* occur on the cliff-top terraces and gentler slopes, and are thought to represent the topographic climax of these habitats. The vegetation on the long stretches of seaside cliffs along the Rikuchu-kaigan National Park is similar to that just mentioned. However, it is further characterized by the occurrence of *Chrysanthemum nipponicum*, frequent growths of dwarf trees of *Pinus densiflora*, and the occurrence of *Puccinellia nipponica* on shore rocks in southern areas.

The seaside scree vegetation in the warm-temperate zone of southern Japan proper is characterized by growths of *Miscanthus condensatus*, *Peucedanum japonicum*, *Carex oahuensis* var. *robusta*, *Rosa Wichuraiana*, *Pittosporum Tobira*, etc. Various species of *Chrysanthemum* are also distributed allopatrically in this community. An association characterized by *Chrysanthemum pacificum* and *Miscanthus condensatus* has been recorded from the Pacific side of central Honshu and the Izu Peninsula.[13)]

On unstable seaside cliffs in the warm-temperate zone, broad-leaved evergreen shrubs such as *Pittosporum Tobira*, *Euonymus japonica* var. *radicifer*, *Eurya emarginata*, *Rhaphiolepis umbellata*, *Ternstroemia gymnanthera* and *Elaegnus macrophylla* grow with herbaceous species such as *Cyrtomium falcatum*, *Peucedanum japonicum*, *Farfugium japonicum*, *Boehmeria biloba*, etc. The more stable slopes of seaside escarpments are occupied by *Quercus phillyraeoides*, *Pittosporum Tobira*, *Alnus Sieboldiana* or *Daphniphyllum Teijsmanni*.

A very characteristic vegetational type is found on the seaside limestone flats of raised coral reefs along the coasts of the Ryukyus.[13] A *Statice arbuscula–Crossostephium chinense* association is found near shorelines receiving abundant spray. It is replaced successively inland by a *Zoysia tenuifolia–Hedyotis biflora* association, and then an *Osteomeles subrotunda–Aster Miyagii* association. The landward side of the raised coral reefs is covered by thickets and forests coinciding with those on fixed dunes.

On some sea cliffs or small rocky islets, vast numbers of seabirds congregate for breeding. Their influence on plant growth may be considerable, often leading to the formation of distinct communities composed of nitrophilous plants. Among the many rookeries of the black-tailed gull (*Larus crassirostris*) found on the coasts of Japan proper and Hokkaido,[20] the densest colony on Kyojima Is. (Shimane Pref.) is utterly devoid of vegetation during the breeding season from January to late July (see Fig. 12), and an autumnal annual weed community of *Portulaca oleracea* and *Digitaria adscendens* appears thereafter. Other rookeries of the gull are dominated by either annual or perennial nitrophilous plants such as *Chenopodium virgatum*, *Dactylis glomerata*, *Brassica* sp., *Poa annua*, *Agropyron tsukushiense* and *Festuca rubra*.

FIG. 12. Black-tailed gulls flocking on Kyojima Is. (Shimane Pref.). The ground is entirely devoid of vegetation during the breeding season.

References

1. M. Numata, Ecological interpretation of vegetational zonation of high mountains, particularly in Japan and Taiwan, *Erdwissenschaftliche Forschung Bd. IV. Landschaftsökologie der Hochgebirge Eurasiens*, p. 288–99, Franz Steiner, Wiesbaden, 1971.
2. K. Ito, Study on the vegetation of the salt marshes in eastern Hokkaido, Japan, *Sapporo Bull. Bot. Gard. Hokkaido Univ.* (Japanese; English summ.), no. 1, 1–102 (1963).
3. A. Miyawaki and T. Ohba, Studien über Strandsalzwiesengesellschaften auf Ost-Hokkaido (Japan), *Sci. Rept. Yokohama Nat. Univ. Sect. II*, no. 12, 1–25 (1965).
4. A. Miyawaki and T. Ohba, Studien über die Strandsalzwiesengesellschaften auf Honshu, Shikoku und Kyushu (Japan), *ibid.*, no. 15, 1–23 (1969).
5. K. Kurachi, Dynamics of coastal vegetation, especially relating to typhoon damage, *D. Sc. Thesis, Osaka City Univ.* (Japanese), 1964.
6. I. Miyata and N. Otani, The vegetation of Iriomote-Jima, Yaeyama Group, the Ryukyus, *First Rept. Kyushu Univ. Exped. to the Yaeyama Group, Ryukyus* (Japanese; English summ.), p. 23–42, 1963.
7. M. Tatewaki, Tatewaki's iconography of the vegetation of the natural forest in Japan, II. Island of Yakushima, *Res. Bull. Coll. Exptl. Forest, Coll. Agr. Hokkaido Univ.* (Japanese; English summ.), **18**, 53–148 (1957).
8. M. Tatewaki and T. Misumi, Tatewaki's iconography of the vegetation of the natural forest in Japan, III. Forest vegetation of southern Kyushu, *ibid.*, **18**, 149–208 (1957).
9. K. Ishizuka, Ecological studies on the vegetation of coastal sand bars, I. An analysis of vegetation on a recently formed sand bar, *Ann. Rept. Gakugei Fac. Iwate Univ.*, **19** (pt. 3), 37–64 (1961).
10. K. Ishizuka, Ecological studies on the vegetation of coastal sand bars, II. Succession in vegetation and developmental processes of dunes, *ibid.*, **20** (pt. 3), 139–68 (1962).
11. W. Lohmeyer and A. Miyawaki, Zur Kenntnis der ephemeren Meeresstrand- und Flussufer-Vegetation in Japan, *Mitt. flor.-soz. Arbeitsgem. N.F.*, **9**, 78–84 (1962).
12. H. Nobuhara, Analysis of coastal vegetation on sandy shores by biological types in Japan, *Jap. J. Bot.*, **19**, 325–51 (1967).
13. *Ed.* A. Miyawaki, Vegetation of Japan compared with other regions of the world, *Encyclop. Sci. Technol.* (Japanese), vol. 3, pp. 535, Gakken, Tokyo, 1967.
14. M. Tatewaki, Forest ecology of the islands of the North Pacific Ocean, *J. Fac. Agr. Hokkaido Univ.*, **50** (pt. 4), 371–486 (1958).
15. K. Ito, Ecological studies of the Notsukezaki sand spit, Hokkaido, II. The forest communities, *Res. Bull. Coll. Exptl. Forest, Coll. Agr. Hokkaido Univ.* (Japanese; English summ.), **27**, 1–48 (1970).

16. K. Saito, K. Yoshioka and K. Ishizuka, Ecological studies on the vegetation of dunes near Sarugamori, Aomori Pref., *Ecol. Rev.,* **16**, 163–80 (1965).
17. S. Okuda, K. Fujiwara and A. Miyawaki, Pflanzensoziologische Studien über die Vegetation des Tsugaru-Halbinsel, des Berges Iwaki und des Juniko Sees, *Sci. Rept. Tsugaru Peninsula, Mt. Iwaki Nat. Park* (Japanese; German summ.), p. 1–40, Nature Conservation Society of Japan, Tokyo, 1970.
18. K. Yoshioka, *Ecological Studies of Japanese Pine Forests* (Japanese), pp. 198, Nippon Ringyo Gijutsu Kyokai, Tokyo, 1958.
19. *Ed.* K. Yoshioka, *JCT(P) Handbook*, I. Materials for the preparation of IBT-CT check sheets (Japanese), pp. 135, JCT(P), Sendai, 1969.
20. K. Ishizuka, Ecology of the ornithocoprophilous plant communities on breeding places of the black-tailed gull, *Larus crassirostris*, along the coast of Japan, I. Vegetation analysis, *Ecol. Rev.*, **16**, 229–44 (1966).

Additional References

a. E. Iwata and K. Ishizuka, Plant succession in Hachirogata polder, *Ecol. Rev.*, **17**, 37–46 (1967).
b. K. Kurachi, Salt spray damage to the coastal forests, *Jap. J. Ecol.*, **5**, 123–27 (1956).
c. M. Numata, A. Miyawaki and S. Itow, Natural and semi-natural vegetation in Japan, *Blumea*, **20**, 435–96 (1972).
d. M. Tatewaki, Phytosociological studies on the forests of *Picea Glehnii, Res. Bull. Coll. Exptl. Forest, Coll. Agr. Hokkaido Univ.* (Japanese), **13,** 1-181 (1944).
e. M. Tsuda, Studies on the halophilic characters of the strand dune plants and of the halophytes in Japan, *Jap. J. Bot.*, **13**, 332–70 (1961).
f. N. Yano, The subterranean organ of sand dune plants in Japan, *J. Sci. Hiroshima Univ. Ser. B, Div. 2 (Bot.)*, **9**, 139–84 (1962).

6

Mountain Vegetation

Kazuo ISHIZUKA*

1. Altitudinal Vegetation Zones
    1.1. Altitudinal vegetation zones and the warmth index
    1.2. The Sea of Japan and Pacific vegetational types
2. The Montane Zone
    2.1. *Fagus crenata* forests
        2.1.1. Beech forests of the Sea of Japan side
        2.1.2. Beech forests of the Pacific Ocean side
    2.2. Distribution of vegetation and topograply of the montane zone
        2.2.1. Xerophytic forests growing on convex land surfaces
        2.2.2. Valley forests
        2.2.3. Swamp forests
    2.3. Special needle-leaved forests
3. The Subalpine Zone
    3.1. Coniferous forests
        3.1.1. Forests of Honshu and Shikoku
        3.1.2. Forests of Hokkaido
        3.1.3. Riverside forests
    3.2. Subalpine vegetation of the Sea of Japan side
4. The Alpine Zone
    4.1. Effects of wind and snow on the vegetation
    4.2. Scrub communities
        4.2.1. Pinus pumila scrub
        4.2.2. Other types of alpine scrub
    4.3. Snow-bed grassland
        4.3.1. Season-hygrophilous types
        4.3.2. Permanently hygrophilous types
    4.4. Dwarf-shrub heath and windward grassland
    4.5. Alpine desert

**Faculty of Liberal Arts, Yamagata University, 1-4-12 Koshirakawa-cho, Yamagata-shi 990, Japan*

As outlined in Chapter 1 of this book, the islands of Japan form part of the circum-Pacific orogenic/volcanic belt. Although in general the heights of the mountains above sea-level are not extremely great, they do rise by more than 10,000 m when measured from the floor of the Japan Trench. In fact, about 70% of the total land surface is occupied by mountains and hills. The activities and ejecta of the volcanoes have exerted a strong effect on the vegetation, particularly the mountain vegetation, as described in Chapter 8. Also, based on meteorological and geographic factors, two distinct climatic types can be recognized in Japan proper, viz. the Sea of Japan type and the Pacific type, which are separated by the mountainous axis (see also Chapter 1). The former climatic type is characterized by heavy snowfall in winter, due to the cold winds blowing across the Sea of Japan from the Asian continent. These circumstances show a marked influence on both the composition and distribution of plant communities in mountainous areas.

This chapter describes the general features of the mountain vegetation of Japan on the basis of altitudinal vegetation zones, and is thus complementary to Chapter 3, where the discussion is based on the corresponding horizontal forest zones.

1. Altitudinal Vegetation Zones

The altitudinal vegetation zones of Japan's mountains have been distinguished and defined by Takeda[1)] in central Japan. His division has gained wide acceptance, and is now used by most Japanese botanists and geographers. The essential details are as follows.

(1) *The Hilly Zone*. This is the area from sea-level up to an altitude of 500–600 m in central Japan. The climatic climax community is assumed to be warm-temperate broad-leaved evergreen forest. However, the original forests have been so seriously damaged and transformed by human activity that they remain only as small stands in sanctuaries around temples and shrines, and on steep, inaccessible mountain slopes and gorges. In their place, paddy fields now dominate the lowland areas, and deciduous coppice forests of *Quercus serrata*, *Q. acutissima*, etc., plantations of *Cryptomeria japonica*, *Chamaecyparis obtusa*, etc. and grasslands (either for grazing or mowing) of *Miscanthus sinensis*, *Zoysia japonica*, etc. occupy the hills and lower mountainsides.

(2) *The Montane Zone*. This zone lies between 500 and 1500 m a.s.l. in central Japan, and is dominated by cool-temperate broad-leaved deciduous forests. The main components of the climatic climax forest are *Fagus crenata* and *Quercus mongolica* var. *grosseserrata*, although cool-temperate coniferous forests containing *Abies homolepis*, *Chamaecyparis obtusa*, *C. pisifera*, etc. are also

commonly found. *Quercus mongolica* var. *grosseserrata* forms the most extensive secondary or coppice forests of the montane zone.

(3) *The Subalpine Zone*. This zone lies between 1500 and 2500 m a.s.l. in central Japan, and is characterized by evergreen coniferous forests of *Abies Mariesii, A. Veitchii, Tsuga diversifolia*, etc., which form the climatic climax. *Betula Ermanii* is usually present admixed with the conifers in this climax forest. In fact, secondary or pioneer forests of *B. Ermanii* and/or *Larix leptolepis* are also characteristic of the subalpine zone. As will be shown later, the coniferous forests are only rarely found in this zone in mountains close to the coast of the Sea of Japan, since they directly face the prevailing northerly and northwesterly winds of winter.

(4) *The Alpine Zone*. The subalpine coniferous forests occur up to a normal forest limit of about 2500 m a.s.l. or so in central Japan. The alpine zone thus embraces all areas extending from the forest limit to the summits of the mountains, and *Pinus pumila* scrub is believed to represent the climatic climax of this zone, even though it is absent from some recent volcanoes at such altitudes.

1.1. Altitudinal vegetation zones and the warmth index

The above-mentioned altitudinal zones of Takeda correspond well with the horizontal zones of forest vegetation described in Chapter 3, including the type of climatic climax found under different conditions. Therefore, it may be expected that the altitudinal zones also correlate with climatic factors such as temperature, and specifically with the "warmth index" (for definition, see Chapter 1, section 2). Kira[2,3] examined this possible correlation by constructing isotherms for his warmth index from local records of monthly mean temperatures, assuming the average altitudinal rate of change of temperature to be fixed at a value of 0.55°C/100 m. The results are illustrated in Fig. 1, and clearly demonstrate that his assumption was well sustained in its broad outline.

The upper limit of the hilly zone, or the zone of warm-temperate broad-leaved evergreen forest, is defined by the +85°C isotherm. Also, due to the coldness of winter, warm-temperate deciduous forests (including mixed forests with *Abies firma* and/or *Tsuga Sieboldii*) occasionally occur in the upper parts of this zone. The boundary between the montane and subalpine zones corresponds generally to the +45°C isotherm, although the boundary between the cool-temperate and subarctic forests of lowland Hokkaido fits better with the +55°C isotherm. The upper forest limit of the subalpine zone coincides roughly with the +15°C isotherm, although there are some mountains where the alpine zone commences well below the level of this isotherm due to otherwise extreme or adverse conditions at the mountaintops.

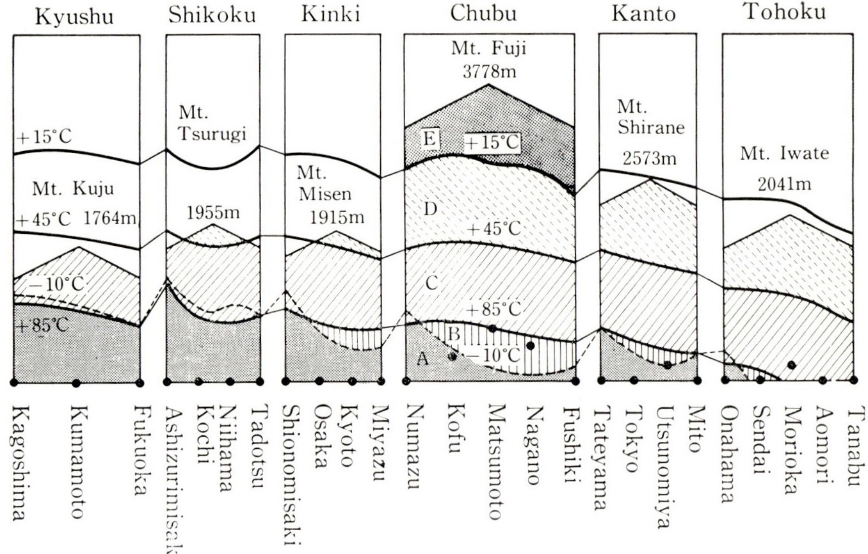

FIG. 1. Relationships between the vertical distribution of forest zones and the warmth and cold indices (after Kira). A, Warm-temperate evergreen forest zone; B, warm-temperate deciduous forest zone; C, cool-temperate deciduous forest zone; D, subalpine coniferous zone; E, alpine zone. The diagram shows the + 15, + 45 and +85°C warmth indices, and the −10°C cold index. Principal meteorological observation sites are indicated by black dots.

1.2. The Sea of Japan and Pacific vegetational types

The altitudinal zonation of vegetation is strongly affected by the climatic differentiation of mainland Japan into Pacific and Sea of Japan types. Imanishi[4)] first reported this based on studies of vegetation zones in the northern Japanese Alps (see Fig. 2). On the Pacific Ocean side, a mixed transitional forest of *Abies firma* and/or *Tsuga Sieboldii* occurs between the hilly and montane zones. In the montane zone, the cool-temperate deciduous forests of *Fagus crenata* are for the most part replaced by cool-temperate coniferous forests of *Abies homolepis*, *Chamaecyparis obtusa*, *C. pisifera*, etc. In the subalpine zone, the dominant conifers are *Abies Mariesii*, *A. Veitchii* and *Tsuga diversifolia*, with occasional occurrence of *Picea jezoensis* var. *hondoensis*. On the Sea of Japan side, *Fagus crenata* forests predominate in the montane zone, abutting directly against the warm-temperate broad-leaved evergreen forests of the hilly zone. The dominant conifer species of the subalpine forests are here confined to *Abies Mariesii* and *Tsuga diversifolia*.

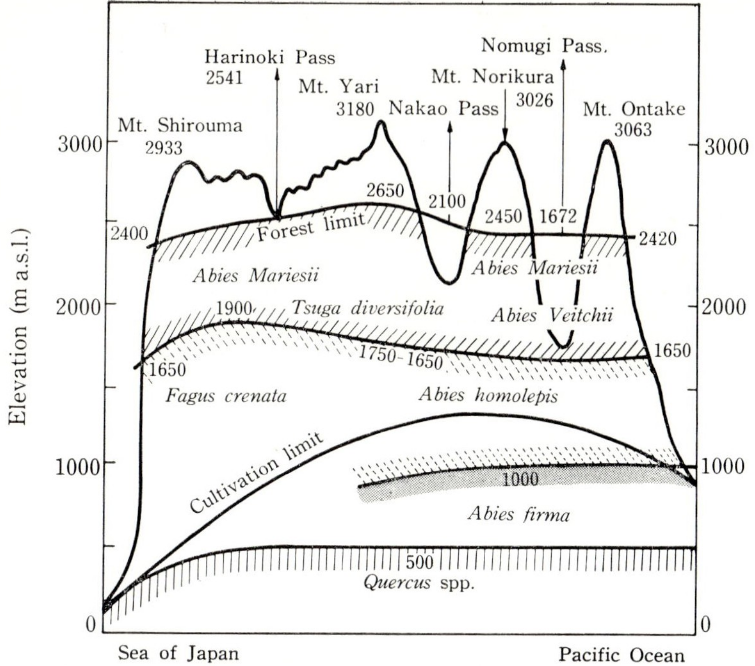

FIG. 2. Profile showing the forest zones across the northern Japanese Alps (after Imanishi).

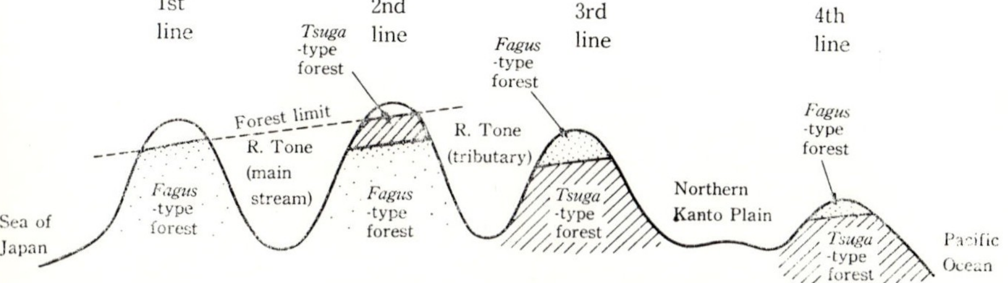

FIG. 3. Profile showing the variation of forest types between the Sea of Japan and Pacific Ocean sides of Honshu (section taken through the northern Kanto district) (after Suzuki).

Suzuki[5] has also confirmed these general trends in the case of the mountains in the northern part of the Kanto district. He and Shidei[6] further pointed out that, in northern Honshu, subalpine coniferous forests are only rarely found, or are entirely lacking from the mountains close to the Sea of Japan coast, i.e. that facing directly the prevailing winds of winter (see Fig. 3). The subalpine zone here is occupied chiefly by deciduous scrub containing *Quercus mongolica* var. *undulatifolia*, *Alnus Maximowiczii*, *Acer Tschonoskii*, etc., various types of snow-bed grasslands and *Sasa kurilensis* scrub. This zone was termed the "quasi-alpine zone" by Shidei.

2. The Montane Zone

2.1. Fagus crenata forests[7]

Fagus crenata forest is the most important climax forest type of the montane zone. It occurs between the "Kuromatsunai line" in southwest Hokkaido (its northern limit), over most of Honshu, Shikoku and Kyushu as far as the Osumi Peninsula in southern Kyushu (its southern limit). From southern Hokkaido to northern Honshu (southwards to a latitude of *ca.* 39°N), it is distributed over the coastal areas up to an altitude of 700 to 1000 m a.s.l (Fig. 4). Further south, it occupies mountainous districts only, starting at 500 m or so and extending up to about 1500 m a.s.l. in central Honshu, and ranging from 1000–1100 m up to 1800–1900 m a.s.l. in Shikoku and Kyushu.

The wide distribution area of this forest embraces various climatic regions, and accordingly a number of community types has been distinguished and recorded, although the most important division is again the segregation of

FIG. 4. *Fagus crenata–Sasa kurilensis* forest at Ohata (Shimokita Peninsula).

vegetational types between the Sea of Japan and Pacific Ocean sides of the islands. These community types were first described in phytosociological terms by Suzuki[5] as different associations, but later, both Miyawaki *et al.*[8,9] and Sasaki[7] raised them to the level of different alliances.

Sasaki's division of Japanese beech forests (his order, Saso–Fagetalia crenatae) into two alliances and seven associations, is consistent with the horizontal distribution of forest types as follows:

Order: Saso–Fagetalia crenatae: occurs in Hokkaido, Honshu, Shikoku and Kyushu.

Alliance: Saso–Fagion crenatae: Hokkaido and the Sea of Japan side of Honshu.

Ass. 1. Abieto sachalinensis var. Mayrianae–Fagetum crenatae: at the northern limit of the community, viz. on Mt. Yurappu in southwest Hokkaido.

Ass. 2. Lindero membranaceae–Fagetum crenatae: from southwest Hokkaido through northern and central Honshu to the Hokuriku region, forming the main part of the alliance.

Ass. 3. (Symploco-) Lindero membranaceae–Fagetum crenatae: southwest Honshu, i.e. northern Kinki and the mountains of the Chugoku region.

Alliance: Sasamorpho–Fagion crenatae: Pacific Ocean side of Honshu, Shikoku and Kyushu.

Ass. 4. Fagetum crenato–japonicae: Pacific Ocean side of northern Honshu.

Ass. 5. Saso nipponicae var. pycnotricae–Fagetum crenatae: northern Kanto region of central Honshu.

Ass. 6. (Sasomorpho-) Abelio–Fagetum crenatae: Tokai and Kinki regions of central Honshu.

Ass. 7. Sapio japonicae–Fagetum crenatae: Shikoku and Kyushu.

2.1.1. Beech forests of the Sea of Japan side (*Fagus crenata–Sasa kurilensis* community (Saso–Fagetum crenatae or Saso–Fagion crenatae))[9–13]

The general physiognomy of the beech forests on the Sea of Japan side of the Japanese islands is characterized best by the dense growths of *Sasa kurilensis* in the shrub layer. The tree layer contains (apart from *Fagus crenata*) other tree species such as *Quercus mongolica* var. *grosseserrata*, *Magnolia obovata*, *Acer mono*, *Kalopanax pictus*, etc. The upper shrub layer is also conspicuous, and consists principally of deciduous shrubs such as *Viburnum furcatum*, *Acer japonicum*, *Magnolia salicifolia*, *Hamamelis japonica* var. *obtusata*, *Rhododendron Albrechti*, *Lindera membranacea*, etc. The forest is further characterized by its

FIG. 5. *Camellia rusticana* as an undergrowth species of *Fagus crenata–Sasa kurilensis* forest (Saruiwa, Iwate Pref.).

more or less constant complement of evergreen shrub species, such as *Aucuba japonica* var. *borealis*, *Camellia rusticana* (Fig. 5), *Daphniphyllum macropodum* var. *humile*, *Ilex leucoclada*, *Skimmia japonica* var. *radicans* and *Torreya nucifera* var. *radicans*.

The occurrence of such a *Fagus crenata–Sasa kurilensis* community is, as mentioned, restricted to the Sea of Japan side where the snow-cover in winter is characteristically heavy. The boundary of this community with the *Fagus crenata–Sasamorpha purpurascens* community coincides approximately with the 50 cm isogram for mean yearly maximum snow-depth. The boundary is often referred to phytogeographically as the "Crassinodi line". In connection with such ecological conditions, the origin of the evergreen shrub species is worthy of special note. Most of the evergreen shrubs mentioned above have their vicarious or mother species (which show an erect growth form) in the warm-temperate evergreen forest zone, and therefore it is generally considered that they have probably attained their prostrate or dwarf growth form in cool-temperate forests under the protection of the heavy snow-cover of winter.

2.1.2. Beech forests of the Pacific Ocean side (*Fagus crenata–Sasamorpha purpurascens* community (Sasamorpho–Fagetum crenatae or Sasamorpho–Fagion crenatae))[8,14–18]

The largest areas of beech forest in Japan are those found on the Sea of Japan side; those of the Pacific Ocean side are rather marginal in character. Here, the tree layer consists of *Fagus crenata* admixed with other tree species

such as *Quercus mongolica* var. *grosseserrata*, *Abies homolepis*, *Fagus japonica*, *Stewartia pseudo-Camellia*, *Carpinus Tschonoskii*, *C. japonica*, *Betula grossa*, *Abies firma*, *Tsuga Sieboldii*, *Picea polita*, etc. Of these species, *Fagus japonica* occurs throughout northern Honshu, while *Abies homolepis* is abundant in central Honshu. *Abies firma* and *Tsuga Sieboldii* are generally found in the lower parts of the beech forests, i.e. in the proximity of the hilly zone.

The shrub layer is again dominated in most areas by dwarf-bamboo species, although *Sasa kurilensis* is here replaced by *Sasamorpha purpurascens* or *Sasa nipponica*. The community is further characterized by the occurrence of various shrubs such as *Stewartia monadelpha*, *Acer Sieboldianum*, *A. micranthum*, *Symplocos coreana*, *Parabenzoin trilobum*, *Lindera umbellata*, etc.

2.2. Distribution of vegetation and topography of the montane zone[8,9,19]

In certain parts of the montane zone, a distinct pattern in the distribution of forest types is apparent, which is closely related to the surface topography. Although *Fagus crenata* forest is generally the climatic climax occupying the greater part of the montane zone, some other forest types are also found locally under special conditions. Three types of habitats, covered by distinctly different forests, may be distinguished as follows: (i) relatively dry, convex land surfaces on rocky slopes or ridges, (ii) moister, concave slopes and valley bottoms, and (iii) flat areas abutting against moors or lakes where the soil is more or less water-logged.

2.2.1. Xerophytic forests growing on convex land surfaces[14]

On rocky ridges or steep slopes in the montane zone are found xerophytic forests which consist generally of *Pinus densiflora*, *P. pentaphylla*, *P. pentaphylla* var. *Himekomatsu*, *Thuja Standishii*, *Quercus mongolica* var. *grosseserrata*, *Tsuga diversifolia*, *Sciadopitys verticillata*, etc. The common occurrence of cool-temperate coniferous forests in this habitat is also to be noted. *Pinus densiflora* forest is found in the lower parts of the montane zone. *Sciadopitys* forest is restricted to central and eastern Honshu, Shikoku and Kyushu. *Pinus pentaphylla* forest (Fig. 6) is found between southern Hokkaido in the north and over northern and central Honshu in the south, while the variety *Pinus pentaphylla* var. *Himekomatsu* is found in the Tokai region and in eastern Honshu, Shikoku and Kyushu.

The undergrowth, although relatively poor in number of species, is luxuriant in growth. The *Sasa* species are rather meagre, but the growth of shrubs such as *Ilex Sugeroki* var. *brevipedunculata*, *Acer Tschonoskii*, *Sorbus commixta*, *Hamamelis japonica* var. *obtusata*, etc., and various ericaceous species such as *Rhododendron* spp., *Vaccinium* spp., etc., is conspicuous.

FIG. 6. *Pinus pentaphylla* forest on a ridge in Mt. Yakeishi (Iwate Pref.).

Several associations have been reported from phytosociological studies of these forests. According to Suzuki,[20] they can be arranged as follows:

Order: Pinetalia pentaphyllae
  Alliance: Pinion pentaphyllae
    Ass.: Rhododendro–Thujetum Standishii: consists of the *Pinus pentaphylla–Thuja Standishii* forests of Hokkaido and Honshu, and the *Tsuga diversifolia* forests of the Sea of Japan side of Honshu.
    Ass.: Ilici–Thujetum Standishii
Order: Pinetalia densiflorae
  Alliance: Pinion densiflorae
    Ass.: Rhododendro–Pinetum azumanum: occurs in eastern Japan and central Kyushu.
    Ass.: Rhododendro–Pinetum kinkianum: Kinki region and the vicinity of the Seto Inland Sea.
    Ass.: Rhododendro–Pinetum Weyrichii: the Pacific Ocean side of southwest Japan.
    Ass.: Rhododendro–Pinetum kiushiani: Mt. Kirishima in southern Kyushu.

2.2.2. *Valley forests*[12,14]

Valley forests are formed on rather moist concave slopes and on relatively stable colluvial or flooded soils of valley bottoms. Representative species of the tree layer are *Pterocarya rhoifolia*, *Magnolia obovata*, *Fraxinus Spaethiana*, *Acer mono*, *Cercidiphyllum japonicum*, *Ulmus Davidiana*, etc. The valley forests of *Fraxinus Spaethiana* are confined to the Pacific Ocean side of central and west-

ern Honshu, Shikoku and Kyushu.

Development of the shrub layer of dwarf-bamboo and deciduous shrubs is suppressed due to the dense growth of the tree and subtree layers. Instead, the herb layer is extremely luxuriant and rich in species. Most important among the flora are the various shade ferns such as *Polystichum tripteron*, *P. retroso-paleaceum*, *Polystichopsis Standishii*, *Dryopteris monticola*, *D. polylepis*, *D. crassirhizoma*, etc. Shade-tolerant shrubs such as *Cephalotaxus Harringtonia* var. *nana* and *Hydrangea macrophylla* var. *acuminata*, tall herbs such as *Rodgersia podophylla*, *Diphylleia Grayi* and *Laportea macrostachya*, and sedges such as *Carex foliosissima*, are also conspicuous.

Associations so far reported from these valley forests may be enumerated as follows:[20]

Alliance: Ulmion Davidianae

Ass.: Ulmetum Davidianae: occurs on alluvial fans in the north Kanto region and Hokkaido.

Alliance: Pterocaryon rhoifoliae

Ass.: Dryopteridi–Fraxinetum commemoralis (=Chrysosplenio–Fraxinetum Spaethianae): central and western Honshu, and Kyushu.

Ass.: Polysticho–Pterocaryetum: Hokkaido and the Sea of Japan side of Honshu and Shikoku.

Ass.: Polysticho–Aesculetum turbinatae: mountains in the Chugoku region.

2.2.3. Swamp forests[21]

Swamp forests of *Fraxinus mandshurica* var. *japonica* and/or *Alnus japonica* occur as small fragments around certain moors and on parts of flat flood-plains in the montane zone. Their distribution is strongly influenced by the level of free ground water and by the degree of development of gley horizons in the soil profile.

Alnus japonica endures water-logged conditions the best, and its forests are found on the peatlands of low moors as well as on gley soils. It is thought that before the birth of human civilization such *Alnus* forests extensively covered the lowland areas of Japan, but that most of them were subsequently destroyed as a result of paddy field cultivation. Nowadays, *Alnus* forests remain only in the lowlands of northeast Hokkaido, at the northern extremity of Honshu (the Shimokita Peninsula), and as small fragments in mountainous areas. A shrub layer of *Rhamnus crenata*, *R. japonica*, etc. often develops, and the herb layer is occupied by moor or hygrophilous species such as *Phragmites communis*, *Lysichiton camtschatense*, *Osmunda cinnamomea* var. *fokiensis*, *Lastrea Thelypteris*, *Carex rhynchophysa*, *C. Augustinowiczii*, etc.

FIG. 7. Swamp forest of *Fraxinus mandshurica* var. *japonica*, associated with *Pterocarya rhoifolia*, near Lake Usori (Shimokita Peninsula).

The *Fraxinus mandshurica* var. *japonica* forests (Fig. 7) are restricted to gley soils around moors and parts of flat flood-plains of valley bottoms. *Pterocarya rhoifolia*, *Aesculus turbinata* and *Ulmus Davidiana* often occur as subordinate species in the tree layer, while *Ilex crenata* var. *paludosa*, *Hydrangea macrophylla* var. *megacarpa*, *Cephalotaxus Harringtonia* var. *nana*, *Syringa reticulata*, etc. are conspicuous in the shrub layer. The herb layer shows differentiation according to the level of free ground water, i.e. from hygrophilous species such as *Lysichiton camtschatense*, *Carex rhynchophysa*, *Trautvetteria japonica* and *Veratrum stamineum*, etc. to tall herbs such as *Filipendula kamtschatica*, *Cacalia hastata* var. *orientalis*, etc.

The following list enumerates associations so far described from Japanese swamp forests:[20]

Order: Fraxino–Ulmetalia

Alliance: Ulmion Davidianae

Ass.: Syringo–Fraxinetum mandshuricae: occurs in lowland areas of Hokkaido.

Ass.: Schisandro–Syringetum: flood-plains in mountains of central Honshu.

Ass.: *Alnus japonica–Stellaria longifolia* association: lowland moors of Hokkaido.

Ass.: *Alnus japonica–Fraxinus mandshurica* var. *japonica* association: lowland areas of Hokkaido and the montane zone of central Honshu.

Ass.: *Fraxinus mandshurica* var. *japonica–Trillium kamtschaticum* association: lowland areas of Hokkaido.

Ass.: *Fraxinus mandshurica* var. *japonica–Meehania urticifolia* association: the montane zone of central Honshu.

2.3. SPECIAL NEEDLE-LEAVED FORESTS[16,22]

The montane/cool-temperate forest zone of Japan is also characterized by the rather sporadic occurrence of coniferous forests. Apart from the comparatively wide distribution of *Abies homolepis* in central and southwest Japan, the most prominent forests consist of conifers belonging to the Taxodiaceae and Cupressaceae. They are of considerable interest from the viewpoint of their phytogeography. At the generic level, *Cryptomeria*, *Sciadopitys* and *Thujopsis* are all endemic to Japan, while both *Chamaecyparis* and *Thuja* are confined to Japan and North America.

Species of *Cryptomeria* (*C. japonica*), *Chamaecyparis* (*C. obtusa* and *C. pisifera*), *Thujopsis* (*T. dolabrata* and *T. dolabrata* var. *Hondai*), *Thuja* (*T. Standishii*) and *Sciadopitys* (*S. verticillata*) have all been utilized widely as timber trees, and constitute the most important species from the viewpoint of Japanese forestry. In particular, the *Cryptomeria* forests of Akita Pref., the *Thujopsis dolabrata* var. *Hondai* forests of Shimokita and Tsugaru (Aomori Pref.), and the *Chamaecyparis obtusa* forests of Kiso (Nagano Pref.) have long been renowned as excellent natural examples of such forests in Japan, although it is generally considered that the pure stands of conifers in them were originally induced artificially from mixed forests of conifers and broad-leaved deciduous trees by the cutting back of the latter. *Cryptomeria japonica*, *Chamaecyparis obtusa* and *Chamaecyparis pisifera* are now the most widely used species for afforestation in Japan.

The "sugi", *Cryptomeria japonica*, is often divided into two varieties, var. *japonica* and *radicans*. The latter is found on the Sea of Japan side of Honshu, and distinguishable from the former by its persistent lower branches which establish new individuals by layering. Indigenous growths of *Cryptomeria japonica* are known from northern Honshu, through Shikoku and Kyushu, to the island of Yaku (south of Kyushu; see Fig. 8), although they are somewhat more conspicuous on the Sea of Japan side of the islands. The native *Cryptomeria* forests of Akita Pref. mentioned above are the most famous, but have recently been cut down to a great extent, resulting in the loss of many wide, pure stands of this conifer from the montane zone. Also, the former mixed forests of *Fagus crenata–Cryptomeria japonica* now remain only as remnants in remote localities of difficult access, where *Cryptomeria* tends in general to occupy drier habitats than *Fagus*. The sugi forest of Yanase (Kochi Pref.) is also well-known, although here *Cryptomeria* is admixed with *Tsuga Sieboldii*.

FIG. 8. *Cryptomeria japonica–Trochodendron aralioides* forest in the lower montane zone of Yaku Is. (Photograph by courtesy of K. Okutomi)

Pseudotsuga japonica, a species endemic to Japan, is also found on steeper slopes, again mixed with *Tsuga Sieboldii*.

The *Cryptomeria* forests on Yaku Is. are used for the production of big timber (*Yaku-sugi*). Huge, wild *Cryptomeria* trees, more than 1000 yr old, grow in the upper hilly zone between 600–700 m and 1200 m a.s.l., as well as in the lower montane region, between 1200 m and 1600–1700 m a.s.l. In the hilly zone, they are admixed with broad-leaved evergreen trees such as *Distylium racemosum*, *Cyclobalanopsis salicina* and *C. acuta*, and conifers such as *Tsuga Sieboldii* and *Abies firma*. The lower montane zone is occupied by *Abies firma*, *Tsuga Sieboldii*, *Chamaecyparis obtusa* and *Trochodendron aralioides*, i.e. in addition to *Cryptomeria japonica*.

Thujopsis dolabrata, or the hiba arbor-vitae, consists of two types, southern and northern. Of these, the southern type (var. *dolabrata*) is widely distributed in the montane zone of Honshu, Shikoku and Kyushu, and tends to admix with *Chamaecyparis obtusa*, *Thuja Standishii*, *Cryptomeria japonica*, *Fagus crenata*, *Tsuga Sieboldii*, *Abies firma*, etc. The northern type (var. *Hondai*) is restricted to southwestern parts of Hokkaido and northern Honshu. It is generally admixed with *Fagus crenata*, *Thuja Standishii*, *Tsuga diversifolia*, *Cryptomeria japonica*, etc. (Fig. 9), although fine, pure forests are not uncommon, especially on the Shimokita and Tsugaru Peninsulas, northern Honshu. Here, the hiba arbor-vitae forms pure stands in the lower parts of the montane zone (i.e. $<$300–400 m in altitude). The upper parts of the montane zone are occupied by *Fagus crenata* forests or mixed forests of *Thujopsis* and *Fagus*. It has been suggested that the pure stands at lower altitudes were, for the most

FIG. 9. Mixed forest of *Thujopsis dolabrata* var. *Hondai* and *Fagus crenata* at Ohata (Shimokita Peninsula).

part, induced from *Fagus–Thujopsis* forests by cutting back of the beech trees. Also, it should be noted that the pure forests of *Thujopsis* tend to cause podsolization of the soil, due to the marked acid reaction of their leaf-fall. On the northern slopes of Mt. Hayachine (Iwate Pref.), *Thujopsis* occurs throughout the montane zone, abutting onto subalpine forests composed chiefly of *Tsuga diversifolia*, *Abies Mariesii* and *Pinus pentaphylla* at altitudes of 1000 m a.s.l. or so.

Chamaecyparis-type forests occur in central and western Honshu, Shikoku, and Kyushu, and tend to form complicated mixtures of various conifers belonging to the Cupressaceae. They are generally regarded as the topographic or edaphic climax forests of steep slopes and ridges. The Kiso district (Nagano Pref.) is especially famous for its fine forests of timber conifers such as *Chamaecyparis obtusa*, *C. pisifera*, *Thujopsis dolabrata*, *Thuja Standishii*, *Sciadopitys verticillata*, *Pinus pentaphylla* and *Tsuga diversifolia*. Among them, the most abundant species are the two cypresses, *Chamaecyparis obtusa* and *C. pisifera*. The "hinoki cypress", *Chamaecyparis obtusa*, grows abundantly on ridges and upslopes, while the "sawara cypress", *C. pisifera*, tends to occupy rather moister habitats in the vicinity of streams, etc.

Towards the Sea of Japan side of Honshu from Kiso, the *Chamaecyparis*-type forests become dominated more by *Thuja Standishii* and also contain various Sea of Japan side elements such as *Ilex Sugeroki* var. *brevipedunculata*, *Pinus pentaphylla*, *Rhododendron Albrechti*, etc. On the other hand, the *Chamaecyparis obtusa* forests of southwestern Japan are characterized by the occurrence of Pacific Ocean side elements such as *Tsuga Sieboldii*, *Pinus parviflora*, *Ilex Sugeroki* var. *longipedunculata*, *Rhododendron Metternichii*, *Pieris japonica*, etc.

Certain phytosociologists[23] have reported a number of distinct associations among the *Chamaecyparis*-type forests of Japan, as follows:

1. Southwest Japan, especially the Pacific Ocean side
   Ass.: Rhododendro–Chamaecyparidetum obtusae: Shikoku and western Honshu.
   Ass.: Pieri–Tsugetum Sieboldii: Shikoku.
   Ass.: Ilici–Tsugetum Sieboldii: western Honshu.
   Ass.: Patrinio–Chamaecyparidetum obtusae: on Mt. Fuji, central Honshu.
   Ass.: *Chamaecyparis obtusa–Rumohra mutica* association, *Pieris japonica* subassociation: on Mt. Fuji and Mt. Chichibu, central Honshu.
2. Northeast Japan, especially the Sea of Japan side
   Ass.: Rhododendro–Thujetum Standishii: *Thuja Standishii–Pinus pentaphylla* forests of Hokkaido and Honshu, and *Tsuga diversifolia* forests of the Sea of Japan side of Honshu.
   Ass.: *Chamaecyparis obtusa–Rumohra mutica* association, *Ilex Sugeroki* var. *brevipedunculata* subassociation: Hida and Kiso districts, central Honshu.
   Ass.: Saseto–Chamaecyparidetum: on Mt. Ontake in the Kiso district.

3. The Subalpine Zone

In the mountains of central Honshu, subalpine vegetation is found between altitudes of about 1500 m and 2500 m a.s.l. The altitude of occurrence drops towards the north, i.e. to between 1000–1300 m and 2000 m a.s.l. in northern Honshu, and to between 0–500 m and 1000–1500 m a.s.l. in Hokkaido. In the mountains of Shikoku and the Kii Peninsula, southwest Japan, the lower limit of the subalpine zone rises to an altitude of about 1700 m a.s.l. or so. The subalpine zone is completely absent from the mountains of Kyushu and the Chugoku district of southwest Japan.

3.1. Coniferous forests

On the basis of the present distribution of subalpine conifers, two distinct areas of occurrence apparently exist; namely, Hokkaido, and the area comprised by Honshu and Shikoku. However, studies of the fossil record by Miki[24] and others have revealed that a marked shifting in the distribution ranges of conifers occurred during the Pleistocene, across the Tsugaru Straits between Honshu and Hokkaido. In Honshu, for example, there are fossil remains of such conifers as *Abies sachalinensis* and *Picea Glehnii*, but these species are now confined almost exclusively to Hokkaido. There is also some evidence

of such transmigration in the present distribution patterns of certain conifers in Japan. For example, *Picea jezoensis* var. *jezoensis*, one of the most important species of the subalpine forests of Hokkaido, has a close relative (var. *hondoensis*) in the mountains on the Pacific Ocean side of central Honshu, again as a member of subalpine forests. Also, *Picea Glehnii*, which is found principally in Hokkaido and in the southernmost parts of Sakhalin and the Kuril Islands, is known to form a small relict stand on a serpentinitic area of Mt. Hayachine, northern Honshu.[25]

3.1.1. Forests of Honshu and Shikoku[9,11,16,26–30]

In the mountains on the Pacific Ocean side of central Honshu, subalpine coniferous forests show their richest floristic composition in Japan. The climax forests are dominated by *Abies Veitchii* and *A. Mariesii*, with *Picea jezoensis* var. *hondoensis* also occurring rather commonly. *A. Veitchii* tends to occupy the lower parts of the subalpine region, i.e. when compared with *A. Mariesii*. *Tsuga diversifolia* grows in rather xerophytic habitats, such as on ridges, hill-crests and steep slopes, as well as on serpentinitic soils, and is often accompanied by *Pinus pentaphylla*, *P. pentaphylla* var. *Himekomatsu*, *P. koraiensis* and *Thuja Standishii*. The distribution ranges of the four rare endemic spruces, *Picea Maximowiczii*, *P. Koyamai*, *P. Shirasawae* and *P. bicolor* var. *reflexa*, are all concentrated in this area, centering on Mt. Yatsugatake and/or the Akaishi mountain range (southern Japanese Alps).

The climax fir forests are often accompanied by *Betula Ermanii*, *B. corylifolia* and *Tsuga diversifolia*. The shrub layer is poorly developed in dense forests, and the forest floor is usually covered by a thick layer of raw, acidic humus. The following species are found on mossy parts of the forest

FIG. 10. Subalpine coniferous forest of *Abies Mariesii* and *Betula Ermanii* (partly in the left distance) on Mt. Zao (Miyagi Pref.). (Photograph by courtesy of K. Saito)

floor: *Oxalis Acetosella*, *Streptopus streptopoides* var. *japonicus*, *Cornus canadensis*, *Ephippianthus Schmidtii*, *Listera nipponica*, *Platanthera ophrydioides*, *Rubus ikenoensis*, *Pyrola renifolia*, *Lycopodium serratum*, etc. In lower-altitude forests, where conditions create moister habitats, tall herbs are often found, such as *Cacalia adenostyloides*, *Diphylleia Grayi*, *Trautvetteria japonica*, *Pteridophyllum racemosum*, etc., together with *Oplopanax japonicus*. The undergrowth of *Tsuga diversifolia* forests growing on ridges and steep slopes contains various ericaceous species, which are characterized by *Ilex Sugeroki* var. *brevipedunculata* on the Sea of Japan side and by *I. Sugeroki* var. *longipedunculata* on the Pacific Ocean side.

The subalpine coniferous forests tend to be poorer in floristic composition both towards the northeast and the southwest, i.e. away from the Pacific Ocean side of central Honshu. On mountains in the northern part of central Honshu, as well as on those of northeast Honshu, the climax forests are thus dominated by *Abies Mariesii* only (Fig. 10), while xerophytic habitats are occupied by either *Tsuga diversifolia*, *Pinus pentaphylla* or *Thuja Standishii*. *Thuja Standishii* often forms dense forests of dwarf trees in moister habitats associated with moors. The density of *Abies Mariesii* in the climax forest tends to decrease in the subalpine zone of northern Honshu, where sparse forests of this fir often occur among the dwarf-bamboo thickets of *Sasa kurilensis*.

The narrow subalpine zones found on the mountains of Shikoku are occupied by sparse forests of *Abies shikokiana*, a species very closely related to *A. Veitchii*. They are situated above the upper montane zone and its predominant *Abies homolepis*. The dwarf-bamboo thickets in this subalpine zone are composed of several *Sasa* species other than *S. kurilensis*, such as *S. ishizuchiensis*, *S. hirtella*, etc.

Native forests of the Japanese larch, *Larix leptolepis*, are confined to subalpine regions of central Honshu, although this tree is most widely used for afforestation of the upper montane region throughout Japan. The native forest[28] occurs as pioneer communities in barren volcanic areas, on landslips, and in drier habitats associated with flood-plains, and also as secondary communities in areas clear-cut of the climax forest. Prominent examples of native larch forests are found on new volcanoes such as Mt. Fuji and Mt. Asama, and especially on Mt. Fuji the growth form is conspicuously dwarf and procumbent above the forest limit.

Betula Ermanii occupies habitats similar to *Larix leptolepis*, although it is found throughout the subalpine regions of Hokkaido, Honshu and Shikoku. It is often the dominant species close to the timber-line for subalpine conifers, and is sometimes found in a shrubby form at even higher altitudes.

3.1.2. *Forests of Hokkaido*[31–33]

In Hokkaido, subalpine conifers form high and fine forests at low altitudes,

extending even to coastal areas in the northeast. However, the forests established on hills and plains have now, for the most part, been cut down. The principal species of the climatic climax forests are *Abies sachalinensis* and *Picea jezoensis*, although in these *Abies–Picea* forests the relative percentages of the two genera are variable. In general, *Picea jezoensis* increases with altitude or with increasing distance northwards.

Abies sachalinensis, a close relative of the *A. sibirica* of Siberia and northern Russia, is distributed widely across Hokkaido, the southern Kuril Islands and Sakhalin. Two varieties, var. *sachalinensis* and var. *Mayriana*, are often distinguished. The latter is found mainly in southwest Hokkaido, while the former is distributed over northeast Hokkaido, the Kuril Islands and Sakhalin. *Picea jezoensis* is a species closely related to the Sitka spruce, *P. sitchensis*, of western North America, and occurs widely in Hokkaido, the southern Kuril Islands, Siberia, Manchuria and Korea.

The lower parts of the subalpine region of Hokkaido are generally covered by *Abies sachalinensis*-dominated forests, with only sparse occurrence of *Picea jezoensis*. *Abies sachalinensis* forms mixed forests with cool-temperate broad-leaved deciduous trees such as *Quercus mongolica* var. *grosseserrata*, *Betula Maximowicziana*, *Carpinus cordata*, *Ulmus laciniata*, *Magnolia obovata*, *Tilia japonica*, etc. The dwarf-bamboo present in the forests is mainly *Sasa paniculata*. The area is further characterized by the frequent occurrence of broad-leaved deciduous forests of cool-temperate trees, such as *Quercus mongolica* var. *grosseserrata*, *Q. dentata*, *Acer mono* var. *glabrum*, *Alnus japonica*, *Fraxinus mandshurica* var. *japonica*, etc. These features have been described by Tatewaki,[32] who assigned the lowland forests of Hokkaido (except those in the southwest) to the so-called "Pan-mixed forest zone", a zone intermediate between the cool-temperate and subarctic zones and found in eastern Asia and northern Europe.

Picea jezoensis–Abies sachalinensis forests with sparsely distributed *Betula Ermanii* are the climatic climax of the upper parts of the subalpine region in Hokkaido. Pure stands of either *Abies sachalinensis* or *Picea jezoensis* occur only rarely. Moreover, the *Picea–Abies* forests may be differentiated according to the type of undergrowth as follows:

(1) *Sasa* type: mainly consisting of *Sasa kurilensis*.
(2) Fern type: *Dryopteris austriaca*, *D. amurensis*, *D. crassirhizoma*, etc., occurring under humid climatic conditions.
(3) Shrub type: *Rhododendron Fauriae* (commonly found on rocky slopes), *Menziesia pentandra*, *Vaccinium axillare*, etc.
(4) *Carex* type: *Carex sachalinensis*, etc.
(5) Moss type: found rarely on rocky slopes, restricted in extent, with

Lycopodium and *Cornus canadensis* characteristic of the ground flora.

(6) Non-plant cover type.

Glehn's spruce, *Picea Glehnii*, plays an important role in the subalpine regions of Hokkaido, rather similar to that of *Tsuga diversifolia*, *Larix leptolepis*, *Thuja Standishii*, etc. in central Honshu. As mentioned above, this spruce is found principally in Hokkaido, but also occurs in restricted localities in the southernmost parts of Sakhalin and the Kuril Islands. *Picea Glehni* often forms fine, pure, uniform forests, which occur in a number of varied but distinctive habitats such as swamps and bogs, serpentinitic areas, coastal sand-dunes, volcanic sands and gravels, and in the neighbourhood of hot springs or rocky slopes. Such native forests are considered to represent the topographic or edaphic climax forests appropriate to these varied habitats. *P. Glehnii* forest is also found as secondary vegetation in areas burnt-out by forest fires in the mountains of eastern Hokkaido.

3.1.3. Riverside forests[33]

Broad-leaved deciduous forests composed of several species of *Populus*, *Salix*, *Alnus*, etc. are known to occur in flood-plains and valleys along riversides, i.e. within regions occupied by the coniferous forests of the subalpine zone. The deciduous forests are generally represented by the pioneer vegetation, which is maintained by the frequent flooding or erosive action of the rivers. Wherever the soil is stable, the forests tend to develop towards the edaphic climax of typical valley forest.

The most conspicuous trees making up these subalpine riverside forests are *Populus Maximowiczii*, *Salix jessoensis*, *S. sachalinensis*, *S. petsusu*, *Toisusu Urbaniana*, *Alnus hirsuta*, etc. The *Populus Maximowiczii–Toisusu Urbaniana* forests are most prominent along the upper parts of rivers, both in central and northern Honshu, as well as in Hokkaido. They stand on rich alluvial soil, and often attain a height of 30 m or so. The underlayer is largely dominated by tall herbs such as *Cacalia hastata* var. *orientalis*, *Petasites japonicus* var. *giganteus*, *Urtica platyphylla*, *Helacleum dulce*, *Angelica anomala*, etc., and sometimes by *Sasa* spp.

Chosenia, a monotypic genus of the Salicaceae, is represented by *C. bracteosa*, which is distributed eastwards from Lake Baikal and extends as far as Kamchatka, Sakhalin and Japan. In Japan, it exhibits a distinctly disjunctive distribution, being confined to two isolated areas, viz. Kamikochi, on the upper reaches of the River Azusa in central Honshu, and along several rivers in the province of Tokachi, central Hokkaido. At these localities, *Chosenia* sometimes forms pure forests, but it is usually found in association with *Toi-*

susu Urbaniana and *Populus Maximowiczii*. It reaches as much as 30 m in height and 1 m in diameter at breast-height.

Several distinct associations have been proposed by different phytosociologists for subalpine coniferous forests. The following list shows a tentative enumeration,[23] including the *Betula Ermanii* forests and broad-leaved deciduous forests of riversides:

1. Subalpine coniferous forests of Honshu and Shikoku
   Ass.: Abietum Mariesii: central and northeast Honshu.
   Ass.: Tsugetum diversifoliae: the Pacific Ocean side of central Japan.
   Ass.: Saseto–Abietum Veitchii: on Mt. Ontake, central Honshu.
   Ass.: Rhododendro–Thujetum: *Tsuga diversifolia* forests of the Sea of Japan side of Honshu.
   Ass.: Abietum shikokianae: Shikoku.
2. Subalpine coniferous forests of Hokkaido
   Ass.: Saso–Piceetum jezoensis: areas of higher elevation.
   Ass.: Dryopteridi–Abietum Mayrianae: areas of lower elevation.
   Ass.: Piceetum Glehnii
3. *Betula Ermani* forests
   Ass.: Alno–Betuletum Ermanii: central and northeastern Honshu and Hokkaido.
   Ass.: Betuletum Ermanii: the Kurobe district on the Sea of Japan side, and the Kanto mountains of central Honshu.
   Ass.: Saussureo–Betuletum Ermanii: on Mt. Fuji, central Honshu.
4. Riverside forests
   Ass.: *Toisusu Urbaniana–Populus Maximowiczii* association: central and northern Honshu, and Hokkaido.
   Ass.: Rubeto–Toisusetum urbanianae: northern parts of the Akaishi mountain range, central Honshu.

3.2. Subalpine vegetation of the Sea of Japan side[9,11,12,26,34]

As mentioned above, subalpine coniferous forests are only rarely found or entirely lacking in mountains near the Sea of Japan coast that face directly the prevailing winds of winter. The heavy snows and strong winds, together with the long duration of snow-cover, prevent the growth of subalpine conifers in this area, and a zone devoid of high forest appears at lower altitudes. The forest limit, at the upper limit of the montane zone, consists of dense, dwarf forests of *Fagus crenata*, while the subalpine regions are occupied by areas of broad-leaved deciduous scrub containing *Quercus mongolica* var. *undulatifolia*, *Alnus Maximowiczii*, *A. Matsumurae*, *Acer Tschonoskii*, *Sorbus commixta*, *Betula Ermanii*, etc., dwarf-bamboo scrub (mainly *Sasa kurilensis*),

FIG. 11. View of the alpine and quasi-alpine zones of the eastern slope of Mt. Iide (Yamagata Pref.). Open stands of *Betula Ermanii*, deciduous thickets, and *Pinus pumila* thickets occur on the ridges, while the convex ground surface under the main ridge is occupied by snow-bed grassland. (Photograph by courtesy of the Yamagata Prefectural Museum)

and various types of extensive snow-bed grasslands. *Betula Ermanii* also forms low, sparse forests near the forest limit (Fig. 11).

Such a pattern in the vertical distribution of vegetation is characteristic of the mountains on the Sea of Japan side of northern Honshu, i.e. of the mountains of the Dewa range and those of the Echigo range. Identical patterns are also found in the central part of the Ou mountain range, from Mt. Akita-koma through Mt. Yakeishi and Mt. Kurikoma to Mt. Funagata. Moreover, it has been reported that a similar situation exists in the mountains near the Sea of Japan coast of Hokkaido, especially in southwestern areas. A coniferous scrub of *Taxus cuspidata* var. *nana* often predominates in subalpine regions of mountains on the Sea of Japan side of southwest Honshu, notably on Mt. Daisen in the Chugoku district. It is commonly thought that this community replaces the broad-leaved deciduous scrub found in northern Japan.

A survey of the horizontal distribution of vegetation on the islands of the north Pacific Ocean reveals a closely resemblant counterpart of the quasi-alpine zone in the vegetation north of the so-called Miyabe Line at the Etorof Straits in the Kuril Islands. Tatewaki[32,35] pointed out the preponderance of *Betula Ermanii* forests, *Alnus Maximowiczii* scrub and *Pinus pumila* scrub in the central Kuril Islands. There is also a conspicuous absence of genera such as *Abies*, *Picea*, *Larix*, *Populus*, *Ulmus* and *Quercus*, whereas *Taxus*, *Acer*, *Sasa*, etc. exhibit some northward extension to the central Kurils. Tatewaki

thus concluded from his analysis of the floristic composition identified that the vegetation of the central and northern Kurils, together with that of the Aleutian Islands, including the Commander Islands, should be assigned to the subarctic zone in spite of the lack of any typical subarctic coniferous forests.

It has recently been revealed, by pollen-analytical studies[36] of material gathered from a number of moors situated on mountains having a quasi-alpine zone, that the growth of subalpine conifers such as *Abies*, *Tsuga* and *Picea* (associated with that of *Betula*) was more extensive in the Upper Pleistocene than it is now. In this context, the present occurrence of *Abies Mariesii* as small stands on Mt. Kurikoma and Mt. Gassan, as well as of *Tsuga diversifolia* on Mt. Iwaki, Mt. Yakeishi and Mt. Iide, is significant. It seems reasonable to regard this present scanty occurrence of *Abies* and *Tsuga* in the quasi-alpine region as relict vegetation from the colder period at the end of the Pleistocene.

Thickets of *Quercus mongolica* var. *undulatifolia* (a dwarf, procumbent form of *Q. mongolica* var. *grosseserrata*) generally predominate in the lower parts of the quasi-alpine zone of northern Honshu. This species is also often dominant on flat and moist habitats around moors and bogs. *Sasa kurilensis* occurs constantly as the undergrowth.

The upper parts of the quasi-alpine region are occupied by various types of vegetation according to the topography and duration of snow-cover in the growing season. Snow remains longest in ravines and depressions on the upper parts of slopes facing south to east (i.e. lying leeward to the prevailing winds of winter), where the growth of deciduous shrubs and dwarf-bamboos is retarded, and snow-bed grasslands flourish over wide areas. *Sasa kurilensis* scrub, as well as broad-leaved deciduous scrub containing *Acer Tschonoskii*, *Sorbus commixta* and occasionally *S. Matsumurana*, occur on the periphery of these snow-bed grasslands.

Alnus Maximowiczii or *A. Matsumurae* scrub is found on steep and rocky slopes and on moister slopes facing north to northeast. It is generally characterized by the absence of *Sasa* spp., and by the occurrence of herbs and grasses such as *Tilingia holopetala*, *Peucedanum multivittatum*, *Smilacina yezoensis*, *Glyceria alnasteretum*, *Calamagrostis Langsdorffii*, etc.

Coniferous scrub containing *Pinus pumila* is found in the quasi-alpine zone in more xerophytic habitats on convex land surfaces and slopes facing south, and *Sasa kurilensis* is also conspicuous here as the undergrowth. Windward grasslands of *Calamagrostis Langsdorffii*, *C. sachalinensis*, etc. are often found on ridges and hilltops exposed directly to strong winds.

Phytosociological studies of the vegetation of the quasi-alpine region in

Japan are by no means complete. However, the following associations have so far been reported by various authors:[23]

Ass.: Nanoquercetum: *Quercus mongolica* var. *undulatifolia* scrub in northern Honshu.

Ass.: Carici–Sasetum: *Sasa kurilensis* scrub and broad-leaved deciduous scrub containing *Acer Tschonoskii* and *Sorbus commixta*, on Mt. Gassan in northern Honshu.

Ass.: Alno–Betuletum Ermanii: *Alnus Maximowiczii* and *A. Matsumurae* scrub in northern Honshu.

Ass.: Melampyro–Rhododendretum trinervis: on Mt. Echigo-sanzan in the Echigo mountain range.

Ass.: Rhododendretum nipponicae: in the same locality.

Ass.: *Taxus cuspidata* var. *nana* association: on Mt. Daisen in southwest Honshu.

4. The Alpine Zone

As stated in the first part of this chapter, the term "alpine zone" is generally applied in Japan to the region above the forest limit of the subalpine coniferous forest zone. It is widely recognized that, in this region, the macroclimate permits the development of scrub vegetation, especially of *Pinus pumila* as the climatic climax, although the summits of a few mountains above 3000 m in central Japan are apparently exceptions. Alpine heaths, grasslands and deserts are found, either as the edaphic or topographic climax induced by some local factor such as exposure to wind, late thawing of the local snow-cover, presence of serpentinitic soil, etc., or as pioneer communities in barren volcanic areas, on landslips, scree, etc.

4.1. Effects of wind and snow on the vegetation

The local distribution of snow is the most important factor governing the distribution of different vegetational types in the alpine zone. In general, snow is swept away from ridges, hill-crests and peaks, as well as from the upper parts of windward slopes, by the strong northerlies or northwesterlies prevailing in winter, and is left as snow-drifts in depressions and valleys on the south- or southeast-facing leeward slopes. On steep mountains, avalanches also play an important role in determining the pattern of snow accumulation in the valleys.

In habitats that become snow-free early, the vegetation is comprised of species which can withstand low temperatures and, in part, heavy wind erosion as well. The dominant plants are mostly xerophilous, since these areas are typically subject to strong desiccation in summer months. The alpine

FIG. 12. The alpine and quasi-alpine zones of Mt. Chokai (Yamagata Pref.). Snow-bed grassland (light in colour) occupies the convex ground surfaces, while *Pinus pumila* thicket (dark) is found on the ridges. In the foreground, *Hemerocallis Middendorffii* var. *esculenta* can be seen in full bloom.

deserts and grasslands found in wind-exposed habitats, alpine heath, and *Pinus pumila* scrub are examples of the so-called "chinophobous" vegetation occurring in these localities (Fig. 12).

On the other hand, the presence of snow-drifts results in a shortening of the growing season, creating other habitats suitable for "chinophilous" vegetation. The extent of the shortening, however, varies considerably within a comparatively small area. Species occurring in snow-bed grasslands can withstand a more or less heavily reduced vegetative season, which retards the growth of shrubs and dwarf-bamboos. Dwarf-bamboo thickets, mainly of *Sasa kurilensis*, as well as scrub comprised of the various broad-leaved deciduous shrubs enumerated in the preceding section, are found in areas peripheral to the snow-bed grasslands with a less reduced growing season (see Fig. 13).

In mountainous areas, the relative extent of chinophilous and chinophobous vegetation naturally differs according to the location of the mountains with respect to the coast of the Sea of Japan or the Pacific Ocean. Thus, there is a preponderance of chinophilous vegetation, and a correspondingly meagre development of chinophobous vegetation, in the quasi-alpine and alpine regions of mountains on the Sea of Japan side of northern Honshu, as mentioned in the preceding section. On the other hand, chinophilous vegetation is found only locally in mountains close to the Pacific coast, such as in the Akaishi mountain range.

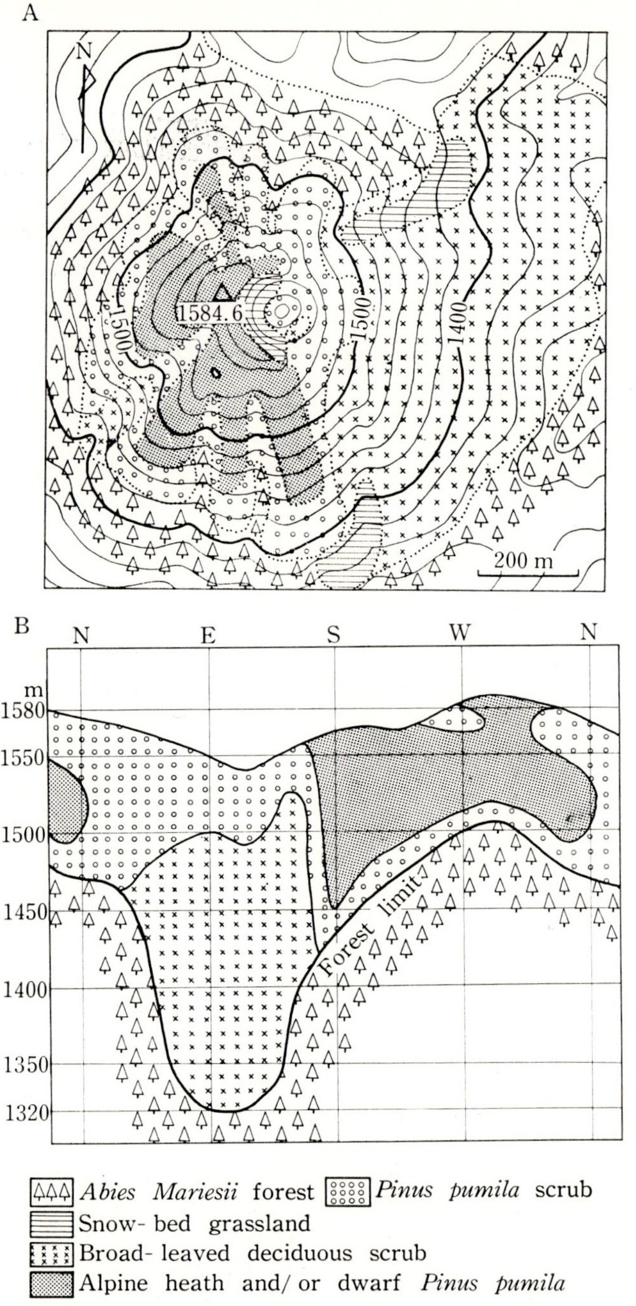

FIG. 13. Vegetation map (A) and distribution pattern of vegetation types in relation to altitude and exposure (B), showing snow-bed grasslands, scrub, heath and forest zones (Mt. Odake, highest point on Mt. Hakkoda, northern Honshu).

4.2. Scrub communities

4.2.1. Pinus pumila scrub[11,12,37]

The alpine regions of Japan's mountains are characterized best by the carpet-like stands of *Pinus pumila*, the Japanese stone pine. This species is widely distributed in the alpine zone from central Honshu northwards. Its southern limit lies on Mt. Tekari-dake, towards the southern end of the Akaishi mountain range, while its western limit is represented by Mt. Hakusan, Ishikawa Pref. In the north, it often drops down almost to sea-level, such as on Rebun Is., northern Hokkaido, as well as in the central and northern Kuril Islands and northernmost part of Sakhalin. It is widely distributed outside Japan in continental northeast Asia, from northern Korea, through Manchuria, the Amur district to eastern Siberia and Kamchatka, and westwards to Lake Baikal. It has not yet been found on either the Commander or Aleutian Islands.

The Japanese stone pine tends to occur as dense, pure communities, often creating wide, impenetrable thickets in the alpine zone (see Fig. 14). It attains a height of more than 1 m at lower altitudes near the forest limit where subalpine conifers such as *Abies Mariesii* and *Tsuga diversifolia* are found admixed with it in less exposed habitats. Under moister conditions, such as in concave depressions, *Pinus pumila* is often admixed with broad-leaved deciduous shrubs such as *Alnus Maximowiczii*, *Sorbus commixta*, *S. sambucifolia*, *S. Matsumurana*, *Acer Tschonoskii*, etc. In windswept habitats such as on ridges and at higher altitudes, the height of the stone pine is greatly reduced, and it is here associated with components of alpine heaths such as *Empetrum nigrum* var. *japonicum*, *Loiseleuria procumbens*, *Arcteria nana*, *Diapensia lapponica* var. *obovata*, etc.

FIG. 14. *Pinus pumila* thickets on a rocky slope of Mt. Hayachine (Iwate Pref.).

The undergrowth of the stone pine thicket is usually rather poorly developed. Among the phanerogams *Vaccinium Vitis-Idaea*, *Cornus canadensis*, *Rubus pedatus*, *Ilex rugosa*, etc. are conspicuous, while the moss layer is composed of species such as *Pleurozium Schreberi*, *Dicranum fuscescens*, *D. majus*, *Ptilium crista-castrensis*, *Hylocomium splendens* and *Hypnum plicatulum*, which are found with high abundance and dominance.

In mountains on the Sea of Japan side of northern Honshu, the occurrence of two coniferous shrubs, *Juniperus communis* var. *nipponica* and *Taxus cuspidata* var. *nana*, is also conspicuous in *Pinus pumila* thickets. These are further characterized here by the occurrence of *Sasa kurilensis* in the undergrowth.

The majority of all stone pine thickets in Japan has been recognized by most Japanese phytosociologists to comprise a single association, Vaccinio–Pinetum pumilae. However, Kobayashi[37] has recently concluded from his studies throughout Japan, including those on cryptogamic plants, that this association should be raised to the rank of an alliance, and that four associations should be distinguished within this alliance, as follows:

Alliance: Vaccinio–Pinion pumilae

- *Ass.*: Ledo–Pinetum pumilae: characterized by the occurrence of *Ledum palustre* var. *diversipilosum*, found in Hokkaido and north-eastern Honshu, and apparently also widely distributed in Siberia.
- *Ass.*: Cetrario–Pinetum pumilae: characterized by fruticose lichens such as *Cetraria crispa* var. *japonica*, *Cladonia rangiferina* and *C. alpestris*, and found on the upper slopes and ridges of the Hida, Kiso and Akaishi mountain ranges, central Honshu, and in the Hidaka mountain range, Hokkaido.
- *Ass.*: Rhodoro–Pinetum pumilae: characterized by *Rhododendron brachycarpum*, and found mainly in mountains on the Sea of Japan side of central and northern Honshu.
- *Ass.*: Rubo–Pinetum pumilae: characterized by *Rubus pedatus*, *Vaccinium ovalifolium*, *Tripetaleia bracteata* and *Gaultheria Miqueliana*, and distributed in central Honshu and Hokkaido.

4.2.2. Other types of alpine scrub

In most alpine regions of Japan, deciduous broad-leaved scrub is also found, either as the topographic or edaphic climax vegetation of the outer margins of bogs, snow-bed grasslands, areas of solfatara vegetation, etc., or as pioneer communities invading landslips or barren volcanic regions. The principal species are *Alnus Maximowiczii*, *Sorbus commixta*, *S. sambucifolia*, *S. Matsumurana*, *Salix Reinii*, *Acer Tschonoskii*, *Ilex Sugeroki* var. *brevipedunculata*, *Juniperus communis* var. *montana*, *J. communis* var. *hondoensis*, *Tripetaleia bracteata*,

Vaccinium ovalifolium, *V. Smallii*, etc. On Mt. Hakkoda in northern Honshu, for example, *Alnus Maximowiczii* scrub and *Salix Reinii* scrub grow on the moister slopes that face east to southeast, whereas *Tripetaleia bracteata–Vaccinium ovalifolium* scrub is characteristic of the immediately surrounding snow-bed grassland. The outer limits of *Sphagnum* bogs and areas of solfatara vegetation in the subalpine zone may also be occupied by mixed thickets of *Acer Tschonoskii*, *Ilex Sugeroki* var. *brevipedunculata*, *Quercus mongolica* var. *undulatifolia*, *Pinus pumila*, etc., in association with dwarf growths of *Abies Mariesii.*

4.3. SNOW-BED GRASSLAND[9,11,12,26,38–40]

Variations in snow-bed vegetation are due partly to the annual exposure time, and partly to the moisture content and condition of the soil. After the snow has melted, the moisture content of the soil is apt to vary considerably. For example, in habitats on the upper slopes of snow-drifts, as well as those at the centre of snow-patches, where soil formation is retarded due to the long-lasting snow-cover, the soil dries up rapidly. Gravelly soils are generally formed in such places, and the prevalent conditions may be termed "season-hygrophilous". On the other hand, soil occurring in flat or gently sloping areas, especially those situated on the lower sides of long-lasting snow-drifts, are moistened or irrigated for longer periods, and thus become peaty through the accumulation of dead organic matter. Such "permanently hygrophilous" conditions usually lead to the formation of bog-like or meadow-like vegetation.

4.3.1. Season-hygrophilous types

Phyllodoce aleutica is the most prominent plant in the vegetation of season-hygrophilous habitats, although *Geum pentapetalum* and various species of *Calamagrostis* (*C. longiseta*, *C. Fauriei*, *C. Langsdorffii*, *C. Matsumurana*, etc.) often become dominant in the outer parts. *Lastrea quelpaertensis*, *Carex pyrenaica*, *C. hakkodensis*, *Arnica unalaschensis*, *Anaphalis alpicola*, *Parnassia alpicola*, *Boykinia lycoctonifolia*, *Lycopodium sitchense* var. *nikoense*, etc. are also found more or less constantly.

At the centre of snow-patches, where the snow-drift remains until late summer or even until late autumn, open communities dominated by various bryophytes occur; these are associated with flowering plants such as *Saxifraga Merkii*, *Carex pyrenaica*, *C. hakkodensis*, *Cardamine nipponica*, etc. Among the bryophytes, there appears to be a regional variation in species composition. *Dicranum falcatum*, *Plagiothecium roeseanum* and *Moerckia nivalis* have been identified on Mt. Hakkoda, *Andreaea nivalis*, *Dicranella secunda*, *Anthelia juratzkana*, etc. on Mt. Gassan, *Dicranum falcatum*, *Marsupella sphacelata*, *Rhacomitrium heterostichum* and *R. anomondontoides* on Mt. Chokai, and *Pogonatum alpinum*, *Arctoa fulvella*, etc. on Mt. Shirouma.

4.3.2. *Permanently hygrophilous types*

In the permanently hygrophilous habitats of snow-bed grasslands, bog-like communities composed chiefly of *Moliniopsis japonica*, *Carex blepharicarpa* and *Fauria crista-galli* are the most widespread. Organic soils showing a textural variation from pure peat to muck are formed, often to a depth of 20 cm or so. They are further characterized by the presence of tiny primroses (*Primula nipponica*) in the mountains of northern Honshu; these are replaced by several types of *P. cuneifolia* in the mountains of central Honshu and Hokkaido. *Narthecium asiaticum*, *Plantago hakusanensis*, *Selaginella selaginoides*, *Parnassia alpicola*, *Tofieldia Okuboi*, etc. are also prominent in these habitats.

In gravelly soils with mobile soil water situated on the lower sides of snow-drifts, various types of meadow-like tall herb communities often flourish. They are composed of *Ranunculus acris* var. *nipponicus*, *Trollius Riederianus* var. *japonicus*, *Trautvetteria japonica*, *Lastrea quelpaertensis*, *Peucedanum multivittatum*, *Tilingia holopetala*, *Mimulus sessilifolius*, etc.

A number of distinct associations have so far been recorded by Japanese phytosociologists[23] from the snow-bed grasslands, as follows:

(1) Season-hygrophilous types

Ass.: Cetralio–Phyllodocetum: the Akaishi mountains, central Honshu.

Ass.: Anaphalio–Phyllodocetum: central and northern Honshu.

Ass.: Calamagrostideto–Phyllodocetum: the Kurobe district, central Honshu, and on Mt. Gassan, northern Honshu.

Ass.: *Salix yezoalpina–Phyllodoce aleutica* association: Hokkaido.

Ass.: *Salix Nakamurana* association: the Akaishi mountains.

Ass.: Sieverseto–Vaccinietum ovalifoliae: on Mt. Ettchu-Asahi, central Honshu.

(1a) Desert types at the centre of snow-patches

Ass.: Cardaminetum nipponicae: central and northern Honshu and Hokkaido.

Ass.: Saxifragetum Merkii: on Mt. Shirouma and Mt. Hotaka, central Honshu.

Ass.: Junceto–Saxifragetum lacinatae: on Mt. Daisetsu, Hokkaido.

Ass.: Carici–Saxifragetum Merkii: on Mt. Daisetsu, Hokkaido.

Ass.: Papaveretum Fauriei: on Mt. Rishiri, Hokkaido.

(1b) Desert types on scree or talus around snow-drifts

Ass.: Veronico–Polygonetum Weyrichii: on Mt. Shirouma and Mt. Hakusan, central Honshu.

Ass.: Violetum brevistipulatae: mountains on the Sea of Japan side of northern Honshu.

(2) Permanently hygrophilous types

Ass.: Primulo–Faurietum crista-galli: on Mt. Iide, northern Honshu.

Ass.: Faurieto–Molinietum: the Oze district and Echigo and Dewa mountains, central and northern Honshu.

Ass.: Aletri–Primuletum hakusanensis: on Mt. Hakusan.

Ass.: Primulo–Caricetum blepharicarpae: the Echigo and Dewa mountains, central and northern Honshu.

Ass.: *Carex blepharicarpa–Fauria crista-galli* association: on Mt. Hakusan and the Echigo mountains, central Honshu.

Ass.: *Plantago hakusanensis* association: the Hida, Echigo and Dewa mountains, central and northern Honshu.

Ass.: *Scirpus caespitosus–Scirpus Maximowiczii* association: Hokkaido.

Ass.: Sphagnetum compacti: the Oze district, central Honshu.

Ass.: *Carex omiana–Sphagnum compactum* association: the Echigo mountains, central Honshu.

Ass.: *Juncus beringensis–Deschampsia caespitosa* association: on the lower sides of snow-drifts irrigated by thawing water; the Akaishi, Hida and Echigo mountains, central Honshu, and the Daisestu and Hidaka mountains, Hokkaido.

4.4. Dwarf-shrub heath and windward grassland[11,12,40,41]

The hilltops, ridges and windblown slopes of the alpine region are the early snow-free areas, and are exposed to the strong effects of wind and desiccation. Alpine desert and windward grassland develop under the most exposed conditions, while alpine dwarf-shrub heath vegetation, having the appearance of a dark-greenish mat, flourishes in more protected and relatively stable habitats.

In the mountains of Japan, the most conspicuous element of alpine heaths is *Empetrum nigrum* var. *japonicum*. Dwarf shrubs such as *Arcteria nana*, *Loiseleuria procumbens*, *Vaccinium uliginosum* and *Diapensia lapponica* var. *obovata* are also characteristic, together with some species of *Cetraria*.

The principal part of the alpine dwarf-shrub heaths of Japan comprises the association Arcterico-Loiseleurietum, although several other associations have been reported by different authors:[23]

Ass.: Arcterico–Loiseleurietum: the Hida and Akaishi mountains, central Honshu.

Ass.: Arctoo–Vaccinietum uliginosi: the Akaishi mountains and on Mt. Ettchu-Asahi, central Honshu.

Ass.: Caricetum Doenitzii: on Mt. Gassan, northern Honshu.

Ass.: Leontopodio–Arcterietum: on Mt. Iide, northern Honshu.

Windward grassland develops in the most severely windswept areas around

mountain summits and ridges, replacing the alpine dwarf-shrub heath. Strong erosion and desiccation by wind are the prime factors in its development. The typical habitats are naturally most common at higher altitudes, although they are also met with at lower altitudes in the subalpine and montane zones.[42]

Although there are regional differences of species, the alpine windward grasslands are generally characterized by the occurrence of leguminous genera such as *Oxytropis*, *Astragalus* and *Hedysarum*, together with certain species of *Leontopodium*, *Euphrasia*, etc. *Oxytropis japonica*, *Minuartia hondoensis*, *Gentiana algida* and *Tilingia Tachiroei* are among the plants occurring more or less commonly in the mountains of central Honshu, while in Hokkaido *Bupleurum triradiatum*, *Patrinia sibirica*, *Minuartia arctica*, etc. are commonly found. Regional speciation of *Oxytropis* is prominent. On Rebun Is. (off the northwest coast of Hokkaido), there is a remarkable example of an alpine windward grassland at lower altitudes.

1. Windward grassland in the alpine region[23]

    Ass.: Kobresio–Oxytropidetum japonicae: the Akaishi mountains and on Mt. Yatsugatake, central Honshu.

    Ass.: Hedysaro–Astragaletum membranaceae: the Akaishi mountains.

    Ass.: Saussureo–Potentilletum Matsumurae: the Akaishi mountains.

    Ass.: *Callianthemum insigne* association: on Mt. Kitadake and Mt. Sammaidake in the Akaishi mountains.

    Ass.: *Oxytropis japonica–Leontopodium Fauriei* association: northern Honshu.

    Ass.: *Oxytropis japonica–Leontopodium hayachinense* association: on Mt. Hayachine, northern Honshu.

    Ass.: *Oxytropis japonica–Leontopodium shinanense* association: the Kiso mountains, central Honshu.

    Ass.: *Oxytropis japonica* var. *sericea* association: in the mountains of central Hokkaido.

    Ass.: *Oxytropis shokanbetsuensis* association: on Mt. Shokanbetsu, western Hokkaido.

    Ass.: *Oxytropis Kudoana* association: the Hidaka mountains, southern Hokkaido.

    Ass.: *Saxifraga Nishidae–Oxytropis rishiriensis* association: on Mt. Rishiri and Mt. Yubari, Hokkaido.

    Ass.: *Oxytropis megalantha–Carex tenuiformis* association: on Rebun Is., northern Hokkaido.

    Ass.: Hypochoreo–Caricetum tenuiformi: in serpentinitic areas of Mt. Apoi, southern Hokkaido.

2. Windward grassland at lower altitudes[23]

Ass.: Orchi–Rhododendretum tsusiophyllae: on Mt. Hakone, central Honshu.

Ass.: *Patrinia triloba* var. *kozushimensis–Calamagrostis autumnalis* association: the Izu Islands.

Ass.: Astilbo–Filipenduletum multijugae: on Mt. Tanzawa and Mt. Hakone, central Honshu.

Ass.: Maianthemo–Rhododendretum kiushiani: on Mts. Kuju, Kirishima, etc., Kyushu.

Ass.: Cacalio–Weigeletum: on Mt. Kirishima, Kyushu.

Ass.: *Rhododendron Metternichii* var. *yakushimanum–Sasa owatarii* association: on Yaku Is., south of Kyushu.

Ass.: *Rhododendron Metternichii* var. *yakushimanum–Juniperus chinensis* var. *Sargentii* association: on Yaku Is.

Ass.: *Shortia soldanelloides* var. *minima–Angelica longeradiata* association: on Yaku Is.

Ass.: *Juniperus chinensis* var. *Sargentii–Carex pisiformis* subsp. *alternifolia* association: on Yaku Is.

4.5. Alpine desert[43,44]

In the alpine regions of the mountains of Japan, alpine desert is restricted to those habitats where the progress of plant successions is retarded by several causes other than the macroclimate. One of the most prominent factors is the repeated accumulation of volcanic ejecta by the many active volcanoes, which provide distinctive habitats for alpine plants (see Chapter 8). Soil instability, due to gravity, wind or water erosion and solifluction, is also important in habitats such as scree or talus, and on escarpments, windswept mountain summits and ridges, etc. The chemical properties of the soil are responsible for the development of deserts around solfataras, etc. in volcanic districts. As stated above, reduction of the growing season by long-lasting snow-drifts may also cause the development of an open vegetation of bryophytes at the centre of snow-bed grasslands.

The floristic composition of alpine deserts varies considerably according to the particular habitat and especially according to the particular region. One of the most widespread types is the Dicentro-Violetum crassae (*Dicentra peregrina–Viola crassa* association) on unstable soil of barren volcanic regions and windswept rocky slopes.

Alpine deserts develop most widely on mountains in serpentinitic areas, where they are very rich in endemic species. Representative serpentinitic mountains in Japan, together with their endemic species, may be enumerated as follows:[43]

Mt. Apoi (southern Hokkaido): *Hypochoeris crepidioides*, *Erigeron Thunbergii* var. *angustifolius*, *Cirsium apoiense*, *Saussurea Riederi* var. *yezoensis*, *Primula hidakana*, *Bupleurum nipponicum* var. *yezoense*, *Viola hidakana*, *Hypericum samaniense*, *Aruncus dioicus* var. *subrotundus*, *Callianthemum Miyabeanum*, *Arenaria Katoana* var. *lanceolata*, *Betula apoiensis*, *Tofieldia coccinea* var. *Kondoi* and *Allium Schoenoprasum* var. *yezomonticola*.

Mt. Yubari (central Hokkaido): *Lagotis glauca* var. *Takedana*, *Primula yuparensis* and *Viola yubariana*.

Mt. Hayachine (northern Honshu): *Leontopodium hayachinense*, *Primula macrocarpa*, *Sanguisorba obtusa* and *Bistorta hayachinensis*.

Mt. Shibutsu and Mt. Tanigawa (central Honshu): *Leontopodium Fauriei* var. *angustifolia* and *Erigeron Thunbergii* var. *heterotrichus*.

Mt. Shirouma and its vicinity (central Honshu): *Cerastium schizopetalum* var. *bifidum* and *Dianthus superbus* var. *amoenus*.

The alpine desert vegetation of Japan has been described in detail by Ohba,[43,44] and the following list is thus prepared according to his classification:

1. Alpine desert excepting serpentinitic areas

Alliance: Stellarion nipponicae: most unstable soils on scree and in barren volcanic areas of Honshu.

- *Ass.*: Melandrio–Cerastetum schizopetalae: the Akaishi mountains, central Honshu.
- *Ass.*: Arabido–Cerastetum schizopetalae: on Mt. Norikura, central Honshu.
- *Ass.*: Arabido–Polygonetum Weyrichii alpinae: volcanic deserts on Mt. Fuji.
- *Ass.*: Deschampsio–Stellarietum nipponicae: on Mt. Norikura.
- *Ass.*: Veronico–Polygonetum Weyrichii: on scree surrounding snowdrifts on the Hida mountains, central Honshu.

Alliance: Violo–Polygonion ajanensis: relatively stable soil of flow-earths (*Strukturboden*), windswept rocky slopes and barren volcanic areas in central and northern Honshu and Hokkaido.

- *Ass.*: Dicentro–Violetum crassae: widely found from central Honshu to Hokkaido, especially in barren volcanic areas.
- *Ass.*: Arenaretum Merckioideae chokaiensis: on old volcanic ash of Mt. Chokai, northern Honshu.
- *Ass.*: Arenaretum Merckioideae: in barren volcanic areas of Mt. Meakan, Hokkaido.
- *Ass.*: Stellario–Polygonetum ajanensis: on wet volcanic ash of Mt. Daisetsu, Hokkaido.

Ass.: Stellario–Pentestemonetum frutescens: the Hidaka mountains, southern Hokkaido.

Ass.: Thlaspi–Polygonetum ajanensis: on rocky slopes of Rebun Is., northern Hokkaido.

Alliance: Saxifrago–Cardaminion nipponicae: at the centre of snow-bed grasslands where the snow-drifts last longest. (Five associations, enumerated in the section on snow-bed grasslands, belong to this alliance.)

2. Alpine desert in serpentinitic mountains

Alliance: Cerasteo–Minuartion vernae japonicae

Ass.: Cerasteo–Minuartetum vernae japonicae: on Mt. Shirouma and in its vicinity, central Honshu.

Alliance: Drabo–Arenarion Katoanae

Ass.: Saussuretum chionophyllae: on Mt. Yubari, Mt. Tottabetsu and Mt. Apoi, Hokkaido.

Ass.: Violetum yubarianae: at the centre of snow-patches on Mt. Yubari, Hokkaido.

Ass.: Arenaretum Katoanae lanceolatae: on Mt. Apoi, Hokkaido.

Ass.: Sanguisorbo–Minuartetum vernae japonicae: on Mt. Hayachine, northern Honshu.

Ass.: Leontopodetum Fauriei angustifoliae: on Mt. Tanigawa and Mt. Shibutsu, central Honshu.

References

1. H. Takeda, Alpine plants, *Saikin Kagaku Koza* (Japanese), **4**, 1–126 (1926).
2. T. Kira, On the altitudinal arrangement of climatic zones in Japan, *Kanchi Nogaku* (Japanese), **2**, 143–73 (1948).
3. T. Kira, Forest zones in Japan, *Ringyo Kaisetsu Series* (Japanese), no. 17, Nippon Ringyo Gijutsu Kyokai, Tokyo, 1949.
4. K. Imanishi, The altitudinal vegetation zones of the Japanese North Alps, *Sangaku* (Japanese), **32**, 269–364 (1937).
5. T. Suzuki, *The Forest Vegetation of East Asia* (Japanese), pp. 137, Kokin-Shoin, Tokyo, 1952.
6. T. Shidei, The forest zones of Ou district, *J. Tohoku Branch Jap. Forestry Soc.* (Japanese), **2**, 2–8, (1952).
7. Y. Sasaki, Versuch zur systematischen und geographischen Gliederung der japanischen Buchenwaldgesellschaften, *Vegetatio*, **22**, 214–49 (1970).
8. A. Miyawaki, T. Ohba and N. Murase, Pflanzensoziologische Studien über die Vegetation in Tanzawa, Prov. Kanagawa, *Sci. Res. Tanzawa Mts.* (Japanese; German summ.), p. 54–102, Kanagawa Prefectural Office, Yokohama, 1964.

9. A. Miyawaki, T. Ohba *et al.*, Pflanzensoziologische Studien über die Vegetation der Umgebung von Echigo-Sanzan und Okutadami, *Sci. Rept. Echigo-Sanzan, Okutadami and Vicinity, Niigata and Fukushima Prefs.* (Japanese; German summ.), p. 57–152, Nature Conservation Soc. Japan, Tokyo, 1968.
10. T. Kashimura, Ecological study of natural forest vegetation in the snowy district along the lower Tadami valley, *Ecol. Rev.*, **17**, 153–70 (1969).
11. T. Suzuki, Die Pflanzengesellschaften und die vertikale Vegetationsstufe vom Hakusan-Gebirge, *Sci. Stud. Hakusan National Park* (Japanese; German summ.), p. 114–56, Ishikawa Prefectural Office, Kanazawa, 1970.
12. T. Suzuki *et al.*, Die Pflanzengesellschaften des Berges Gassan, *Sci. Res. Gassan and Asahi Mts. 1955* (Japanese), p. 144–99, Yamagata Prefectural Office, Yamagata, 1956.
13. K. Yoshioka, Montane forest of Mt. Hakkoda, I. Fagetum, *Ecol. Rev.* (Japanese), **3**, 187–205, 322–30 (1937); **4**, 27–38 (1938).
14. Y. Horikawa and Y. Sasaki, Phytosociological studies on the vegetation of Geihoku district (the Sandankyo Gorge and its vicinity), Hiroshima Pref., *Sci. Rept. Sandankyo Gorge and Yawata Highland* (Japanese; English summ.), p. 85–107, Board of Education, Hiroshima Pref., Hiroshima, 1959.
15. T. Kashimura, Natural forest communities in the Abukuma Mts., *Ecol. Rev.*, **17**, 75–85 (1968).
16. A. Miyawaki *et al.*, Pflanzensoziologische Studien über die Vegetation auf dem Südhang des Berges Fuji, *Sci. Stud. South Slope Mt. Fuji* (Japanese; German summ.), p. 1–59, Shizuoka Prefectural Office, Shizuoka, 1967.
17. T. Yamanaka, Deciduous forests in the cool-temperate zone of Shikoku, *Res. Rept. Kochi Univ.*, **11**, Nat. Sci. I, no. 2, 9–14 (1962).
18. K. Yoshioka, Sociological studies of the forests in the Tohoku district, II. Forests in the *Fagus crenata* climax zone in the suburbs of Sendai, *Bull. Soc. Pl. Ecol.* (Japanese; English summ.), **2**, 69–75 (1952).
19. K. Saito, Ecological approaches to the study of forest distribution in Mt. Hakkoda, Northeast Japan, with special reference to the soil condition, *Ecol. Rev.*, **17**, 217–71 (1971).
20. T. Suzuki, Preliminary system of the Japanese natural forest communities, *Shinrin Ritchi* (Japanese; English summ.), **8**, 1–12 (1966).
21. M. Tatewaki, Tatewaki's iconography of the vegetation of the natural forest in Japan, VI. Vegetation of the deciduous broad-leaved forest along the Okhotsk Sea, Prov. Kitami, Hokkaido (Japanese; English summ.), Kitami Forest Office, Kitami, 1961.
22. T. Yamanaka, Studies on the forest vegetation of Yanaze district, Shikoku, Japan, *Res. Rept. Kochi Univ.* (Japanese; English summ.), **3** (10), 1–12 (1954).
23. *Ed.* K. Yoshioka, *JCT(P) Handbook*, I. Materials for the preparation of IBP-CT check sheets (Japanese), pp. 135, JCT(P), Sendai, 1969.
24. S. Miki, Pinaceae of Japan, with special reference to its remains, *J. Inst. Polytech. Osaka City Univ. Ser. D.*, **8**, 221–72 (1957).

25. K. Ishizuka, A relict stand of *Picea Glehnii* Masters on Mt. Hayachine, Iwate Pref., *Ecol. Rev.*, **15,** 155–62 (1961).
26. A. Mitawaki, S. Itow and S. Okuda, Pflanzensoziologische Studien über die Vegetation der Umgebung von Aizukomagadake und Tashiroyama (Fukushima Pref.), *Sci. Rept. Mt. Aizukoma, Tashiroyama and its vicinity, Fukushima Pref.* (Japanese; German summ.), p. 15–43, Nature Conservation Soc. Japan, Tokyo, 1967.
27. M. Tatewaki, K. Ito and M. Tohyama, Phytosociological study on the forests of Japanese hemlock (*Tsuga diversifolia*), *Res. Bull. Coll. Exptl. Forest, Coll. Agr. Hokkaido Univ.* (Japanese; English summ.), **23**, 83–146 (1963).
28. M. Tatewaki, K. Ito and M. Tohyama, Phytosociological study on the forests of Japanese larch (*Larix leptolepis* Gordon), *ibid.*, **24**, 1–176 (1965).
29. T. Yamanaka, On the subalpine forest vegetation in Shikoku, Japan, *Bot. Mag. Tokyo* (Japanese; English summ.), **72**, 120–25 (1959).
30. K. Yoshioka, Montane forest of Mt. Hakkoda, II. Abietum, *Ecol. Rev.* (Japanese), **4**, 150–58, 227–49, 253–64 (1938).
31. M. Tatewaki, Phytosociological studies on the forests of *Picea Glehnii, Res. Bull. Coll. Exptl. Forest, Coll. Agr. Hokkaido Univ.* (Japanese), **13**, 1–181 (1944).
32. M. Tatewaki, Forest ecology of the islands of the north Pacific Ocean, *J. Fac. Agr. Hokkaido Univ.*, **50**, 369–486 (1958).
33. M. Tatewaki *et al., A Memoir of the Scientific Investigation of the Primeval Forests in the Headwaters of the River Ishikari, Hokkaido, Japan* (Japanese; English summ.), Sapporo, 1955.
34. K. Ishizuka, H. Tachibana and K. Saito, Vegetation of Mt. Chokai, *Sci. Rept. Mt. Chokai and Tobishima Is.* (Japanese), p. 52–88, Sci. Res. Ass. Yamagata Pref., Yamagata, 1972.
35. M. Tatewaki, Geobotanical studies on the Kuril Islands, *Acta Horti Gotoburgensis*, **21**, 43–123 (1957).
36. M. Yamanaka, K. Saito and K. Ishizuka, Historical and ecological studies of *Abies Mariesii* on Mt. Gassan, the Dewa Mts., northeast Japan, *Jap. J. Ecol.*, **23**, in press.
37. K. Kobayashi, Phytosociological studies on the scrub of dwarf pine (*Pinus pumila*) in Japan, *J. Sci. Hiroshima Univ. Ser. B, Div. 2* (*Bot.*), **14**, 1–52, 1971.
38. K. Ishizuka, Plant communities developed in the place of snow-patches on Mt. Hakkoda, *Sci. Rept. Tohoku Univ. Ser. 4* (*Biol.*), **18**, 494–506 (1950).
39. K. Ogasawara, T. Suzuki and Y. Yuki, The effects of snow on vegetation, with special reference to the snow-patch vegetation on Mt. Gassan, *Sci. Res. Gassan and Asahi Mts. 1956* (Japanese), p. 214–40, Yamagata Prefectural Office, Yamagata, 1957.
40. H. Shimizu, Phytosociological studies of alpine herbaceous and dwarf shrub communities on the Iide Mts., *Jap. J. Ecol.* (Japanese; English summ.), **17**, 149–56 (1967).
41. K. Asano and T. Suzuki, Pflanzensoziologische Behandlung über die alpine Vegetation des Akaishi-Gebirges und der Krautartigen alpinen Heide, *ibid.*, **17**, 251–62 (1967).

42. A. Miyawaki *et al.*, *Vegetationskundliche Studien auf den Berg Hakone* (Japanese; German summ.), pp. 60, Yokohama, 1969.
43. T. Ohba, Über die Serpentine-Pflanzengesellschaften der alpinen Stufe Japans, *Bull. Kanagawa Pref. Mus.* (Japanese; German summ.), **1**, 37–64 (1968).
44. T. Ohba, Eine pflanzensoziologische Gliederung über die Wüstenpflanzengesellschaften auf alpinen Stufen Japans, *ibid.*, **1**, 23–70 (1970).

Additional References

a. Agency for Cultural Affairs, *Vegetation Maps and Maps of Natural Monuments* (published separately for each prefecture), Ministry of Education, Tokyo, 1969–.
b. Y. Hayashi, The natural distribution of important trees indigenous to Japan. Conifers, reports I, II and III, *Bull. Gov. Forest Exptl. Sta.* (Japanese; English summ.), no. 48, 1–240; no. 55, 1–251; no. 75, 1–173, 1951–54.
c. E. Iwata, Ecological studies on the secondary vegetation of Kitakami mountain range, with special reference to grassland vegetation, *Res. Bull. Fac. Agr. Gifu Univ.* (Japanese; English summ.), **30**, 288–430 (1971).
d. A. Miyawaki and S. Itow, Phytosociological approach to the conservation of nature and natural resources in Japan (paper presented at the Divisional Meeting of Conservation), *11th Pacific Sci. Congr.*, p. 1–5, Tokyo, 1966.
e. *Ed.* A. Miyawaki, Vegetation of Japan compared with other regions of the world, *Encyclop. Sci. Technol.* (Japanese), vol. 3, pp. 535, Gakken, Tokyo, 1967.
f. M. Numata, A. Miyawaki and S. Itow, Natural and semi-natural vegetation in Japan, *Blumea*, **20**, 435–96 (1972).
g. Society of Forest Environment, *Forest Environment Map of Japan: Soils, Vegetation, Precipitation, Snow Depth and Warmth Index* (Japanese; English summ.), Tokyo, 1972.
h. K. Takahashi, Studies on vertical distribution of the forests in Middle Honshu, *Bull. Gov. Forest Exptl. Sta.* (Japanese; English summ.), no. 142, 1–171 (1962.)
i. Vegetation Research Society of Nagano Pref., *Scientific Studies on the Vegetation of Nagano Pref.* (Japanese), pp. 75, Nagano Prefectural Office, Nagano, 1971.
j. T. Yamanaka, Sociological studies on the serpentine vegetation, VII. General consideration on the serpentine vegetation in Shikoku, *Bull. Fac. Educ. Kochi Univ.* (Japanese; English summ.), no. 11, 87–104 (1959).
k. T. Yamanaka, The forest and scrub vegetation in limestone areas of Shikoku, Japan, *Vegetatio*, **19**, 286–307 (1969).
l. *Ed.* K. Yoshioka, Systematics and dynamics of the terrestrial plant communities in Japan, *Ann. Rept. JIBP-CT(P)* (Japanese), *Fiscal 1967*, pp. 210 (1968).
m. *Ed.* K. Yoshioka, Types and conservation of terrestrial plant communities in Japan, *ibid.*, *Fiscal 1968*, pp. 100; *Fiscal 1969*, pp. 148; *Fiscal 1970*, pp. 178; *Fiscal 1971*, pp. 143; *Fiscal 1972*, in press (1969–73).

7

Aquatic and Wetland Vegetation

Kuniji YOSHIOKA*

1. Aquatic Vegetation of Lakes and Ponds
    1.1. Lake Suwa
    1.2. Lake Biwa
    1.3. Ogura Pond
    1.4. Lake Towada
    1.5. Penke Pond and Panke Pond
    1.6. The Goshiki ponds
2. Aquatic Vegetation of Rivers
3. Swamp Forest
4. Marshes
    4.1. Tajimagahara
    4.2. Marshes of the Byobuyama dunes
    4.3. Other remnant marshes
    4.4. The Enoto marsh
5. Paddy Fields and Irrigation Ponds
    5.1. Paddy fields
    5.2. Marshes of irrigation ponds
6. Fens and Bogs
    6.1. The Kushiro and Sarobetsu moors
    6.2. Akaiyachi Moor
    6.3. Ozegahara Moor
        6.3.1. Gallery forests
        6.3.2. Bog ponds
        6.3.3. Swamps
        6.3.4. Fens
        6.3.5. Bogs
    6.4. Blanket bog

**Biological Institute, Tohoku University, Aramaki Aza Aoba, Sendai-shi 980, Japan*

As described in Chapter 1 of this book, the islands of Japan lie across a broad range of latitudes, and are influenced by a humid climate and high rainfall. Moreover, the many volcanoes and widespread volcanic terrain have given rise to large numbers of lakes and areas of low inclination. Such flat and gently sloping regions, covered with deposits of volcanic ash, commonly lead into areas of impeded water drainage. The country, therefore, abounds in aquatic and wetland vegetation of widely different characteristics.

Japan is a mountainous country with a comparatively narrow lowland strip along the coast. This lowland, for the most part, originates in alluvial deposits that are often underlain by diluvium. Formerly, sluggish streams passed across a comparatively wider lowland area, giving rise to lakes, ponds, swamps, marshes, and occasionally bogs. However, most of the lowland waters apart from rivers and wetlands have now been reclaimed for use as rice paddy, and have so lost most of their original natural features. Nevertheless, their general characteristics can still be recognized from the remnants that do remain after reclamation. Fig. 1 shows the location of the principal lakes and wetlands discussed in this chapter.

FIG. 1. Map showing the location of principal lakes and wetlands discussed. ○, Lake or pond; ×, wetland; ⊗, lake and wetland.

1. Aquatic Vegetation of Lakes and Ponds

Among the many lakes and ponds in Japan, eutrophic, oligotrophic, dystrophic and acidotrophic types can be distinguished.[1] These support a great variety of aquatic plants, which in turn show marked ecological differences according to the trophic type and geographical location.

Most of the rivers are swift-flowing and exhibit very little silt deposition; however, a few are sluggish, with marked silting. The sluggish rivers have in general been encroached upon for use as rice paddy or for housing, and are so deprived of their associated wetland and aquatic vegetation.

The following discussion of aquatic (and swamp) vegetation in Japan is therefore limited chiefly to lakes and ponds.

1.1. Lake Suwa

Lake Suwa lies in the center of Nagano Pref., central Japan, and is a typical eutrophic lake of the temperate zone. The surface area is 14.108 km², and the maximum depth 7 m. The transparency, as measured by a Secchi disc, is of the order of 1 m.

Several places along the lake-side are overgrown with emersed plants such as *Phragmites communis*, *Zizania latifolia* and *Scirpus triqueter*, resulting in the formation of swamps. At depths of 1–2 m, this swamp vegetation is replaced by floating leaf communities consisting of *Nymphaea tetragona*, *Nuphar japonicum*, *Nymphoides peltata*, etc. In deeper water (2–3 m) there is a submerged community composed of *Potamogeton perfoliatus*, *P. oxyphyllus*, *P. Maackianus*, *Najus marina*, *Vallisneria asiatica*, *Myriophyllum spicatum*, *Ceratophyllum demersum*, etc., which fades away gradually at depths of between 3 and 4 m. Characeous plants are not found in the lake, even at depths below 4 m. This is probably a result of the high turbidity and consequent extremely weak light intensity in deeper regions, insufficient to support their photosynthesis.[2]

1.2. Lake Biwa

Lake Biwa, Japan's largest lake, is located in the Kinki district of central Japan. The surface area is 679.50 km², and the maximum depth 102 m. It is essentially an oligotrophic lake and has a transparency of from 2 to 10 m.

Paddy fields and housing have been advanced close to the water's edge, and only narrow remnants of former marshes and swamps now remain in such regions. Floating leaf communities are rarely found in the lake, except in calm inlets, due to heavy waves produced by strong prevailing winds. Therefore, submerged communities constitute the main aquatic flora.

In inlets sheltered from the heavy waves, the following zonation of plant

communities can be distinguished:[3] water of depth 0–2 m is occupied by *Hydrilla verticillata*, *Ceratophyllum demersum*, *Vallisneria asiatica*, *Potamogeton perfoliatus* and *P. malaianus*; water of depth 2–4 m is occupied chiefly by *Najus marina*, *Myriophyllum spicatum*, *Hydrilla verticillata*, *Potamogeton Maackianus*, *Vallisneria denseserrulata*, and *V. asiatica* associated with *Nitella flexilis* and *Chara Braunii*; deeper water (4–6 m) supports a very sparse flora with few species, including *Najus marina* and *Potamogeton Maackianus*. Also, *Potamogeton biwaensis* and *Vallisneria asiatica* var. *biwaensis* are endemic to Lake Biwa, while *Elodea canadensis*, which flourishes in the lake, was introduced from North America.

Table 1 shows average figures for vertical plant zonation in deep, clear lakes (of the Biwa type) and shallow, turbid lakes (of the Suwa type). However, the water of Lake Biwa, as well as that of Lake Suwa, has recently become increasing polluted (especially eutrophicated), giving rise to increased turbidity. This situation has probably raised the lower limits of occurrence of the submerged plant communities.

TABLE 1. Vertical zonation of aquatic vegetation in two different types of lakes in Japan

Vegetation type	Depth range (m)	
	Deep, clear lakes	Shallow, turbid lakes
Emersed	0–1	0–1
Floating leaf	1–2	1–3
Submerged	2–8	2–4
Characeous	3–18	2–4

1.3. OGURA POND

Ogura Pond is situated in the suburbs of Kyoto City, central Honshu. It is of the eutrophic type, has a surface area of 1 km^2 and a maximum depth of only 2m.[4]

The outer margin of the pond, which is less than 1 m in depth, is densely overgrown with reed swamp species such as *Zizania latifolia*, *Phragmites communis*, *Typha latifolia* and *T. angustifolia*. Towards the center there is abundant *Nelumbo nucifera*, which is intolerant to water of greater depth than 2 m. This plant was introduced to Japan from India, via China, and is cultivated in many ponds and pools throughout the country apart from Hokkaido.

The center of the pond is occupied by two types of communities. One occurs on the stable bottom, and comprises such specific plants as *Euryale ferox*, *Nymphaea tetragona*, *Nuphar japonicum*, *Nymphoides indica*, *Trapella sinensis* and *Trapa natans*. The other is seen mainly on the unstable bottom, and consists

FIG. 2. Floating leaf (*Brasenia Schreberi*) and emersed plant (*Zizania latifolia*) communities in a shallow, eutrophic pond.

chiefly of submerged species such as *Potamogeton crispus*, *P. Maackianus*, *P. malaianus*, *Vallisneria asiatica*, *Myriophyllum spicatum*, *Ceratophyllum demersum*, etc. Macrophytoplankton such as *Hydrocharis asiatica*, *Azolla imbricata*, *Lemna minor* and *Spirodela polyrhiza* are frequently found in the open water between the swamp and floating leaf communities. Free-living insectivorous species such as *Aldrovanda vesiculosa*, *Utricularia japonica* and *U. exoleta* also occur in the open water.

Ogura Pond is thus a typical example of eutrophic waters of the warm-temperate zone which abound in aquatic species related closely to tropical or subtropical types, but which are remarkably different from those of the temperate zone. Fig. 2 illustrates a shallow, eutrophic pond.

1.4. Lake Towada

Lake Towada is located near the northern extremity of the main island of Japan (Honshu) at an altitude of 400 m a.s.l. It is a caldera lake, having a surface area of 60 km^2 and a maximum depth of 330 m. It belongs to the oligotrophic type, with a transparency of the order of 20 m.[5]

The lake is almost devoid of swamp and floating communities, due to heavy waves caused by strong winds and to non-silting. Submerged plant communities occur only at the inner bay sheltered from heavy winds. The shallow water, 1–8 m in depth, is dominated by *Myriophyllum spicatum* in association with *Potamogeton Maackianus*, *P. perfoliatus*, *P. compressus* and *P. heterophylla*. Among the phanerogams, only *Potamogeton pectinatus* attains depths of the order of 10 m. Stoneworts such as *Chara globularis* and *Nitella flexilis* also appear in the shallow water below 2 m in depth, mixed with the phanero-

gams. They further extend down to a depth of 16 m in the case of *Chara globularis*, and to a maximum depth of about 26 m in the case of *Nitella flexilis*. The extreme transparency of the water permits the occurrence of these characeous plants at such deep levels in the lake.

1.5. Penke Pond and Panke Pond

The Penke and Panke ponds are small dystrophic bodies of water lying in the Sarobetsu moor at the northwestern end of Hokkaido, and are subject to a subarctic climate. They are bordered by swamps composed of *Zizania latifolia*, *Phragmites communis* and *Scirpus tabernaemontani*, associated and interspersed with *Trapa japonica*. Towards the centre of these ponds floating leaf communities dominated by *Trapa japonica* occur, followed by a *Nymphaea tetragona* community and then a *Nuphar pumilum* community comprised of *Nuphar pumilum* and *N. pumilum* var. *ozeanum*. Submerged communities are rarely developed owing to the stagnant and turbid nature of the water, and to the consequent weak light intensity at depth.

Nuphar pumilum and *N. pumilum* var. *ozeana* are typical dystrophic water plants of boreal climates, and extend from Hokkaido to the mountainous regions of northern Honshu.[6,7]

1.6. The Goshiki ponds

The Goshiki ponds are located at the northern foot of Mt. Bandai, Fukushima Pref., northeast Japan. They are a chain of small ponds which came into existence following the 1800 eruption of Mt. Bandai, and include eutrophic as well as acidotrophic types.

The acidotrophic ponds are fed with mineral water of pH 3.8–6.4 and are heavily overgrown with the aquatic moss *Drepanocladus fluitans*, which extends from the shore to depths of 2–3 m. *Drepanocladus fluitans* gives rise to a dense moss mat, upon which *Phragmites communis* or *Equisetum ramosissimum* often occur (see Fig. 3). Dead moss is not decomposed because of the cool, acidic conditions and accumulates as a peat layer in the same way as sphagnum mosses. Also, a bog situated near the Sukawa spa on the northern slopes of Mt. Kurikoma, Akita Pref., is underlain with *Drepanocladus* peat of more than 5 m in thickness.

The Goshiki ponds include eutrophic types with pH values of 6.6–7.0. In contrast to the more acidic ponds, these are without an overgrowth of *Drepanocladus fluitans*, but instead phanerogams such as *Potamogeton perfoliatus* and *Myriophyllum spicatum* (i.e. normal components of the flora of eutrophic ponds) occur. Along the shores, occasional swampy areas consist solely of *Phragmites communis*, a species in common with the acidotrophic ponds.

The Goshiki ponds were formed about 85 yr ago by a mudflow which

FIG. 3. Aquatic moss (*Drepanocladus fluitans*) mat bearing *Phragmites communis*. (Urabandai, Fukushima Pref.).

dammed up brooks formerly flowing through the area, and the present aquatic vegetation has developed since that time. These ponds are thus an interesting site for studies on the earlier stages of hydrarch plant successions.

2. Aquatic Vegetation of Rivers

As mentioned above, the Japanese islands are conspicuously mountainous and are traversed by swift-flowing rivers in general. The bed material of these rivers is mostly shingle, gravel and sand, which are unable to support the typical aquatic communities. However, some sluggish rivers, bedded mainly with silt and clay, sustain aquatic vegetation, as for example the River Oppa, Miyagi Pref., northeast Honshu. Along the shores of this river, reed swamp composed of emersed plants such as *Phragmites communis*, *Miscanthus sacchariflorus*, *Zizania latifolia*, *Typha latifolia*, *T. orientalis* and *T. congesta* develops. Deeper water (0.5–1.5 m) is inhabited by submerged plants such as *Potamogeton oxyphyllus*, *P. malaianus*, *P. crispus*, *Hydrilla verticillata*, etc.

The aquatic vegetation of rivers, where developed, is thus strikingly similar to that of lakes and ponds. One exceptional, curious aquatic plant is *Cladopus japonicus*, a phanerogam of tropical origin and resembling a moss, which adheres to stones at the bottom of rapid-flowing streams in Kagoshima Pref., southernmost Kyushu.

3. Swamp Forest

Swamps established in lowland areas along rivers and lakes, and in damp

FIG. 4. *Alnus japonica* swamp forest with reedswamp dominated by *Phragmites communis* in the foreground.

depressions, often support woods with dominant *Alnus japonica* or *Fraxinus mandshurica*.

Alnus japonica grows on tussocks, stools or stoney patches in reed swamp at the edge of open water. The swampy woods themselves usually contain wide stretches of open water between the trees. In Hokkaido, some swampy alder woods still remain undisturbed from land reclamation, and exhibit the following typical undergrowth species: *Phragmites communis*, *Sanguisorba albiflora*, *Senecio palmatus*, *Lysichiton camtschatense*, *Lycopus parviflorus*, *L. lucidus*, *Lysimachia vulgaris* var. *dahurica*, *Impatiens Textori*, *I. noli-tangere*, etc. (see Fig. 4).

On less moist soil *Alnus japonica* is replaced by *Fraxinus mandshurica*, often associated with *Ulmus Davidiana* var. *japonica*. Such mixed wetland forest has a dense undergrowth comprised of *Hydrangea macrophylla* subsp. *serrata*, *Trillium kamtschaticum*, *T. Tschonoskii*, *Anemone flaccida*, *Calamagrostis Langsdorffii*, *Osmunda japonica*, *Athyrium multifidum*, *Dryopteris tokyoensis*, *D. crassirhizoma*, etc. The *Fraxinus mandshurica* forest gradually merges into forest dominated by *Ulmus Davidiana* var. *japonica* (belonging to mesic type forest) at the wet-dryland boundary.[8)]

As an example of such vegetational transition, Lake Usoriyama of the Shimokita Peninsula (northernmost Honshu) will next be described. The lake itself is acidotrophic (pH 3.4) and inhabited by the bottom-dwelling aquatic moss *Leptodictyum riparium*, but along the southern shore there is a clear floral transition from aquatic through wetland vegetation to a climactic forest community.[9)]

A carpet of *Leptodictyum riparium* covers the bottom in deeper parts of the lake (1.5–10 m), but it is succeeded by a floating leaf-moss community com-

FIG. 5. Reedswamp dominated by *Phragmites communis* on the shores of Lake Usoriyama.

FIG. 6. *Fraxinus mandshurica* swamp forest established on wetland near the western shores of Lake Usoriyama.

posed of *Potamogeton alpinus*, *P. oxyphyllus* and *Nuphar japonicum*, together with *Leptodictyum riparium*. Water of 0.3–0.5 m in depth is dominated by *Scirpus lacustris* var. *Tabernaemontani*, while regions shallower than 0.3 m support *Phragmites communis* (see Fig. 5).

The reed swamp borders onto a narrow belt of *Salix subfragilis* thickets at the edge of the water, and these willow swamp thickets are in turn replaced by alder woods with an undergrowth of swamp plants such as *Calla palustris*, *Lycopus Maackianus*, *Carex dispalata* and *Lysichiton camtschatense*. The alder woods in general contain a single species, *Alnus japonica*, and the individual trees tend to be of even age. Next comes a rather broad belt of *Fraxinus mandshurica* woods. These are dominated by *Fraxinus mandshurica*, associated occasionally with *Alnus japonica* and *Pterocarya rhoifolia*, have an undergrowth of *Hydrangea paniculata*, *Carex dispalata*, *Lysichiton camtschatense*, *Athyrium deltoidofrons* and *Dryopteris crassirhizoma*. The *Fraxinus mandshurica* swamp forests (see Fig. 6) are to some extent interspersed with open water, but they are in general not so extensively water-logged as the alder swamp forests.

The *Fraxinus mandshurica* swamp forests finally pass into a climatic climax forest of *Fagus crenata–Thujopsis dolabrata* at higher levels which are never under water throughout the year.

In other areas of Honshu, most of the swamp forests have been largely reclaimed for use as rice paddy, but those remnants which do remain are worthy of future preservation and study.

4. Marshes

The term "marsh" is applied to a soil-vegetation type developing in water-logged soil with a summer water level close to or conforming with the ground surface, as in the "sedge meadow" of U.S. terminology, to which type it is broadly equivalent.

In Japan marshes occur in association with the wet, mineral-rich soils of dried-up lakes and stream beds, around the shores of extant lakes and streams, and in depressions. They support various sedges, rushes, grasses and flowering herbs. However, these marshes have in the majority of cases been partially converted into paddy fields, due to their remarkable suitability as sites for rice production.

4.1. Tajimagahara

Tajimagahara is a remnant marsh on the River Arakawa, Urawa City, near Tokyo. It is noted for the abundant occurrence of *Primula Sieboldi*, one of the most beautiful of Japan's wild flowers, and the area is protected as a national monument by the Ministry of Education. Tajimagahara is located

on a small alluvial plain of wet, silty sand which is occasionally flooded by rising water.[10)]

The marsh displays an excellent vernal array of flowering herbs such as *Primula Sieboldi*, *Euphorbia adenochlora*, *Ranunculus ternatus* and *Corydalis decumbens*. The aestival community, which succeeds the vernal one, is a luxuriant growth of *Miscanthus sinensis*, *Thalictrum simplex* and *Rumex Acetosa*.

4.2. Marshes of the Byobuyama dunes

Along the west coast of the Tsugaru Peninsula (northern Honshu) runs a belt of dunes called Byobuyama, in which marshes occur around ponds and in depressions.[11–13)]

On a damp cover of silty sand, marsh dominated by *Carex Thunbergii* (associated frequently with *Lythrum salicina* and *Lysimackia vulgaris* var. *davurica*) is found. It includes, moreover, *Carex pseudocuraica*, *C. vesicaria*, *Lycopus Maackianus*, *Sanguisorba tenuifolia* var. *alba*, *Triadenum japonicum*, *Lastrea Thelypteris* and *Onoclea sensibilis* var. *interrupta*.

Soils containing substantial amounts of organic material are occupied by a marsh community different from that just mentioned. It is characterized by the presence of *Juncus yokoscensis* and *Ischaemum aristatum* var. *glaucum*, which are associated with *Sanguisorba tenuifolia* var. *alba*, *Iris ensata* var. *spontanea*, *Hemerocallis Middendorffii* var. *esculenta*, *Hosta rectifolia*, *Cirsium tenuifolia*, etc.

4.3. Other remnant marshes

In Honshu, small marsh remnants are often preserved and designated as national monuments because of their dense growth of *Iris laevigata* or *Iris ensata* var. *spontanea*, which provides a tourist attraction during the flowering season. Various other species occur, as follows: *Eriocaulon Sieboldtianum*, *E. nudicupse*, *Fimbristylis dichotoma*, *Isachne globosa*, *Habenaria radiata*, *Platanthera sachalinensis* and *Hosta longissima* var. *brevifolia*. Also, there is a frequent abundance of insectivorous species, such as *Utricularia bifida*, *U. yakushimensis*, *Drosera rotundifolia*, *D. spathulata* and *D. peltata* var. *nipponica*.[10)]

4.4. The Enoto marsh

Near Beppu City (northeastern Kyushu), at an altitude of 700 m a.s.l., lies the Enoto moor. The main portion of the moor is occupied by marsh, which has developed on secondarily deposited volcanic ash and is fed by spring water issuing from the surrounding slopes.[14)]

The marsh is located at the margin between swamp and grassland areas and supports, in the main, the following species of sedges: *Rhynchospora Fujiiana*, *Rh. Umemurae* var. *Hattoriana* and *Fimbristylis complanata* f. *exalata*,

associated frequently with *Carex heterolepis*, *C. cinerascens*, *C. forficula*, *Rhynchospora chinensis* var. *fauria* and *Sesleria tesselata*. Various kinds of flowering herbs also occur, as follows: *Cirsium Sieboldi*, *Inula britannica* subsp. *japonica*, *Lobelia sessilifolia*, *Triadenum japonicum*, *Primula Sieboldi*, *Platanthera hologlottis*, *Iris ensata* var. *spontanea* and *Hemerocallis Thunbergii*, together with occasional insectivorous species such as *Drosera rotundifolia*, *Utricularia yakushimensis*, *U. racemosa* and *U. bifida*.

These marshy areas are among the most interesting sites in Japan from the viewpoint of floral and ecological studies. Their remnants thus deserve careful preservation, especially in view of the fact that they have to a large extent already been obliterated for reclamation as paddy fields.

5. Paddy Fields and Irrigation Ponds

5.1. Paddy fields

As mentioned in the introduction to this chapter, Japan has narrow plains along its coast, and a mountainous hinterland. The plains have mostly been converted to paddy fields and partly to urban and industrial areas. Regions for use as paddy fields were reclaimed mainly by drainage of inland waters, swamps, marshes, fens and bogs, or by irrigation of upland areas. They range throughout the lowlands up to the sub-montane zone, although here they are restricted to valley bottoms, other flat areas or gentle slopes.

Since rice forms the staple diet of the Japanese people, paddy fields now spread widely over the plains, and give rise to a very characteristic landscape form. The fields are divided into blocks, each of about 10 areas, enclosed by narrow, low banks which serve to retain the water in the fields during the rice-growing season from early June to mid-October.

A specific aquatic-weed community is found in the paddy fields at the time of rice cultivation. However, recently this has been heavily, though not entirely, suppressed by the use of agricultural chemicals. The emersed weed community is made up of *Cyperus difformis*, *Monochoria vaginalis*, *Echinochloa crus-galli*, *Dopatrium junceum*, *Limnophila sessiliflora*, *Sagittaria pygmaea*, *Alisma canaliculatum*, *Blyxa ceratosperma* and *B. caulescens*. Megaphytoplankton such as *Salvinia natans* and *Spirodela polyrhiza* are often found in the open water between rice plants.[15]

Following the rice harvest, the paddy fields are drained by suspension of irrigation and lain fallow or planted with upland winter crops for the winter-spring period from November to May. Fields left fallow soon become dominated by biennial weeds such as *Alopecurus aequalis* var. *amurensis*, *Beckmannia erucaeformis* and *Ranunculus sceleratus*. *Astragalus sinensis*, introduced from China,

is often cultivated during the winter-spring season as green manure, and its deep pink flowers provide a characteristic element in the lowland landscape in mid-spring. Thus, the weed communities of paddy fields display an annual cycle, varying markedly from the irrigated to the drained periods.

Some paddy fields are kept filled with water throughout the year and cultivated with *Juncus effusus* var. *decipiens* or *Nelumbo nucifera*. The former is grown as material for use in making the "tatami" (straw mats) typical of Japanese houses, while the latter (introduced from India, via China) is cultivated for use as an ornamental plant or for food.

5.2. Marshes of irrigation ponds

Those parts of southwest Japan without heavy rainfall abound in artificial irrigation ponds, which provide water for the paddy fields during rice plant cultivation. The ponds remain full of water during the resting season but their level gradually falls during the rice growing season, partially exposing the bottom to the air. They are eutrophic ponds, inhabited by various native, aquatic plants. The submerged community consists of *Hydrilla verticillata* and *Potamogeton capiscus*, and the floating leaf community of *Potamogeton distinctus*, *Nymphoides indica*, *N. peltata*, *Trapa natans*, etc. Macrophytoplankton such as *Lemna paucicostata*, *L. minor*. *L. trisulca*, *Azolla imbricata* and *Salvinia natans* (and often *Eichhornia crassipes*, introduced from tropical America) float on the open water. A similar flora is also present in the irrigation canals that run between the ponds and paddy fields.[16]

The littoral zone of the irrigation ponds is exposed to the air every year during the growing season for rice plants. This exposed bottom supports annuals such as *Lindernia verbenaefolia*, *Centipeda minima*, *Eleocharis pellucida*, *Hedyotis prostrata*, *Persicaria Sieboldi* and *Cyperus polystachyos*, and perennials such as *Jussiaea prostrata*, *Fimbristylis diphylloides*, *Rhynchospora chinensis*, *Lysimachia Fortunei*, *Lobelia chinensis* and *Ludwigia ovalis*. Thus, the vegetation is characterized by annual plants of short life and amphibious perennial plants.

6. Fens and Bogs

The term "fen" is applied to places where the soil consists of peat but is alkaline, neutral or rarely somewhat acidic. It is used mainly in the United Kingdom, to some extent in North America, and corresponds to the German, "Niedermoor". In Japan, however, even though the floristic composition in general resembles that of Europe, fens occur not on alkaline or neutral peat but rather on weak acidic peat.

Bogs develop on peat under highly acidic conditions, bearing vegetation remarkably different from that of fens. They correspond in general to the

"Hochmoor" of German. In the United Kingdom, bogs are divided into raised, valley, and blanket types, and examples roughly equivalent to each are also found in Japan.[17]

Fens and bogs occur only rarely in the southern half of Japan, i.e. in regions influenced by a warm-temperate climate, but are common in the northern, cool-temperate and subarctic regions. Under such cool, humid conditions, fens are gradually replaced by bogs (usually raised bogs) at higher or colder places, often through a transitory community dominated by *Moliniopsis japonica*, which is vicarious to the *Molinia coerulea* of Europe, and fen thicket. Occasionally, raised bogs are established on top of valley bogs, which form over the upper parts of valley fens fed by water that crosses terrain underlain by acidic rock formations.

Blanket bog is the typical climatic vegetation of Ireland, where the precipitation is high and the air moist and cool. It has not necessarily arisen from fen-lands, but rather developed directly on flat or gently sloping terrain. In the alpine and subalpine zones of Japan's mountainous regions, which are subject to high precipitation and cool, humid atmospheric conditions, bogs roughly equivalent to blanket bog occur in flat or gently sloping areas of poor drainage.

Raised bog usually originates in association with lakes, estuaries and valley bogs. It arises through a number of stages; open water, swamp, marsh, and fen or valley bog. It is therefore possible to observe two or more types of aquatic and wetland vegetation in a single moor, when it is of comparatively large extent. In Japan, however, bog- and fen-lands have often been reclaimed, especially in the lowlands, although they remain intact in northern Hokkaido.

6.1. The Kushiro and Sarobetsu moors

Kushiro moor is located along the Pacific coast of southeast Hokkaido, and is famous as the only known breeding ground of the Japanese crane, *Grus japonensis*.[18] Sarobetsu moor is situated on the coast of the Japan Sea at the northwestern end of Hokkaido.[6,7] These moors are presumed to have been established on estuary fen-lands.

Two types of fen communities occur; the *Phragmites communis–Carex Miyabei* and *Carex Augustinowiczii–Phragmites communis* communities. The former is found at the moor periphery, which is frequently covered by water; the latter is found in the central zone, which is saturated but not covered with water. *Carex Augustinowiczii* plays an important role in the hydrarch succession, forming sufficiently large tussocks to cause considerable irregularity in the ground surface. The tussocks themselves are inhabited by *Thelypteris palustris*, *Artemisia vulgaris* var. *kamtschatica* and *Adenophora triphylla* var. *tetra-*

FIG. 7. Alder thicket comprised of *Alnus japonica* (foliated) and *Fraxinus mandshurica* (defoliated).

phylla, while *Sanguisorba tenuifolia* var. *alba*, *Solidago Virga-aurea* var. *yezoensis*, *Hosta rectifolia* and *Calamagrostis Langsdorffii* occur in the spaces between tussocks.

With increasing accumulation of acidic peat and rise of ground level above the water surface, a *Moliniopsis japonica* community, or alder thicket, transitory from fen to bog, develops. As mentioned, *Moliniopsis japonica* is quite similar in ecological properties to the *Molinia coerulea* found in Europe. It builds communities on acidic peat without silt, whereas the alder thicket occurs on peat which contains silt deposited by flood water.

The *Moliniopsis japonica* community is dominated by *Moliniopsis japonica* in association with *Myrica Gale* var. *tomentosa*, *Eriophorum gracile*, *Hemerocallis Middendorffii* and *Osmunda cinnamomea*. Occasionally, *Drosera rotundifolia*, *Pogonia japonica*, *Eleocharis japonica*, *Oxycoccus quadripetalus* and *Sphagnum squarrosum* are found locally. The community as a whole is thus similar to the *Molinia coerulea* community of Europe, both in floristic composition and in ecological properties.

The alder thicket, which corresponds to the so-called "carr" of England, is composed of wetland trees, shrubs and herbs (see Fig. 7). It is dominated by stunted *Alnus japonica* trees associated frequently with *Fraxinus mandshurica*, and occasionally *Picea Glehnii*, and an undergrowth of *Moliniopsis japonica*, *Eriophorum gracile*, *Hydrangea paniculata* and *Sasa palmata*.

These transitory communities are succeeded by bog at higher altitudes, where typical bog plants grow on the accumulated acidic peat. The bogs in the Kushiro and Sarobetsu moors are not so markedly raised as those of Ozegahara and Yashimagahara in Honshu. Their ground surface has a comparatively even appearance, although it does rise sufficiently to avoid the direct influence of ground water. The dominant floral component is

bog moss, including species such as *Sphagnum papillosum*, *Sph. compactum* and *Sph. fuscum*. On hummocks, dwarf shrubs such as *Empetrum nigrum*, *Vaccinium Vitis-Idaea*, *V. quadrifolium*, *Andromeda polifolia* var. *grandifolia*, *Ledum palustre* var. *yezoense*, *Camaedaphne calyculata* and *Myrica Gale* var. *tomentosa* occur, and between the hummocks, *Eriophorum vaginatum* and *Carex Middendorffii* are found. Also, and especially in hollows, *Rhynchospora alba*, *Lycopodium inundatum* and *Drosera anglica* commonly occur.

It appears that bog pools are rare on the moors of Hokkaido, except for a few, larger ponds that remain as a result of wetland formation in the vicinity of lakes. These ponds are dystrophic, with a submerged community of *Potamogeton pectinatus* and *P. oxyphyllus*, succeeded by a floating leaf community of *Nymphaea tetragona*, *Nuphar pumilum*, *N. pumilum* var. *ozeana* and *Trapa japonica*. The pond periphery is occupied by an emersed community of *Scirpus lacustris* var. *Tabernaemontani* and *Equisetum fluviale* and a floating mat of *Menyanthes trifoliata* and *Comarum palustre*.

6.2. Akaiyachi Moor

Akaiyachi Moor is situated on the western bank of Lake Inawashiro, Fukushima Pref., Honshu. It probably originated as a shallow valley sloping down into the lake, surrounded by low hills covered with *Pinus densiflora*, *Quercus serrata*, *Castanea crenata* and *Prunus verecunda*, i.e. except the western end which was formerly a bog but has more recently been converted into rice paddy.[19]

It is generally presumed that the floral succession of the region began from swamp forest. There is a peat layer 3–5 m in thickness, which commonly contains dead stumps of *Fraxinus mandshurica*, so demonstrating the existence of swamp forest at least at an early stage of the hydrarch succession.

At present the moor lacks fen, but does exhibit transitory and bog communities. Two transitory communities are recognized, alder thicket and an *Osmunda cinnamomea* community.

The alder thicket, or carr, is composed of stunted trees of the species *Alnus japonica*, *Fraxinus mandshurica* and *Magnolia Kobus*, and shrubs such as *Ilex Nemotoi*, *I. crenata* var. *radicans*, *Rhamnus crenata*, *Hydrangea paniculata* and *Viburnum Sargentii*. *Miscanthus sinensis*, *Moliniopsis japonica* and *Lastrea Thelypteris* var. *pubescens* frequently occur as undergrowth.

The *Osmunda cinnamomea* fern community occurs on soil that is scarcely influenced by underground water. It is dominated by *O. cinnamomea* in association with *Phragmites communis*, *Moliniopsis japonica* and *Ilex crenata* var. *radicans* (see Fig. 8). Bog mosses such as *Sphagnum papillosum*, *Sph. palustre* and *Sph. magellanicum* also occur locally.

With increasing accumulation of acidic peat, the transitory communities

FIG. 8. *Osmunda cinnamomea* community intermingled sparsely with dead *Phragmites communis* shoots.

merge into bog consisting of raised and hollow portions. The raised portions are occupied by a *Moliniopsis japonica–Sphagnum papillosum* community characterizd by dominant *Sphagnum papillosum* and by an abundance of *Moliniopsis japonica* and *Oxycoccus quadripetalus*. *Pogonia japonica*, *Spiranthes sinensis*, *Gentiana axillariflora* and *G. Thunbergii* var. *minor* also occur frequently. The hollows are dominated by *Rhynchospora alba* and *Sphagnum cuspidatum* in association with *Scheuchzeria palustris*, *Eleocharis japonica*, *Carex omiana* var. *monticola* and *Lycopodium inundatum*. Insectivorous plants such as *Drosera rotundifolia*, *Utricularia bifida*, *U. racemosa* and *U. intermedia* are also found.

At the centre of the moor, a core was taken through the peat to the basal mineral zone and the horizons shown in Table 2 were identified. The basal

TABLE 2. Horizons identified in a core from the central area of Akaiyachi Moor, Honshu

Depth (cm)	Horizon
0–10	Living peat
10–45	*Sphagnum–Moliniopsis* peat
45–80	*Sphagnum–Rhynchospora* peat
80–120	*Sphagnum–Moliniopsis* peat
120–240	*Sphagnum–Oxycoccus* peat
240–280	Volcanic ejecta with *Phragmites*
280–340	*Phragmites* with wood remains
340–	Gley

gley layer was not a lake deposit but mostly undecomposed mineral crystals. It is therefore presumed that the present moor did not develop from lacus-

trine waters but from water-logged land covered by an alder-ash forest. This swamp forest probably went through a short transitory period, with a transitory community, to bog showing a cycle similar to that involving an alternation of hummocks and hollows.

6.3. Ozegahara Moor

Ozegahara Moor lies at the boundary of Fukushima, Gunma and Niigata Pref., northeast Honshu, and occupies a valley bottom which was formerly a lake, known as the "Old Ozehara Lake". It extends 6 km (NE to SW) by 1–2 km (NW to SE) at an altitude of about 1400 m a.s.l. The moor is surrounded by mountains covered by *Fagus crenata* forest at 1400 to 1600 m, and by *Abies Mariesii* forest at 1600 to 1900 m. However, conifers such as *Abies Mariesii*, *Picea jezoensis* and *Larix leptolepis* also descend to an altitude of about 1400 m, forming a narrow belt between the beech forest above and moor below.[20] Fig. 9 illustrates the general appearance of the moor.

Towards the end of the Pleistocene, the valley was dammed up by a lava flow issuing from Mt. Hiuchi to the north, so forming the "Old Ozehara Lake". Subsequently, the basin was filled by sediments and peat, resulting in the formation of moorland. *Fraxinus mandshurica* forest developed on the wet sediment, as shown by the many stumps buried beneath the 5–6 m thick peat layers. Shallow water, on the other hand, was covered by a floating mat of *Menyanthes trifoliata* and *Comarum palustre*. Carbon(^{14}C)-dating of buried *Fraxinus mandshurica* wood gives an age of 5678$\pm$700 yr, which, together with other evidence, indicates that Ozegahara Moor began its development about 8000–10,000 yr ago.

FIG. 9. Northern part of Ozegahara Moor showing reedswamp (foreground), raised bog (central field) and surrounding highland (background).

Many boreal species that migrated south during the Ice Age are still preserved in Ozegahara Moor.[21] In the case of *Drosera anglica*, *Scheuchzeria palustris* and *Nuphar pumilum* var. *ozeana*, it is their southernmost point of occurrence in Japan. New endemics such as *Cirsium hondoensis*, *Euphorbia togakusensis* var. *ozeanum* and *Inula ciliaris* var. *glandulosa* have also developed in the moor.

6.3.1. *Gallery forests*

Rivers bounded by low banks wind through the moor, and are the only sites of eutrophic water and well-drained mineral soil in the moor. The river banks are occupied by forest which differs markedly from the surrounding bog vegetation. This forest corresponds roughly to the "Gallerie Wälder" which occur along rivers running through bogs. In Ozegahara Moor, the gallery forest is composed of *Ulmus Davidiana*, *Betula Ermanii*, *Aesculus turbinata*, *Tilia japonica*, *Pterocarya rhoifolia*, *Fraxinus mandshurica*, *Quercus crispula*, *Acer mono*, *Prunus Ssiori*, *Acer Miyabei*, etc. The ground below is densely grown with herbs such as *Clinopodium sachalinense*, *Allium Victorialis* ssp. *platyphylla* and *Cacalia hastata* ssp. *orientalis*, and ferns such as *Dryopteris crassirhizoma*, *D. monticola*, *Laportea macrostachya* and *Thelypteris bukoensis*. This gallery forest is the place of highest productivity and standing crop within the moor.

In addition, in sluggish and silted-up portions of the rivers, emerged communities of *Equisetum fluviale* and submerged communities of *Hipparis vulgaris* often occur.

6.3.2. *Bog ponds*

Ozegahara Moor contains a large number of ponds, amounting in total

FIG. 10. Bog pond with floating islets (located in the central area of a raised bog).

to about 400. The majority are filled with water that is brown, acidic, and low in mineral content. The larger and deeper ponds support floating leaf plants such as *Nymphaea japonica* and *Nuphar pumilum* var. *ozeana* in the centre, and *Menyanthes trifoliata* along the periphery. Commonly, there are also islets of peat of different sizes and forms in these ponds, many of which are also floating (see Fig. 10).

6.3.3. Swamps

The border of the moor often receives abundant water from the surrounding mountain slopes, and is thinly covered with water throughout the growing season. The resulting swampland is occupied by a *Comarum palustre–Menyanthes trifoliata* community, which resembles the Comareto–Menyanthemum community of Europe. Besides *Comarum palustre* and *Menyanthes trifoliata*, it includes *Carex limosa*, *Lobelia sessilifolia*, *Iris laevigata*, *Caltha palustris*, *Myrica Gale* var. *tomentosa*, *Equisetum fluviale* and *Lysichiton camtschatense*. On sites covered by stagnant water, a community dominated by *Carex rhynchospora* or *Phragmites communis* occurs.

Such swamp occupies an area of less than 10% of the moor, although it does, as indicated, support a wide variety of flowering plants.

6.3.4. Fens

Fens occur only in narrow, limited areas near the foot of gentle slopes that border the moor. They are fed by the eutrophic and weak acidic water that issues from the nearby mountain slopes. The dominant fen community is *Phragmites communis–Osmunda cinnamomea*, together with a variety of herbs such as *Sanguisorba officinalis* var. *carnea*, *Euphorbia Sieboldiana*, *Ligularia Fischeri*, *Cirsium homolepis*, *Veratrum japonicum*, *V. nigrum* and *Lysichiton camtschatense*. Bog plants such as *Sphagnum amblyphyllum*, *Sph. pulchrum*, *Oxycoccus quadripetalus* and *Eriophorum vaginatum* also appear locally in the fens.

In addition, a narrow belt of *Abies Mariesii–Picea jezoensis* forest occurs along those parts of the moor periphery where the water issuing from the nearby slopes lies in a stagnant condition. This coniferous forest itself degrades into fen dominated by *Phragmites communis*, *Osmunda cinnamomea*, *Moliniopsis japonica* and *Lysichiton camtschatense*, and the fen in turn merges into raised bog. Thus, in Ozegahara Moor, fens can be considered to have arisen as an early stage of developing valley bog, replacing the marginal forests.[22]

6.3.5. Bogs

The greater part of Ozegahara Moor is bogland, which can be divided roughly into flat and raised types, where the ground surface is generally even in flat bog and uneven in raised bog. Areas of flat bog are occupied

principally by a *Moliniopsis japonica–Carex Middendorffii* community that includes *Sphagnum papillosum*, *Sph. amblyphyllum*, *Oxycoccus quadripetalus*, *Drosera rotundifolia*, *Carex omiana*, *Eriophorum vaginatum*, *Andromeda polifolia* and *Narthecium asiaticum*. It occurs on moist, loose peaty soil, suggesting that it has in fact developed from a floating mat community composed of *Comarum palustre* and *Menyanthes trifoliata*.[23)]

Raised bog is found in several places in the moor, and can be further subdivided into three types based on the relief and accompanying vegetation. The first, which corresponds to a regeneration complex composed of hummocks and hollows, usually occurs on gentle slopes. The hummocks are formed of *Carex Middendorffii*, *Moliniopsis japonica*, *Oxycoccus quadripetalus*, *Sphagnum papillosum* and *Sph. fuscum*, while the hollows support *Rhynchospora alba*, *Scheuchzeria palustris*, *Drosera anglica*, *D. rotundifolia*, *Lycopodium inundatum*, *Sphagnum pulchrum* and *Sph. papillosum*.

The second type corresponds to a resting area and/or pond complex that occurs on the moist, flat tops of raised bog, and which typically has larger, deeper ponds. It supports a community of *Rhynchospora alba*, *Eriophorum vaginatum*, *Narthecium asiaticum*, *Sieversia pentapetala*, *Andromeda polifolia*, *Sphagnum papillosum* and *Sph. compactum* (see Fig. 11).

The third type is equivalent to a marginal complex, occupying the marginal slopes of areas of raised bog. The peaty soil shows variable moisture content according to the degree of drainage, which in turn depends on the land surface inclination. This gives rise to a variety of plant communities. The moister margin is dominated by *Moliniopsis japonica* (see Fig. 11) in association with *Carex Middendorffii*, *Eriophorum vaginatum*, *Sanguisorba officinalis* var. *carnea*, *Gentiana Thunbergii* var. *minor*, *Myrica Gale* var. *tomentosa*, etc. *Hemerocallis esculenta* is the dominant plant in some localities, presenting a distinctive scene during the summer flowering season. On somewhat steeper slopes, but still with moist peat, a community dominated by the fern *Osmunda cinnamomea* occurs (also see Fig. 11). The *Osmunda cinnamomea* is associated with *Moliniopsis japonica* and *Carex Middendorffii*, and shrubs such as *Rhododendron japonicum* and *Malus Toringo* var. *Koringo*. At drier sites of greater inclination, the fern community is replaced by scrub dominated by *Prunus Grayana* and *Sorbus commixta*, in association with *Hydrangea paniculata*, *Malus Toringo* var. *Koringo*, *Prunus nipponica*, *Rhododendron japonicum* and *Rhus trichocarpa*. This scrub has an undergrowth of *Osmunda cinnamomea*, *Matteuccia Struthiopteris*, *Pteridium aquilinum* var. *japonicum* and *Sasa ozeana*.

On well-drained slopes, groves of birch trees or conifers often develop. The birch woods are dominated by *Betula platyphylla* var. *japonica* in association with *Betula Ermanii*, *Prunus Grayana* and *Sorbus commixta*. The undergrowth is composed of shrubs such as *Hydrangea paniculata*, *Rhus trichocarpa*, *Rhododendron*

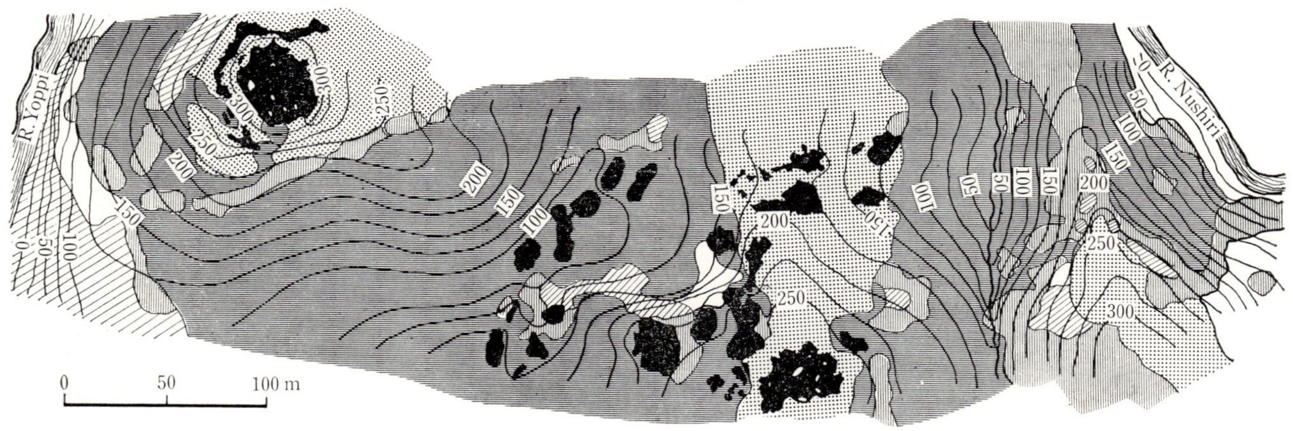

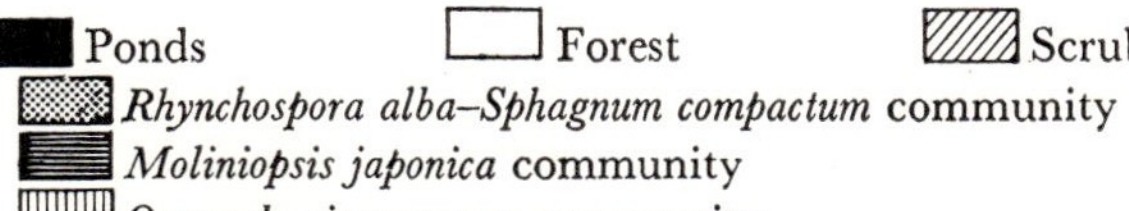

Contour lines in centimeters with respect to zero datum line.

FIG. 11. Map illustrating the topography and vegetation of Ozegahara Moor. (Northern part of Nakatashiro A moor. Yoshioka, 1954).

japonica and *Sasa ozeana*. The forest floor supports a community of *Osmunda cinnamomea*, *Moliniopsis japonica*, *Matteuccia Struthiopteris*, *Majanthemum dilatatum* var. *nipponicum* and *Sanguisorba officinalis* var. *nipponicum*. Groves of *Larix leptolepis* with a dense undergrowth of *Sasa ozeana* are also common, and sometimes include *Tsuga diversifolia*, *Thuja Standishii* and *Pinus pentaphylla*.

Ozegahara Moor is thus a place of great interest from the viewpoint of Japan's flora and vegetation. One other moor (Yashimagahara, in Nagano Pref., central Honshu) is also very famous, but it falls rather behind Ozegahara Moor in both ecological properties and geographical extent, and is not discussed in detail here.

6.4. Blanket bog

As mentioned above, another type of bog, corresponding to blanket bog, occurs in the mountainous regions of northern Japan that are influenced by a cool, moist climate (see Fig. 12). Here, flat or gently sloping areas covered by fine soil are the site of bogs. These occur in place of forest or scrubland which are both intolerant to poor drainage. In Japan, blanket bog usually develops on the slopes of volcanoes where the precipitation, especially snowfall, is high. The volcanoes are in most cases gently sloping and covered with ash, the latter seriously impeding drainage and resulting in bog formation directly on the land surface.

Blanket bogs are found on Mt. Taisetsu in Hokkaido, and on Mt. Hakkoda (see Fig. 13), Mt. Hachimantai, Mt. Gassan, Mt. Komagatake, Mt. Hiuchi, Mt. Naeba and Mt. Tateyama on Honshu. They have almost identical floristic components to normal bog, although the ground surface is, in contrast, generally even. The principal flora consists of *Moliniopsis japonica*, *Rhynchos-*

FIG. 12. Blanket bog established on gentle slopes surrounded by coniferous *Abies Mariesii* forest.

FIG. 13. Bog pond in blanket bog (Mt. Hakkoda). The floating leaf community is composed of *Nymphaea tetragona*; the foreground area is occupied by *Lobelia sessilifolia*.

pora alba, *R. Yasudana*, *Fauria crista-galli*, *Eriophorum vaginatum*, *Narthecium asiaticum*, *Oxycoccus quadripetalus*, *Sieversia pentapetala*, *Sphagnum tenellum*, etc.[8,24]

In Japan, blanket bog often contains islets of woodland or scrub, occupying hillocks within it. The woods are usually dominated by *Abies firma* with an undergrowth of *Sasa kurilensis*; the scrub is usually dominated by *Pinus pumila* and broad-leaved deciduous shrubs in association with *Empetrum nigrum*, *Vaccinium Vitis-Idaea* and *Loiseleuria procumbens*. These islets are gradually replaced by bog with increasing thickness of underlying peat.

References

1. S. Yoshimura, *Limnology* (Japanese), pp. 426, Tokyo, 1932.
2. H. Nakano, Hydrophytes in Lake Suwa, *Bot. Mag. Tokyo* (Japanese), **28**, 65–74, 127–32 (1914).

3. H. Yamaguchi, A survey of the larger aquatic plants in the southern basin of Lake Biwa, *Ecol. Rev.* (Japanese), **4**, 17–26 (1932.)
4. S. Miki, Ökologische Studien über die Sumf und Wassergewächse sowie ihre Formation in Ogura Teich, *Kyoto Pref. Shiseki Shochi Chosakai Hokoku* (Japanese), no. 8, 81–145, 1927.
5. T. Jimbo, M. Takamatsu and H. Kuraishi, Notes on the aquatic vegetation of Lake Towada, *Ecol. Rev.*, **14**, 1–9 (1955).
6. K. Ito, M. Tohyama, K. Ishizuka and T. Tsujii, The mire vegetation of Sarobetsu, *Ann. Rept. JIBP-CT(P)*, p. 1–5, 1968.
7. Hokkaido Kaihatsukyoku, *Ecology of Sarobetsu, VII. Biology* (Japanese), pp. 439, Tokyo, 1959.
8. A. Miyawaki, S. Itow and S. Okuda, Pflanzensoziologische Studien über die Vegetation der Umgebung von Aizukomagatake und Tashiroyama (Fukushima Pref.), *Sci. Res. Mt. Aizukoma and Taishaku National Park* (Japanese; German summ.), p. 15–43, Nature Conservation Soc. Japan, Tokyo, 1967.
9. T. Kashimura *et al.*, Note on the vegetation in and around Lake Usoriyama, *Ecol. Rev.*, **16**, 153–62 (1965).
10. M. Honda, *Plants Designated as National Monuments* (Japanese), pp. 438, Tokyo, 1957.
11. M. Saito and S. Ishikawa, Ecological studies on vegetation of basins and bogs in the northern part of the Byobusan area, *Bull. Educ. Fac. Hirosaki Univ.*, no. 18, 6–15 (1968).
12. M. Saito, S. Hasegawa and T. Kon, Vegetation of the Kakurenuma Moor, *J. Aomori Pref. Biol. Soc.*, **11**, 8–11 (1969).
13. S. Okuda, K. Fujiwara and A. Miyawaki, Pflanzensoziologische Studien über die Vegetation der Tsugaru-Halbinsel, des Berges Iwaki und des Juniko Sees, *Sci. Rept. Tsugaru Peninsula, Mt. Iwaki National Park* (Japanese; German summ.), p. 1–40, 1970.
14. Yubu, Tsurumi Scientific Researches, *Nature of Okubeppu* (Japanese), pp. 40, 1971.
15. *Ed.* A. Miyawaki, Vegetation of Japan compared with other regions of the world, *Encyclop. Sci. Technol.* (Japanese), vol. 3, pp. 535, Gakken, Tokyo, 1967.
16. S. Kamuro, Phytosociological studies on the littoral vegetation of the artificial ponds in the Tokai district, central Japan, *Mem. Fac. Educ. Fukui Univ. Ser. II Nat. Sci.*, no. 17, 53–79 (1967).
17. A. G. Tansley, *The British Islands and their Vegetation*, pp. 930, Cambridge University Press, 1939.
18. M. Tanaka, Phytosociological studies on the moors of eastern Hokkaido, I. Structure of the plant commutities in the Kushiro Moor, *J. Hokkaido Gakugei Univ.*, no. 10, 96–111 (1959).
19. K. Yoshioka, Phytosociological study on the vegetation of Akaiyachi Moor, *Ecol. Rev.*, **15**, 163–75 (1961).
20. T. Suzuki, Forest and bog vegetation within Ozegahara Basin, *Ozegahara*, p. 428–79, Tokyo, 1954.
21. H. Hara and M. Mizushima, List of vascular plants of Ozegahara Moor and

the surrounding districts (Japanese; English summ.), in *ref. 20.*
22. K. Yoshioka, Structure and development of plant communities in the Ozegahara Moor (Japanese; English summ.), in *ref. 20*, p. 170–204.
23. A. Miyawaki and K. Fujiwara, *Vegetationskundliche Untersuchungen in Ozegahara-Moor, Mittel Japan* (Japanese; German summm.), pp. 152, Tokyo, 1970.
24. K. Yoshioka, Ecological studies of the Takadayachi Moor, I. General aspects of the environment and vegetation, *Ecol. Rev.*, **16**, 13–26 (1963).

Additional References

a. J. T. Curtis, *The Vegetation of Wisconsin*, pp. 657, University of Wisconsin Press, 1959.
b. Faculty of Education, Oita University, *Scientific Researches of Kuju* (Japanese), pp. 741, 1968.
c. Y. Horikawa, H. Suzuki, H. Yokokawa and T. Matsumura, Moorland vegetation in the Yawata highlands, *Sci. Res. Sandankyo Gorge and Yawata Highlands, Hiroshima, Japan* (Japanese; English summ.), p. 121–52, 1959.
d. H. Nakano, The vegetation of lakes and swamps in Japan, *Bot. Mag. Tokyo*, **25**, 35–51 (1911); **28**, 65–74, 127–32 (1914); **30**, 31–50 (1916).
e. H. Osvald, *Die Vegetation des Hochmoores Komosse*, pp. 436, Uppsala, 1932.
f. K. Saito, Ecological studies of the Takadayashi Moor, III. Relation between the vegetation and peat, *Ecol. Rev.*, **16**, 33–37 (1963).
g. M. Tanaka, Phytosociological studies on the moors of eastern Hokkaido, II. Structure of the plant communities in the Kiritappu Moor, *J. Hokkaido Gakugei Univ.*, no. 10, 112–25 (1959).
h. M. Tatewaki and K. Ishizuka, Vegetation of the Senjogahara Moor, Nikko, *J. Bot. Garden. Hokkaido Univ.* (Japanese; English summ.), no. 2, 1–72 (1969).
i. K. Urakami and S. Ichimura, Peatland: its characteristics and agriculture, *Bull. Hokkaido Agr. Exptl. Sta.* (Japanese), no. 60, 1–306 (1937).
j. H. Walter, *Einfuhrung in die Pflanzengeographie Deutschland*, pp. 458, Jena, 1927.
k. J. E. Weaver and F. E. Clements, *Plant Ecology*, pp. 601, McGraw-Hill, 1938.
l. M. Yamanaka, Ecological studies of the Takadayachi Moor, III. Pollen analytical study of the Takadayachi Moor, *Ecol. Rev.*, **16**, 27–32 (1963).
m. N. Yano, *Vegetation of Kirigamine* (Japanese; English summ.), pp. 66, Suwa, 1971.

8

Volcanic Vegetation

Kuniji YOSHIOKA*

1. Mt. Usu and Showa-shinzan
2. Mt. Komagatake
3. The Yakebashiri Lava Flow on Mt. Iwate
4. The Sainokawara Lava Flow on Mt. Zao
5. The Urabandai Mud Flow on Mt. Bandai
6. Lava Flows on Mt. Fuji
7. Mt. Asama
8. Izu-Oshima Is.
9. Miyakejima Is.
10. Mt. Aso
11. Sakurajima Is.
12. Solfataras and Vapour Fumaroles
    12.1. Solfataras
        12.1.1. The Osoreyama solfataras
        12.1.2. The Sukayu solfataras
        12.1.3. The Goshogake solfataras
        12.1.4. The Oowakutani solfataras
        12.1.5. The Unzen solfataras
    12.2. Vapour fumaroles
13. Conclusion

**Biological Institute, Tohoku University, Aramaki Aza Aoba, Sendai-shi 980, Japan*

Volcanic vegetation is interesting and important from the viewpoint of plant ecology, and especially from the viewpoint of plant successions. The first intensive studies were those on the vegetation of the volcanic island of Krakatau (Java), following the eruption of August 1883. Revegetation on the new volcanic ejecta spread out over the island was investigated independently by many researchers.[1-7] Inspired by the results, botanists also began studies on the vegetation of other volcanoes of the world, such as of Taal in the Philippines,[8] Katamai in Alaska,[9,10] Mauna Kea and Mauna Loa in Hawaii,[11,12] El Paricutin in Mexico,[13,14] and Mt. Cameroun in W. Equatorial Africa.[15]

In Japan, Yoshii[16,17] first carried out intensive ecological research on the revegetation of the volcano Komagatake in Hokkaido, following the great eruption of 1929. He also studied the volcanic vegetation of several other volcanoes throughout the country. Following Yoshii, many other researchers have investigated different volcanoes, such as Mt. Aso in Kyushu (Asai[18]), the volcanic island of Izu-Oshima (Tezuka[19]), and the volcanic island of Sakurajima in Kyushu (Tagawa[20]). The research on Mt. Komagatake has been extended by Tatewaki *et al.*[21] and by the present author.[22]

The various volcanoes listed above are situated in widely separated parts of the globe, in areas with contrasting climates, and they issue very different kinds of ejecta. Such circumstances are clearly reflected in the nature and patterns of the plant successions, which are correspondingly diverse. Indeed, ecological studies of vegetation developing on volcanic ejecta have contributed much to the resolution of the important problems concerning plant successions. In Japan, there are widespread volcanic zones and areas covered by volcanic debris, the general features of which are reviewed in Chapter 1 of this book. The patterns of plant successions differ markedly from locality to locality, although the final stages of each succession in general converge to a single climax community under similar climatic and orographic conditions. This chapter gives several outstanding examples of volcanic vegetation from different parts of Japan.

1. Mt. Usu and Showa-shinzan

Mt. Usu is situated to the south of Lake Toya in Hokkaido, at 42°10′N. It comprises two volcanoes, Oousu (725 m) and Kousu (611 m), rising side by side. Since 1626 several eruptions have been recorded, and on each occasion mud flows and volcanic ash were produced in large quantities.

In 1943–45, a volcanic dome of quartz andesite, Showa-shinzan, was uplifted in an upland field on the eastern side of Oousu, attaining a height of 150 m above the ground surface (257 m a.s.l.) (see Fig. 1). By 1965, the dome

FIG. 1. Showa-shinzan, showing its general appearance and vegetation in 1965.

itself was only scarcely grown with vascular plants, although the moist rock surface was also occasionally found to support mosses and algae. At the foot of the dome, where detritus, ash and soil have accumulated together, a rather dense vegetation has now been established. At the immediate foot, where the soil condition is rather unstable, a herb community occurs, comprised of the following species: *Reynoutria sachalinensis*, *Oenothera biennis*, *Artemisia montana*, *Petasites japonicus* var. *giganteus*, *Ixeris japonica*, *Taraxacum officinale*, *Aster Glehni*, *Picris japonica*, *Trifolium repens*, *Lotus corniculatus*, *Miscanthus sinensis* and *Poa annua*. A few seedlings belonging to such trees and shrubs as *Populus Maximowiczii*, *Betula platyphylla* var. *japonica*, *Salix Bakko* and *S. integra* are also found. This community is thus a mixture of native and exotic plants, arising because the locality was, before the eruption, a cultivated field overgrown with exotic weeds.

The gently sloping skirt of rather stable soil is occupied by a pioneer forest consisting of *Populus Maximowiczii*, *Betula Ermanii* and *B. platyphylla* var. *japonica* (5–7 m high and 10–15 yr old in 1965). Beneath this tree canopy (see Fig. 2), *Salix sachalinensis*, *S. Bakko* and *S. integra* (2–3 m high) occur, but their growth is restricted under intense shade. The herb layer is made up of species almost identical to those of the herb community mentioned above, although there is a tendency for native species to be dominant over exotic ones.

A violent eruption also took place in 1910, on the northern side of Kousu, and caused the uplift of the new cone, Meiji-shinzan, which remained covered with blown ash and soil after the eruption. In 1965, 55 yr later, an investigation was made of the new vegetation established on the lower slopes

of the cone where the soil was moderately moist and the surface unexposed to strong winds. A dense pioneer forest of 10–15 m tall trees, consisting mainly of *Populus Maximowiczii* and *Alnus hirsuta* var. *siberica*, associated with *Betula Maximowicziana* and *Cornus controversa*, has been identified. The sub-tree layer (2–5 m tall) is comprised of *Acer mono*, *Prunus Sargentii*, *Sorbus alnifolia*, *Quercus mongolica* var. *grosseserrata*, *Salix Bakko* and *Aralia elata*, the former three species growing vigorously while the latter two are more recessive. The shrub layer is 1–2 m in height and includes such plants as *Hydrangea paniculata*, *Morus bombycis* and *Rosa multiflora*. The herb layer attains a maximum height of about 1 m and is comprised of native species only, such as *Petasites japonicus* var. *giganteus*, *Reynoutria sachalinensis*, *Pyrola japonica*, *Asperula odorata*, *Sanicula chinensis*, *Disporum sessile* and *Lilium cordatum* var. *Glehni*.

On the drier soil near the crater a shorter and sparser pioneer forest occurs. This is dominated by *Populus Maximowiczii* and *Betula platyphylla* var. *japonica* and has an undergrowth of *Cornus controversa*, *Prunus verecunda* and *Kalopanax pictus* saplings, plus shrubs such as *Morus bombycis* and *Rosa multiflora*. The herb layer is composed of *Miscanthus sinensis* and *Patrinia villosa* (native species) and *Poa pratensis*, *Trifolium repens* and *Oenothera biennis* (exotic species). The drier soil community thus occupies a lower position in the plant succession than the one on moderately moist soil, although there is progression towards mesophytism with the invasion of *Acer mono*, *Quercus mongolica* var. *grosseserrata*, *Prunus verecunda*, etc.

The main volcano of Oousu underwent its latest eruption (on the southern side) in 1822, producing large mud flows. Observations in 1965 proved the establishment of a mesophytic forest on protected sections of these flows with moderate moisture content in the soil. The forest is composed mainly of tall

FIG. 2. Pioneer forest established near the base of Showa-shinzan. *Populus Maximowiczii* and *Betula platyphylla* var. *japonica* can be seen in the foreground.

mesophytic trees such as *Acer mono*, *Kalopanax pictus*, *Magnolia obovata*, *Cornus controversa* and *Ulmus Davidiana*, associated with pioneer trees such as *Populus Maximowiczii*, *Alnus hirsuta* and *Betula platyphylla* var. *japonica*. Beneath the dense canopy is a sub-tree layer composed of growing saplings of the mesophytic trees plus growth-restricted sun plants such as *Alnus Maximowiczii*, *Salix Bakko*, *S. sachalinensis* and *Hydrangea paniculata*. The shrub layer is scattered with shade-tolerant plants such as *Euonymus Sieboldiana* and *Viburnum Sargentii*, while the herb layer supports a dense growth of shade-tolerant herbs such as *Dryopteris crassirhizoma*, *Trillium Smallii*, *Disporum sessile*, *Lilium cordatum* var. *Glehni*, *Angelica pubescens*, *Pachysandra terminalis*, *Asperula odorata* and *Cacalia hastata*, associated with a few remaining pioneer plants such as *Petasites japonicus* var. *giganteus*. This mesophytic forest is thought to represent closely the climax forest of the area studied.

2. Mt. Komagatake

The volcano Mt. Komagatake rises to an altitude of 1140 m a.s.l., and is situated about 18 km north of Hakodate City, Hokkaido, at 42°4′ N, 140°43′ E. It has erupted repeatedly in recorded history, each time issuing large amounts of pumice. The latest, violent eruption occurred on June 19, 1929, and a tremendous amount of pumice was ejected and deposited in two different ways, i.e. as pumice flows running down the steep slopes and as pumice fields built up of airborne pumice ejected from and dropping around the crater. Both the pumice flows and pumice fields initially formed as aggregates of hot andesitic blocks, reaching a maximum total thickness of about 10 m, and the vegetation covered by them was for the most part completely destroyed.

The revegetation of the pumice flows and fields was extensively investigated by Yoshii[16,17] at intervals of from one to several years, starting shortly after the eruption. Also, more recently, following the work of Yoshii and of Tatewaki *et al.*,[21] the present author[22] has undertaken an extensive study of the vegetation of the Mt. Komagatake volcano.

The pumice flows and fields investigated lie between 200 and 600 m a.s.l., and the climax forest is judged to be *Fagus crenata* forest. However, most of the forests remaining unscathed after the eruption were secondary ones consisting primarily of *Quercus mongolica* var. *grosseserrata*, associated with *Castanea crenata*, *Prunus Ssiori*, *P. verecunda*, *Sorbus commixta*, *Populus Maximowiczii*, *Betula Ermanii*, etc.

In 1931, two years after the eruption, the moss *Polytrichum commune* appeared as the first visible pioneer in small depressions among the pumice blocks, although no other plants apart from the moss were found, even in the

finer pumice deposits. However, investigations in 1933 revealed the scattered occurrence of seedlings of ligneous plants such as *Salix Bakko*, *S. integra*, *Populus Maximowiczii* and *Betula platyphylla*, as well as herbaceous plants such as *Miscanthus sinensis*, *Reynoutria sachalinensis* and *Anaphalis margaritacea*. No vascular plants were seen in the depressions occupied by *Polytrichum commune*. Instead, they grew in narrow interstices between pumice blocks. That is to say, the vascular plants preferentially occupied a biotope different from that of the moss, and were able to establish themselves independently of the presence of mosses or lichens. There were no observable differences in this situation between pumice flows and pumice fields.

In 1935, about 20 taxa were found to be growing on the pumice deposits, in addition to *Polytrichum commune*. However, they were rather scattered and showed only hampered growth. The grasses were *Miscanthus sinensis*, *Calamagrostis hakonensis*, *Phragmites communis* and *Poa annua*; and the sedges, *Carex oxyandra* and *Luzula capitata*. The herbs included exotic weeds such as *Rumex Acetosella* and *Oenothera Lamarckiana* plus native species such as *Reynoutria sachalinensis*, *Epilobium angustifolium*, *Galium japonicum*, *Anaphalis margaritacea*, *Solidago Virga-aurea* var. *leiocarpa*, *Petasites japonicus* ssp. *giganteus* and *Artemisia montana*. The tree seedlings were *Betula Maximowicziana*, *B. platyphylla* var. *japonica* and *Populus Maximowiczii*, and the shrub seedlings, *Salix Bakko*, *S. sachalinensis* and *Aralia cordata*.

By 1942, the following species among those confirmed in 1935 had disappeared: *Calamagrostis hakonensis*, *Phragmites communis*, *Poa annua*, *Carex oxyandra*, *Luzula capitata*, *Rumex Acetosella*, *Oenothera Lamarckiana*, *Epilobium angustifolium* and *Aralia cordata*. They were presumably insufficiently tolerant to the severe conditions prevailing in the pumice flows and pumice fields. *Senecio cannabifolius* was the only new invader.

In 1960, 31 yr after the eruption, a closed pioneer forest was established on a pumice flow more or less protected from strong winds and close to other remaining forests. The trees were 10–12 m high and the dominant species, *Populus Maximowiczii* and *Betula platyphylla* var. *japonica*. Beneath this canopy were saplings of other trees such as *Prunus Sargentii* and *Acer mono*, together with shrubs such as *Salix Bakko*, *S. sachalinensis*, *Hydrangea paniculata* and *Rhus trichocarpa*. Identical species were also found in open places at the centre of the pumice flow, although they were sparser and more stunted. Besides *Polytrichum commune*, another moss, *Rhacomitrium canescens* var. *ericoides*, and the lichen *Stereocaulon vesvianum* were now found for the first time on the pumice flows and fields. Clearly, as mentioned, the vascular plants had successfully established themselves on the pumice deposits quite independently of cryptogams such as mosses and lichens, although the latter are usually regarded as pioneers that precede vascular plants in typical plant successions.

FIG. 3. Southern slopes of Mt. Komagatake covered by a pumice flow at the foot, photographed four years after the eruption of 1929. (By courtesy of Y. Yoshii)

FIG. 4. View of Mt. Komagatake corresponding to that in Fig. 3, photographed again in 1965.

The latest investigations made in the same areas in 1965 revealed no significant changes from the situation in 1960.

In conclusion it can be said that the plant succession developing on the pumice flows of Mt. Komagatake is somewhat different, both in tempo and in floristic composition, from that on the mud flows of Mt. Usu, although the pioneer forests established on the volcanic ejecta in both cases are distinctly similar. The change in general appearance of Mt. Komagatake with the development of its pumice vegetation is illustrated in Fig. 3 and 4.

3. The Yakebashiri Lava Flow on Mt. Iwate

The volcano Mt. Iwate rises to an altitude of 2041 m a.s.l., and is situated to the northwest of Morioka City, northeast Honshu, at about 39°50′ N. It erupted in 1686, producing the Yakebashiri lava flow on its northeastern slopes between 1100 m and 680 m a.s.l. The lava is basaltic in type, with a scoriaceous surface, but the greater part remains almost entirely barren even though almost 300 yr have elapsed since its formation.

The scoriaceous surface of the lava flow is thinly populated by vascular plants, although lichens such as *Stereocaulon vesvianum* and *Cladonia* sp., and the moss *Rhacomitrium canescens* var. *ericoides*, are the dominant plants. The flow contains many small depressions and thin grooves in which fine detritus has collected, and these support plants where they are protected from violent winds and strong sunlight. Such localities are usually occupied by small thickets (2–3 m high) consisting of many different tree and shrub species. The tree species found are *Pinus densiflora*, *Populus Sieboldi*, *Betula Ermanii*, *Alnus Maximowiczii*, *Quercus mongolica* var. *grosseserrata*, *Maackia amurensis* var. *Buergeri*, *Fraxinus Sieboldiana*, *Sorbus alnifolia* and *S. commixta*. Mixed with these gnarled trees are various shrubs such as *Salix Reinii*, *Corylus Sieboldiana*, *Lindera membranacea*, *Weigela hortensis*, *Rhododendron Fauriae*, *R. japonicum*, *Tripetaleia paniculata*, *Prunus nipponica*, *P. Sargentii*, *Acer Tschonoskii* and *Hydrangea paniculata*. Herbaceous species are conspicuously sparse in comparison with ligneous ones but include *Reynoutria sachalinensis*, *Majanthemum bifolium*, *Miscanthus sinensis* and *Solidago Virga-aurea*. Several other kinds of mosses in addition to those mentioned above are also frequently found on the ground surface.

The edge of the lava flow is rich in fine scoriaceous debris and close to the surrounding tree seed source. It is thus often occupied by pioneer forests dominated by *Quercus mongolica* var. *grosseserrata* or *Pinus densiflora*. However, as shown above, the greater part of the lava flow still remains in the moss-lichen stage, and the mild microhabitats that do occur are generally occupied by ligneous communities rather than herbaceous ones. That is to say, the

ligneous species show marked superiority over herbaceous ones in the plant succession that has developed on the Yakebashiri lava flow. Clearly, herbaceous communities are not always necessary for the establishment of ligneous ones in any given environment. The climatic climax of this area is thought to be *Fagus crenata* forest, although the present state of the succession is far from approaching such a plant assemblage. Indeed, as a whole, the tempo of development of the succession on the Yakebashiri lava flow is very slow compared with that on other volcanic ejecta in Japan.

4. The Sainokawara Lava Flow on Mt. Zao

Mt. Zao rises to an altitude of 1840 m a.s.l., and is situated on the border of Miyagi Pref. and Yamagata Pref., northeast Honshu, at about 38°8′ N. Although no reliable records have actually been found, the mountain is believed to have erupted about 1200 yr ago, producing the Sainokawara lava flow on its eastern slopes between 1600 m and 1200 m a.s.l., i.e. in the region primarily occupied by subalpine *Abies Mariesii* forest. The thick flow consists of andesitic lava which is partially weathered in places to yield a thin soil on the surface (see Fig. 5).

The Sainokawara lava flow abuts onto *Abies Mariesii* forest to the south and mixed scrub to the north, across the wide valley of the Nigorikawa. The plant succession exhibits a distinct north-south gradient, probably arising from the southerly situation of the seed source and from exposure to cold northerly prevailing winds in winter. The northern areas are scattered with creeping trees of the species *Pinus pentaphylla*, associated with alpine plants such as *Empetrum nigrum*, *Loiseleuria procumbens*, *Anaphalis margaritacea* var. *angustior* and

FIG. 5. The Sainokawara lava flow in the subalpine zone of Mt. Zao.

FIG. 6. *Dicentra peregrina* on scoria of the Sainokawara lava flow.

FIG. 7. *Pinus pentaphylla* thicket with an undergrowth of *Empetrum nigrum*, occurring on the Sainokawara lava flow as an intermediate stage of the plant succession.

Deschampsia flexuosa. Elevated parts of the flow with scoriaceous deposits on the surface are often inhabited by alpine herbs such as *Dicentra peregrina* (see Fig. 6) and *Hieracium japonicum*. A rather dense thicket also occurs towards the north, consisting mainly of *Pinus pentaphylla* (2–3 m high) associated with *Pinus densiflora*, *Larix leptolepis*, *Maackia amurensis*, *Alnus Maximowiczii* and *Salix Reinii*. Dwarf alpine plants such as *Empetrum nigrum*, *Loiseleuria procumbens*, *Ledum palustre* var. *diversipilosum*, *Shortia soldanelloides*, etc. are commonly found in association (see Fig. 7). The southern limit of the lava flow that adjoins the *Abies Mariesii* forest is occupied by *Pinus pentaphylla* forest (5–10 m high). It contains saplings of *Abies Mariesii*, an indication of progressive succession towards the climax forest. Thus, although the Sainokawara lava flow is situated in the subalpine zone, it is at present inhabited by alpine, montane and submontane species, in addition to subalpine ones.

The lava flow also covers the upper parts of an older extrusion, which extends down to about 900 m a.s.l. The lower part of this older flow supports a pure *Quercus mongolica* var. *grosseserrata* forest at altitudes of 900–1200 m (which correspond to the *Fagus crenata* climax forest zone). However, *Fagus crenata* has not yet invaded this oak forest, probably due to the relative immobility of its seed.

5. The Urabandai Mud Flow on Mt. Bandai

Mt. Bandai rises to an altitude of 1818 m a.s.l., and is situated to the north of Lake Inawashiro, Fukushima Pref., northeast Honshu, at about 37°36′ N. In 1888, it erupted violently and abruptly producing a huge mud flow containing vast amounts of soil and boulders (the Urabandai mud flow) on its northern slopes, between 1100 m and 800 m a.s.l. The lower parts of the mud flow dammed up streams and brooks, giving rise to many lakes and ponds and so to a diversity of landform in the area (see Fig. 8).

The mud flow is divided into two main regions according to the structure of the ground surface, i.e. into rocky fields and muddy fields. The former occupy the greater part of the flow, apart from the lower, narrower areas, and are dominated by *Pinus densiflora* (10–15 m high), frequently associated with broad-leaved deciduous trees such as *Acer mono*, *Populus Sieboldi*, *Alnus firma*, *Betula Ermanii* and *B. Maximowicziana*. The principal shrubs are *Hydrangea paniculata*, *Coriaria japonica*, *Weigela hortensis*, *Gaultheria adenothrix*, etc. Also occurring commonly in the rocky fields are *Quercus mongolica* var. *grosseserrata*, *Viburnum furcatum* and *Sasa kurilensis*, which are components of the (presumed) climax forest of *Fagus crenata*.

The mud fields in the lower reaches of the mud flow are not covered by forest, but mostly support grassland dominated by *Miscanthus sinensis* as-

FIG. 8. Region of the Urabandai mud flow on the northern slopes of Mt. Bandai where many ponds and lakes have formed by the damming up of streams and brooks. *Pinus densiflora* forest occurs on rocky areas of the mud flow.

sociated with shrubs such as *Weigela hortensis*, *Hydrangea paniculata* and *Coriaria japonica* (see Fig. 9). The comparatively sparse occurrence of trees in this grassland is attributed to the dense growth of *Miscanthus sinensis*, which probably inhibits any large-scale invasion of trees.

According to the present author's own studies, the Urabandai mud flow was already occupied by *Pinus densiflora* forest (in rocky fields) and *Miscanthus sinensis* (in muddy fields) in 1933, i.e. about 45 yr after the eruption. The plant succession had thus made rapid progress, although even today it is still rather far from the *Fagus crenata* climax forest stage.

FIG. 9. *Miscanthus sinensis* grassland on fine deposits of the Urabandai mud flow where there is retardation of the plant succession. *Salix serissaefolia* occurs in the moist depressions.

6. Lava Flows on Mt. Fuji

Mt. Fuji is the highest volcano in Japan. Its cone rises from the surrounding plain to an altitude of 3776 m a.s.l., and is situated about 100 km southeast of Tokyo, at 35°22′ N.

Mt. Fuji has erupted repeatedly in the past, alternately issuing lava flows and mixed scoria and ash deposits. The two strongest recorded eruptions took place in 864 and 1707. At present there are five dammed-up lakes, known as Yamanaka-ko, Kawaguchi-ko, Sai-ko, Shoji-ko and Motosu-ko, along the northern foot of the mountain.

In 864 an extensive lava flow, the Aokigahara lava flow, was formed on the northwestern slopes of Mt. Fuji, between 1300 m and 900 m a.s.l. This flow is composed of basalt and is only slightly weathered at the surface even though 1100 yr have elapsed since its formation. Nevertheless, it supports a dense growth of evergreen conifers (hence its name, Aokigahara, "the plain of evergreen trees") (see Fig. 10). The forest consists mainly of evergreen conifers such as *Tsuga Sieboldii*, *Picea polita*, *Picea jezoensis* var. *hondoensis*, *Abies firma*, *Pinus densiflora*, *P. pentaphylla* var. *Himekomatsu* and *Chamaecyparis obtusa*, associated with the deciduous *Larix leptolepis*. Some broad-leaved deciduous trees, such as *Quercus mongolica* var. *grosseserrata*, *Betula grossa* and *B. corylifolia*, also occur mixed with the conifers. The dense undergrowth is made up of small broad-leaved evergreen trees such as *Ilex pedunculosa*, *I. crenata*, *Pieris japonica* and *Skimmia japonica*, together with deciduous shrubs such as *Rhododendron Wadanum*, *Lyonia Neziki* and *Rhus trichocarpa*.

At an altitude of about 1100 m a.s.l., the flow runs round a parasitic cone,

FIG. 10. The Aokigahara lava flow, densely covered by evergreen coniferous trees, on the northwestern slopes of Mt. Fuji.

known as Omuro, which was probably formed long before the Aokigahara lava flow. The exposed surface of this cone has a rather deep, mature soil which supports a dense forest of large broad-leaved deciduous trees. The forest is dominated by *Fagus japonica*, in association with *Quercus mongolica* var. *grosseserrata*, *Acer palmatum*, *A. mono* var. *dissectum* and *Abies homolepis*. The undergrowth is comprised of *Lindera umbellata*, *L. sericea* var. *tenuis*, *Pieris japonica*, etc.

To the immediate north of Mt. Fuji is the Tertiary mountain range, the Misaka Mts., which are covered by broad-leaved deciduous *Fagus crenata* forests. It is clear, therefore, that the Aokigahara lava flow lies at an altitude appropriate to a climax forest of *Fagus japonica* or *F. crenata*. However, as shown, its vegetation remains at an intermediate stage in the plant succession, even though about 1100 yr have elapsed since the eruption. Compared to broad-leaved deciduous trees such as *Fagus crenata* and *F. japonica*, the conifers are more tolerant to the severe conditions that still prevail on the lava flow. This situation, i.e. where a pioneer forest of conifers is first successful, to be followed later by broad-leaved deciduous trees, is typical of plant communities invading this type of volcanic extrusion.

There are some other conspicuous lava flows on Mt. Fuji, although these are rather lesser in extent and somewhat older than the Aokigahara lava flow. Their original volcanic vegetation has, through human interference, been almost completely destroyed, except for some forestland on the Takamarubi lava flow, which occurs on the northeastern slopes of the volcano to the west of Lake Yamanaka.

In total about 70 ha of the Takamarubi lava flow is occupied by an exceptional, pure forest of *Picea polita* at an altitude of 1000 m a.s.l. The undergrowth consists of shrubs such as *Prunus incisa*, *Clethra barbinervis*, *Deutzia crenata*, *Hydrangea paniculata*, etc. This type of volcanic vegetation is of extremely rare occurrence, but at present still remains relatively free from the destructive influence of man. However, *Picea polita* is unable to reproduce and develop effectively beneath a thick canopy, and areas of competitive species such as *Prunus incisa*, *Deutzia polita* and *Clethra barbinervis* are gradually becoming exposed, following the formation of thickets in place of true forest. Clearly, the conservation and preservation of such pioneer and intermediate stages of volcanic plant successions will prove very difficult.

7. Mt. Asama

Mt. Asama rises to an altitude of 2542 m a.s.l., and is situated on the border of Gunma Pref. and Nagano Pref., central Honshu, at 36°25′ N. It has undergone repeated, violent eruptions in recorded history since 685, the

FIG. 11. Lichens and mosses covering the lava surface of the Onino-oshidashi lava flow, Mt. Asama.

latest great eruption occurring in 1783. On this occasion, extensive flows of lava and mud were formed, and a vast amount of pumice was ejected from the crater. The lava flowed almost due north, the mud northeastwards, and the pumice collected mostly on the eastern slopes. These ejecta were deposited at altitudes of 1000–2300 m, the lower half of which corresponds to the montane climax forest zone of *Fagus crenata* and the upper half to the subalpine coniferous forest zone of *Abies Mariesii* and *Tsuga diversifolia*. The Asama volcano is thus particularly important from the viewpoint of studies on plant successions, since many different kinds of volcanic ejecta were produced simultaneously and came to lie in different localities in the same broad zones on the mountain side.

The principal lava flow, known as Onino-oshidashi, runs from the crater at the top of the mountain down the northern slopes to an altitude of 1200 m. It is a thick andesitic flow with a rather porous but scarcely weathered surface. Exposed rock is often thinly covered by a cryptogam community of *Stereocaulon exutum*, *Cladonia sylvatica*, *Dicranum japonicum* and *Rhacomitrium hypnoides* (see Fig. 11), associated with a few alpine plants such as *Loiseleuria procumbens* and *Phyllodoce nipponica*. Several depressions in the lava flow are filled with scoriaceous blocks, lava debris and volcanic ash, and are protected from the effects of strong winds and desiccation. Near the nose of the lava flow, these depressions are occupied by 3–6 m high open thickets of *Pinus densiflora*, *Larix leptolepis*, *Alnus Sieboldiana* and *Betula platyphylla* var. *japonica*, with an undergrowth of alpine plants such as *Rhododendron Fauriae*, *Empetrum nigrum*, *Loiseleuria procumbens*, *Phyllodoce nipponica*, *Arcteria nana* and *Shortia soldanelloides* (see Fig. 12).

FIG. 12. *Pinus densiflora* and *Rhododendron Fauriae* in a depression on the Onino-oshidashi lava flow, Mt. Asama.

The ejected pumice was thickly deposited over the eastern side of the mountain at altitudes of from 1600 m to 1100 m, forming an extensive pumice field. The lower parts of this field (1100–1300 m a.s.l.) are densely covered with an exceptionally well-developed *Pinus densiflora* forest (about 20 m high). Broad-leaved deciduous trees such as *Quercus mongolica* var. *grosseserrata*, *Prunus verecunda*, *Cornus controversa*, etc. occur beneath the canopy. The upper parts of the pumice field (1300–1500 m a.s.l.) are dominated by *Larix leptolepis* (10–15 m high) associated with *Betula platyphylla* var. *japonica*. This larch forest has an undergrowth of *Alnus Sieboldiana*, *Rhododendron japonicum*, *Hydrangea paniculata*, etc. (see Fig. 13).

The mud flow runs from an elevation of about 1300 m, down the valley of the River Azuma, and is composed of volcanic ash, sand, scoria and rock debris. In most places it has been reclaimed for human use, i.e. apart from the upper portions which are overgrown with a variety of plants. Open areas exposed to strong winds are sparsely grown with alpine plants such as *Empetrum nigrum*, *Vaccinium uliginosum*, *Loiseleuria procumbens*, *Deschampsia flexuosa* and *Carex flexuosa*, together with the ubiquitous species *Miscanthus sinensis* and *Reynoutria japonica*. Other areas, more or less protected from severe winds, support a mixed forest of *Pinus densiflora*, *Alnus Sieboldiana* and *Quercus serrata*, with an undergrowth of *Vaccinium uliginosum*, *Empetrum nigrum*, *Deschampsia flexuosa*, *Carex oxyandra* and *Miscanthus sinensis*. However, this forest is rather open and the trees attain a maximum height of only about 5 m or so.

In conclusion it can thus be said that the plant succession on Mt. Asama has, since the 1783 eruption, made conspicuously rapid progress on the

FIG. 13. *Larix leptolepis* forest naturally established on a pumice field that was formed contemporaneously with the lava flow on Mt. Asama.

pumice flow, but has been remarkably slow to develop on the lava flow. Certain intermediate stages can be identified, as for example on those parts of the mud flow which are exposed to strong winds.

8. Izu-Oshima Is.

The islands of the so-called Izu-Shichito extend in a line southwards from the entrance to Tokyo Bay. All are of volcanic origin, and are named (from north to south) Oshima, Toshijima, Niijima, Kozushima, Miyakejima, Mikurajima and Hachijojima, respectively. Among these seven islands, Oshima and Miyakejima are active volcanoes, and have erupted repeatedly during recent history.

Oshima Is., the northernmost island, has formed by the gradual accumulation of volcanic ejecta, etc. from and around the so-called Mihara volcano, which is known to have erupted severely at least seven times since 684. On each occasion, flows of basaltic lava were issued from and around the central crater area.

Tezuka has made an intensive study[19)] of the vegetation present on lava flows of different ages, and has recognized the following sequence of stages: (1) Bare land, (2) Rocky desert, with scattered *Reynoutria hachidyoensis* and *Carex Okuboi*, (3) Scrub, dominated by *Alnus Sieboldiana* and *Weigela coraeensis*, (4) Mixed broad-leaved forest, dominated by *Alnus Sieboldiana*, *Prunus Lannesiana* form. *simpliciflora*, *Cornus controversa*, etc., and (5) Climax broad-leaved evergreen forest, dominated by *Castanopsis cuspidata* var. *Sieboldii* and *Machilus Thunbergii*.

The lichen and moss stages, or lichen-moss stage, usually present as the lithosere (such as on typical lava flows) is indiscernible in this case. Also, the herbaceous stage, typically composed primarily of herbs, is unrepresented. However, a rocky desert stage, comprising scattered herbs and tree seedlings, can be distinguished (stage 2 of the list). That is to say, the important pioneers on the Oshima lava flows are not lichens and mosses, but rather ligneous plants and herbs that have settled in such favourable biotopes as rock crevices and depressions in the lava flows, where there is protection from strong winds and sunlight and some accumulation of soil.

The 1950–51 lava flow (the latest one) was scattered from the start with *Alnus Sieboldiana* and *Reynoutria hachidyoensis*, never supporting a community of mosses or lichens. Such a situation may thus represent the beginning of the rocky desert stage, following which there is formation of a rather dense scrub dominated by *Alnus Sieboldiana* and *Weigela coraeensis* (i.e. there is an increase of shrubs and trees rather than of herbaceous plants). The 1876–77 lava flow is now in such a scrub stage (stage 3 of the list), and this is typically followed by a mixed broad-leaved forest dominated by *Alnus Sieboldiana*, *Prunus Lannesiana* form. *simpliciflora* and *Cornus controversa* (stage 4), which is now developing on the 1777–78 lava flow. The broad-leaved deciduous trees are then gradually replaced by shade-tolerant broad-leaved evergreen trees such as *Machilus Thunbergii* and *Castanopsis cuspidata* var. *Sieboldii*, which constitute two of the main components of the climatic climax forest of the island (i.e. stage 5). The lava flow dating probably from the year 684 is occupied by such a broad-leaved evergreen forest.

Soil formation on the lava flows is first discernible in the scrub stage but it is still confined to a relatively thin A horizon even in the climax forest stage. The chemical properties of the soil, such as its content of total nitrogen, ammonium nitrogen, total phosphorus, soluble calcium, soluble potassium,

TABLE 1. Amounts of organic matter and nutrients (per 1 m^2 of land surface) contained in soils of different successional stages on the Oshima volcanics. (Organic matter and nutrients present in the surface layer of litter, etc. have been excluded.) (After Tezuka, 1961)

Stage / Substance	Bare land	Rocky desert	Scrub	Mixed forest	Climax forest
Organic matter (kg)	1.5	4.7	20.0	32.5	28.8
Total N (g)	2	144	515	990	743
Available N (g)	0	1.5	4.6	4.1	4.7
Total P (g)	5.7	11.3	47.6	46.0	34.0
Available P (mg)	60	258	699	461	474
Available K (g)	0	2.3	12.1	13.4	13.7
Available Ca (g)	6	54	195	255	311

moisture, etc., and the cation-exchange capacity, increase proportionally with the amount of humus in the soil (see Table 1), although other properties such as the content of soluble phosphorus and nitrate nitrogen vary more irregularly. Nevertheless, in general, the nutrient content of the soil formed on these lava flows rises with increase in amount of organic matter, and the value per unit area increases through the successive stages up to the climax forest, where a stable state is reached. Soil conditions, however, are the main controlling factor only in the earlier stages of the succession, light being the dominant factor governing the later stages.

9. Miyakejima Is.

Miyakejima Is. is situated about 80 km to the south of the island of Izu-Oshima. It is known to have undergone significant eruptions in 1643, 1712, 1834, 1874, 1940 and 1962, each time issuing appreciable amounts of lava with some scoria and ash.

The island was surveyed for its new volcanic vegetation by the present author at the end of December 1940, five months after the 1940 eruption, but was found to show no signs of plant invasion into the new lava flow or, for the most part, the scoria fields. However, some thin scoria fields had begun to recover their vegetation.

At that time (i.e. 66 yr after the 1874 eruption), the 1874 volcanics showed different states of revegetation according to the nature and situation of the ejecta. The lava flow was only sparsely grown with plants in coastal areas exposed to strong sea-winds, but supported a dense growth of ligneous species on inland hillsides protected from the winds. The coastal lava flow was occasionally grown with the moss *Rhacomitrium* sp. and epiphytic ferns such as *Pyrrosia lingua*, *Nephrolepis auriculata* and *Pleopeltis Thunbergiana*. Small, open thickets also occurred but were restricted to depressions. They were composed of *Alnus Sieboldiana*, *Hydrangea macrophylla*, *Deutzia scabra*, *Machilus Thunbergii* and *Pinus Thunbergii*, with an undergrowth of *Pyrrosia lingua* and *Nephrolepis auriculata*. The protected hillside lava flow was for the most part covered by dense thickets consisting of broad-leaved deciduous trees and shrubs such as *Alnus Sieboldiana*, *Styrax japonica*, *Prunus Lannesiana* var. *speciosa*, *Deutzia scabra*, *Hydrangea macrophylla*, *Stachyurus praecox* var. *Matsuzakii* and *Callicarpa mollis*. Associated with the deciduous trees and shrubs were evergreen species such as *Machilus Thunbergii*, *Castanopsis cuspidata* var. *Sieboldii*, *Eurya japonica*, *Ligustrum pauciflorum* and *Daphniphyllum macropodum*. The undergrowth consisted of *Pyrrosia lingua*, *Lemmaphyllum microphyllum* and *Psilotum nudum*.

The coastal part of the 1834 lava flow was covered by rather dense, mixed

thickets of broad-leaved deciduous and evergreen species similar to those on the protected hillside lava flow of 1874. The older flows of 1712 and 1643, although their precise location and extent are not accurately known, were clearly covered by broad-leaved evergreen trees such as *Machilus Thunbergii*, *Camellia japonica* and probably *Castanopsis cuspidata* var. *Sieboldii*, which is the dominant tree of the climax forest of the area.

10. Mt. Aso

Mt. Aso lies at the centre of the island of Kyushu and is famous as the world's greatest caldera, the name referring inclusively to the entire volcano incorporating five separate central peaks, Nakadake, Kishimadake, Eboshidake, Takadake and Nekodake. Among these peaks, only Nakadake is at present active.

In recorded history, i.e. since about 1300 yr ago, the peaks have erupted repeatedly and ejected immense amounts of volcanic ash onto the older lava flows. Asai[18] has investigated the volcanic vegetation of the cones, ranging from their summits to the lateral slopes.

Areas in the immediate vicinity of the crater of the active peak, Nakadake, are still almost completely bare due to repeated coverings of hot ash ejected from the crater. However, at distances of about 300–500 m from the crater, *Reynoutria japonica* f. *colorans* (a typical pioneer of volcanic desert) occurs frequently, forming cushions several metres in diameter. *Carex blepharicarpa*, *Miscanthus Matsumurae* and *Calamagrostis autumnalis* are also sparsely intergrown with the *Reynoutria japonica* f. *colorans*.

External to this *Reynoutria* community is a *Rhododendron kiusianum–Alnus firma* scrub community including *Hydrangea paniculata*, *Salix Sieboldiana*, *Miscanthus sinensis*, *M. Matsumurae*, etc. Such scrub normally replaces the *Reynoutria* community in the course of the succession developing on the ash field, shifting next to a broad-leaved deciduous forest dominated by *Quercus mongolica* var. *grosseserrata* and *Magnolia parviflora*. The upper parts of the extinct volcanoes are also occupied by broad-leaved deciduous trees, including *Quercus mongolica* var. *grosseserrata*, *Magnolia parviflora*, *Acer rufinerve*, *Zelkova serrata*, *Stewartia pseudo-Camellia*, *Acanthopanax sciadophylloides*, etc.

Mountain slopes below 1000 m a.s.l. are occupied by grasslands of the *Miscanthus sinensis–M. Matsumurae* and *Zoysia japonica–Imperata cylindrica* types, both of which have developed under the impact of special biotic factors including grazing, mowing and firing. The *Zoysia japonica–Imperata cylindrica* community (a short grassland) in fact came into existence through the intensive grazing of the *Miscanthus sinensis–M. Matsumurae* community (a tall grassland) by cattle.

The climatic climax of the areas concerned is presumed to be *Abies firma* forest with an undergrowth of broad-leaved evergreen trees. However, it is as yet unrepresented in any locality probably due to the repeated eruptions and to human interference.

11. Sakurajima Is.

The volcanic island of Sakurajima is situated in Kagoshima Bay near the southern tip of Kyushu, and is connected with the Osumi Peninsula of the mainland by the lava flow of 1914. Violent eruptions have been recorded four times in recent history, i.e. in 1476, 1779, 1914 and 1946. On each occasion large andesitic lava flows were produced, sometimes with ash and/or pumice. The volcanic vegetation has been extensively studied by Tagawa.[20]

The latest (1946) lava flow is at present commonly inhabited by the lichen *Stereocaulon denudatum*, and also occasionally by the moss *Dieranella tosaensis*, the fern *Nephrolepis auriculata* and the herb *Reynoutria japonica* (see Fig. 14). The lava remains in an unweathered condition, having at most only a thin overlay of volcanic ash and sand. Thus, it remains essentially in the lichen-moss stage.

The 1914 lava flow is situated on the opposite flank of the volcano to the 1946 flow, and at the time of eruption was overlain with appreciable amounts of ash and sand, particularly in areas where there were depressions. The flow is at present inhabited by numerous species at a rather higher density than the 1946 lava flow. The lichens and mosses are greater in both species number and quantity, and there is a marked invasion of vascular species. Herbaceous plants such as *Reynoutria japonica*, *Miscanthus sinensis*, *Nephrolepis auri-*

FIG. 14. *Reynoutria japonica* growing in a small depression on the 1946 lava flow of Sakurajima Is.

FIG. 15. *Nephrolepis auriculata* and *Miscanthus sinensis* growing in a depression on the 1914 lava flow of Sakurajima Is.

culata and *Dryopteris erythrosora* are abundant, and there are many ligneous species such as *Deutzia Sieboldiana* var. *Deppeliana*, *Hydrangea paniculata*, *Rhododendron Kaempferi*, *Eurya japonica*, *Alnus firma*, *Ficus crenata* and *Machilus japonica*.

The vegetation on the 1779 and 1476 lava flows was once cleared for the most part and the areas reforested with *Pinus Thunbergii*. However, in both cases there are some spots which were neither cleared of vegetation nor replanted with *Pinus Thunbergii*. These remain relatively intact, and support dense forests of broad-leaved evergreen trees. On the 1779 lava flow small stands dominated by the evergreen oak *Cyclobalanopsis glauca* occur, with an associated community of *Ligustrum japonicum*, *Vaccinium bracteatum*, *Farfugium japonicum*, *Nephrolepis auriculata*, etc. On the 1476 lava flow there are groves dominated by *Machilus Thunbergii* in association with evergreens such as *Cyclobalanopsis glauca*, *Cinnamomum japonicum*, *Pasania edulis* and *Camellia japonica*. The undergrowth contains herbs such as *Desmodium racemosum*, *Miscanthus sinensis*, *Cyclosorus acuminatus*, etc.

Tagawa[20] has suggested the following five successional stages for the volcanic vegetation of Sakurajima Is.

(1)	Lichen and moss stage	(20 yr after eruption)
(2)	Herb stage	(50 yr after eruption)
(3)	Scrub stage	(100 yr after eruption)
(4)	*Cyclobalanopsis glauca* forest stage	(150 yr after eruption)
(5)	*Machilus Thunbergii* forest stage	(450 yr after eruption)

The plant succession on the Sakurajima lava flows thus begins with the lichen and moss stage (in contrast to the algae found on similar volcanics in N. America). However, along with lichens and mosses, there is invasion by

vascular species such as *Reynoutria japonica*, *Nephrolepis auriculata* and *Pinus Thunbergii*, particularly in depressions and crevices on the lava surface where accumulations of pumice and/or volcanic ash are found. Also, the vascular plants can probably grow independently of the lichens and mosses, a situation which resembles that on the lava flows of Mt. Asama and Izu-Oshima Is. (In fact, conditions on the Sakurajima volcano in many ways resemble those on the Oshima volcano from the viewpoint of both volcanic ejecta and climate, giving rise to several points of similarity between the respective plant successions.) The climatic climax forest on Sakurajima Is. is dominated equally by *Machilus Thunbergii* and prob. *Castanopsis cuspidata* var. *Sieboldii* and is probably fully established within 1000 yr after the formation of a lava flow.

12. Solfataras and Vapour Fumaroles

The term "fumarole" is a general name applied to volcanic vents that issue vapours such as steam, sulfurous gases and carbon dioxide. Solfataras are understood as fumaroles issuing abundant sulfurous gases such as hydrogen sulfide and sulfur dioxide, and frequently steam. Vapour fumaroles are those vents which issue little sulfurous gas but abundant hot aqueous vapours. In Japan, a conspicuously large number of solfataras and some vapour fumaroles occur in association with the volcanics.

12.1. Solfataras

The vicinity of solfataras constitutes a desolate environment, due to the issuance of hot vapours (especially sulfurous gases), the bleached ground and peculiar vegetation. The latter is entirely different from that of surrounding areas, a fact which, according to Faber,[23,24] is attributable more to soil properties than to the direct influence of the sulfurous gases, etc. It was also reported that alpine plants often descend to solfataras located far below their usual habitat, due to the development of soil conditions similar to those of alpine soil.

The most well-known of Japan's solfataras are Atosa-nupuri of the Kawayu spa and Noboribetsu of the Noboribetsu spa in Hokkaido, Osoreyama of the Shimokita Peninsula, Sukayu on the western slopes of Mt. Hakkoda, Goshogake of the Hachimantai Heights and Owakutani of the Hakone spa, Honshu, and Unzen of the Unzen spa, on the southwestern side of Mt. Unzen, Kyushu.

12.1.1. The Osoreyama solfataras

The Osoreyama solfataras are situated close to the northern shore of Lake

Usoriyama, Shimokita Peninsula, Honshu. These solfataras are situated at an altitude of about 220 m a.s.l. and are surrounded by a climatic climax community of *Fagus crenata–Thujopsis dolabrata* var. *Hondai*.

The author and his co-workers[25] have made an ecological study of the Osoreyama solfatara region. The central part is entirely desolate and covered by greyish-white, barren soil bleached by sulfurous gases and strongly acidic vapours escaping from the scattered solfataras and hot springs in the area. Nevertheless, in some limited spots a few plant species are able to grow. For example, *Deschampsia flexuosa* (a well-known acidophilic plant of the boreal and alpine zones of the northern hemisphere) is distributed thinly on certain of the drier sites, while *Carex angustisquama* (a plant restricted to northeastern districts of Honshu) occurs sparsely on wet soil moistened by acidic water of pH 1.8–2.0 issuing from the springs.

The marginal portion of the Osoreyama solfatara region is densely grown with plants and can be divided into two distinct regions according to the vegetation type. One is a heathy community dominated by *Ledum palustre* var. *diversipilosum*, associated with *Moliniopsis japonica*, *Vaccinium Smallii*, *Rhus trichocarpa* and *Ilex crenata* var. *paludosa*. *Ledum palustre* var. *diversipilosum* itself flourishes in many solfatara areas in north and northeastern Japan, and also occurs in the alpine and subalpine zones of mountainous districts. The *Ledum palustre* var. *diversipilosum* community has established itself on solfatara soil exhibiting a fine structure and containing moderate amounts of moisture.

The second community is scrub dominated by *Rhododendron Fauriae* var. *roseum* and *Ilex Sugeroki* ssp. *brevipedunculata*, associated with *Sorbus commixta*, *Rhus trichocarpa*, *Leucothoe Grayana* and *Ledum palustre* var. *diversipilosum*. The majority of these species normally occur in alpine and subalpine zones of mountainous districts in Japan, and the *Rhododendron Fauriae* var. *roseum* community as a whole thus resembles alpine scrub. The community occurs on soils exhibiting a coarse structure and low moisture content, i.e. in direct contrast to the *Ledum palustre* var. *diversipilosum* community.

Towards its outer edge, the scrub is gradually replaced by a *Fagus crenata–Thujopsis dolabrata* climax forest, often through a *Quercus mongolica* var. *grosseserrata* stage. Fig. 16 illustrates typical vegetation of the Osoreyama solfatara region.

According to chemical analyses of soils from the Osoreyama solfataras, they are extremely acidic (pH 1.4–3.8) and remarkably poor in nutrient content, especially in potassium, phosphorus, calcium and nitrate nitrogen (as has already been reported by Faber[23] for Javanese solfataras). Excess aluminium has been considered as one factor controlling the development of solfatara vegetation, i.e. since the suggestion of Faber in his report.[23] However, the levels of aluminium, iron and magnesium were very low, even when

FIG. 16. Solfatara vegetation of Osoreyama, consisting of a *Ledum palustre* var. *diversipilosum* mat fringed by *Carex angustisquama* (central foreground) and climax beech and oak forests (background).

compared with those of the climax forest soils of Osoreyama. Thus, high aluminium content in the soil should not be considered an absolute controlling factor in the establishment of solfatara vegetation.

12.1.2. The Sukayu solfataras

The Sukayu solfataras lie on the western slopes of Mt. Hakkoda, northeast Honshu, at about 900 m a.s.l., and include old parasitic craters, hot springs, areas of barren land, bogs and scrub. The region is for the most part surrounded by a climax montane forest of *Fagus crenata*. The solfatara areas are occupied largely by scrub comprised of alpine shrubs such as *Sorbus commixta*, *Ilex Sugeroki* var. *brevipedunculata*, *Acer Tschonoskii*, *Rhododendron Fauriae*, *Vaccinium axillare* and *Pinus pumila*, with an occasional undergrowth of *Empetrum nigrum*, *Vaccinium Vitis-Idaea* and *Gaultheria Miqueliana*. At the bottom of one of the old parasitic craters, there is a bog dominated by *Sphagnum recurvum* var. *amblyphyllum* and *Eriophorum vaginatum*, associated with *Moliniopsis japonica*, *Oxycoccus quadripetalus*, *Drosera rotundifolia*, etc. pH values for the various soils are 1.4–2.8 in barren areas, 3.8–4.0 in the solfatara communities, and 5.0–6.0 in the surrounding forest area. Most of the solfatara plants also descend from the subalpine and alpine zones into the montane zone (as shown similarly by Faber in Java).

12.1.3. The Goshogake solfataras

The Goshogake solfataras are situated at about 900 m a.s.l. in the Hachimantai Heights, Akita Pref., northeast Honshu. They are located in the montane zone occupied by *Fagus crenata* forest. Near the centre of the solfa-

tara region, there is a heath community of *Ledum palustre* var. *diversipilosum* associated with *Empetrum nigrum*, *Rhododendron Fauriae* and *Vaccinium axillare*. This shifts outwards through *Pinus parviflora* var. *pentaphylla* forest to the climax *Fagus crenata* forest.

Such *Pinus parviflora* var. *pentaphylla* forest often occurs in the solfataras of northeast Honshu, although it is primarily the topographic climax vegetation on low, dry ridges of the climatic climax *Fagus crenata* forest zone. The solfataras are thus probably rather similar in soil properties to ridges occupied normally by *Pinus parviflora* var. *pentaphylla*.

12.1.4. The Oowakutani solfataras

The Oowakutani solfataras lie on the northern slopes of Mt. Hakone in the Hakone spa district, central Honshu. They are situated in a rather wide gorge and emit hot water vapours and sulfurous gases. At other places in the gorge there are several boiling hot springs issuing from the surface. The solfatara area is thought to be situated at about the lower limit of the climax *Fagus crenata* forest. Near the centre of the region, there is a scattered community of *Miscanthus sinensis*, *Reynoutria japonica*, *Alnus firma*, *Hydrangea paniculata* and *Enkianthus campanulatus* var. *Palibinii*, while towards the periphery, *Sorbus commixta*, *Clethra barbinervis*, *Leucothoe Grayana*, *Pieris japonica* and *Ilex crenata* occur. The pH values of the solfatara soils are about 2.8 in barren areas and 3.6–3.8 in places with scattered plants.

12.1.5. The Unzen solfataras

There are numerous solfataras in the Unzen spa region on the southwestern side of Mt. Unzen, Shimabara Peninsula, Kyushu. Among them, the most extensive are the Oojigoku solfataras, which are scattered with *Miscanthus sinensis*, *Fimbristylis dichotomata* var. *Tashiroana*, *Pinus densiflora* and *Enkianthus cernuus*. These solfataras are bordered by a low forest of *Pinus densiflora*, with a dense undergrowth of *Enkianthus cernuus*, *Rhododendron kiusianum* and *Miscanthus sinensis*.

12.2. Vapour fumaroles

The vapour fumaroles discussed here are mainly volcanic vents issuing aqueous vapours at temperatures below 100°C. They are surrounded by a peculiar, characteristic vegetation which is conspicuously different from that of solfataras. In Japan, this peculiar vegetation is found chiefly in the cool-temperate zone, and is usually characterized by the presence of *Lycopodium cernuum* (which itself ranges primarily from tropical/subtropical to warm-temperate regions).

Previously, *Lycopodium cernuum* was known to occur in several isolated

fumaroles situated in the cool-temperate zone where it is reportedly no longer found, such as at the Nakanoyu spa (36°12′ N, 1300 m a.s.l.), Nagano Pref., Hakone (36°16′ N, 1000 m a.s.l.) and Noboribetsu (42°29′ N, 300 m a.s.l.). However, the characteristic fumarole vegetation has recently been found at Katanuma in the Naruko spa, Miyagi Pref., northeast Japan (38°48′ N, 160 m and 480 m a.s.l.). Here, the fumarole vegetation is dominated mainly by *Lycopodium cernuum* in association with *Odontoschisma denudatum* and *Jungermannia infusca*.[26] These fumaroles lie in the cool-temperate zone where conditions (mean annual temperature, *ca.* 9°C) are far removed from those of the proper northern limit of *Lycopodium cernuum* (mean annual temperature, >14–15°C) (see Fig. 17).

The present author[26] has also carried out a study of the detailed conditions prevailing at the Naruko fumaroles, i.e. where they issue sufficient quantities of hot vapour to raise the soil and air temperature near the ground surface appreciably. In places where the *Lycopodium cernuum* community flourished, air and soil temperatures were measured, with the following results. In the proximity of the fumaroles, the air temperature 10 cm above the ground was 25–30°C in late September and 5–8°C in early January, compared with 18–19°C and −3°C, respectively, away from the fumaroles. The soil temperature 2 cm below the ground surface was 30–40°C in late September and 15–30°C in early January, in the vicinity of the fumaroles, compared with 16–18°C and 1–2°C (beneath the snow cover), respectively, away from the fumaroles (see Fig. 18). That is to say, the soil and air temperatures of the fumarole area are conspicuously higher than those away from the fumaroles, especially in winter, and this probably enables *Lycopodium cernuum* to flourish even though the climate is essentially a cool-temperate one.

FIG. 17. Winter aspect of a snow-bound fumarole of the Naruko spa. *Lycopodium cernuum* occurs abundantly around the fumarole.

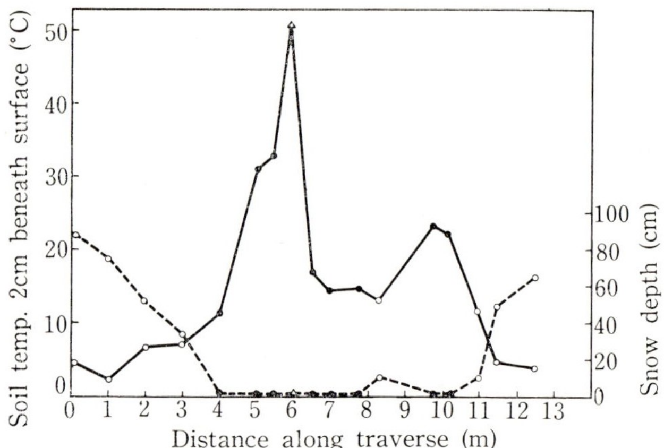

FIG. 18. Winter soil temperatures and snow depth at and in the vicinity of a fumarole of the Naruko spa. (Solid line shows temperature data; broken line, snow depth. △, Fumarole; ●, *Lycopodium* community; ○, scrub.)

13. Conclusion

The general pattern of the xerosere, starting from bare rock, typically approximates to the following sequence: Bare rock – Lichen/moss stage – Herb stage – Shrub stage – Woodland stage – Climax forest stage.

In the case of the xeroseres given here for Japanese volcanic vegetation, the complete series just mentioned was not always represented, although a common general trend from low, sparse vegetation, through intermediate stages, to tall, dense vegetation was easily recognizable in the plant successions. Also, the xeroseres developing on volcanic ejecta were conspicuously diverse because of the rich variety of ejecta located in many different climatic regions, ranging from subarctic through temperate to subtropical areas.

The lichen/moss stage is generally found only on bare lava flows with flat and/or convex surfaces, while depressions containing major or minor deposits of volcanic ash or lava debris, etc. are almost immediately able to support vascular species without any preceding cryptogams. Indeed, invasion of the depressions by vascular plants is commonly earlier than that of the bare lava surface by lichens and mosses.

The plant successions of mud flows, ash and pumice fields typically do not begin from the lichen/moss stage, but rather from the herb or mixed herb and shrub stage. The latter, in particular, readily shifts to the shrub stage and then, under favourable circumstances, develops into the woodland stage.

However, a herb stage dominated by a dense growth of *Miscanthus sinensis*, etc. may conspicuously retard further development of the succession.

The shrub and woodland stages are frequently occupied by broad-leaved deciduous shrubs and trees (esp. sun plants such as pines) throughout widely contrasting climatic zones. However, the later stages (and particularly the climax stage) are made up of relatively few species characteristic to each particular climatic zone.

Climatic climax communities dominated by broad-leaved evergreen trees are generally attained within about 800–1000 yr from the time of eruption in the case of the moist and warm (e.g. subtropical and warm-temperate) zones, compared with 1000 yr (at least) in the case of deciduous beech communities of the cool-temperate zone.

In Japan, there are many solfataras and vapour fumaroles. These typically show a characteristic type of vegetation, which is here included under the general category of "volcanic vegetation". This vegetation, however, is conspicuously different in ecological properties from the ordinary volcanic vegetation that develops to the climatic climax on volcanic ejecta. The solfatara and fumarole vegetation is arrested at a subclimactic stage, due to the effect of emitted gases and the peculiar soil properties, and is thus characterized by plant species entirely different from those of the surrounding areas.

Japanese volcanoes have, through their activity recently brought about many areas of lifeless and rocky desert, which are now gradually regaining their vegetation through the various stages of plant successions appropriate to the prevailing environmental conditions. These volcanoes have produced not only new rock formations but also many new lakes and ponds, and so a wide variety of new and changing landscapes. The vegetation of such regions thus constitutes an important topic for new and detailed research.

References

1. M. Treub, Notice sur la nouvelle flore de Krakatau, *Ann. Jard. Bot. Buitenzorg*, **7**, 213–23 (1888).
2. O. Penzig, Die Fortschritte der Flora des Krakatau, *ibid.*, **3**, 92–113 (1902).
3. A. Ernst, Die neue Flora der Vulkaninsel Krakatau, *Vierteljahrsschr. Natur. Ges. Zurich*, **52**, 289–363 (1907).
4. A. Ernst, Die biologische Krakatauproblem, *ibid.*, **79**, 1–187 (1934).
5. C. A. Backer, *The Problem of Krakatau as Seen by a Botanist*, pp. 299, Hague, 1929.
6. W. M. D. van Leeuwen, The flora and fauna of the islands of the Krakatau group in 1919, *Ann. Jard. Bot. Buitenzorg*, **31**, 103–40 (1921).

7. W. M. D. van Leeuwen, Krakatau 1883 to 1933, *ibid.*, 1–506 (1936).
8. F. C. Gates, The pioneer of Taal volcano in 1911, *Philipp. J. Sci. Bot.*, **9**, 391–434 (1914).
9. R. F. Griggs, The colonization of the Katamai ash, a new and inorganic "soil", *Am. J. Bot.*, **20**, 92–113 (1933).
10. R. F. Griggs and D. Ready, Growth of liverworts from Katamai in nitrogen-free media, *ibid.*, **21**, 265–77 (1934).
11. C. E. Hartt and M. C. Neal, Plant ecology of Mauna Kea, Hawaii, *Ecology*, **21**, 237–66 (1940).
12. F. R. Fosberg, Upper limit of vegetation on Mauna Loa, Hawaii, *ibid.*, **40**, 144–46 (1959).
13. W. A. Eggler, Plant communities in the vicinity of the volcano El Paricutin, Mexico, after two and a half years of eruption, *Ecology*, **29**, 415–36 (1948).
14. W. A. Eggler, Manner of invasion of volcanic deposits by plants, with further evidence from Paricutin and Jorullo, *Ecol. Monogr.*, **29**, 267–84 (1959).
15. R. W. Keay, Lowland vegetation on the 1922 lava flow, Cameroun Mountain, *J. Ecol.*, **47**, 25–38 (1959).
16. Y. Yoshii, Revegetation of the volcano Komagatake after the great eruption in 1929, *Bot. Mag. Tokyo*, **46**, 208–19 (1932).
17. Y. Yoshii, Die Besiederungsverlauf am Vulkan Komagatake nach dem Ausbruch von 1929, *Ecol. Rev.* (Japanese), **8**, 170–220 (1942).
18. T. Asai, Zur Ökologie der Vulkanpflanzen von Asosan, *Kumamoto J. Sci. Ser. B*, no. 1, 33–81 (1952).
19. Y. Tezuka, Development of vegetation in relation to soil formation in the volcanic island of Oshima, Izu, Japan, *Jap. J. Bot.*, **17**, 371–402 (1961).
20. H. Tagawa, A study of the volcanic vegetation of Sakurajima, south-east Japan, I. Dynamics of vegetation, *Mem. Fac. Sci. Kyushu Univ., Ser. E. Biology*, **3**, 165–228 (1964).
21. M. Tatewaki, R. Shibakusa, A. Matsushita and S. Kojima, Vegetation of Mt. Koma, Prov. Oshima, Hokkaido, Japan, *Nippon Shinrinshokusei Kenkyukai* (Japanese), 1–82, 1966.
22. K. Yoshioka, Development and recovery of vegetation since the 1929 eruption of Mt. Komagatake, Hokkaido, I. Akaikawa pumice flow, *Ecol. Rev.*, **17**, 271–92 (1969).
23. F. C. Faber, Untersuchen über das Physiologie der javanischen Solfataren Pflanzen, *Flora N. F.*, **18–19**, 89–110 (1925).
24. F. C. Faber, Die Kraterpflanzen Javas in physiologischen Beziehung, *Arb. Treub-Lab.*, 1–119 (1927).
25. K. Yoshioka, K. Saito and H. Tachibana, Solfatara vegetation at Osoreyama, *Ecol. Rev.*, **16**, 137–51 (1965).
26. K. Yoshioka, *Lycopodium cernuum* community, as a fumarole vegetation in the cool-temperate zone of Japan, *ibid.*, **17**, 115–22 (1968).

ADDITIONAL REFERENCES

a. R. A. Howard, Vulkanism and vegetation in the Lesser Antilles, *J. Arnold Arbor.*, **43**, 229–311 (1962).

b. H. Molisch, *Pflanzenbiologie in Japan auf Grund einiger Beobachtungen*, Gustav Fischer, Jena, 1926.

c. K. Negoro, Studies on solfatara plants, *Seibutsugaku no Shimpo* (Japanese), p. 151–92, Kyoritsu Shuppan, Tokyo, 1943.

9

Conservation of Flora and Vegetation in Japan

Makoto NUMATA*

The conservation of plant life in Japan was first promoted officially under an act passed in 1919, which designated various plants, etc. as "national monuments". However, the most important objects of plant conservation were initially rare plants and large trees, little emphasis being given to the preservation of natural stands of vegetation.[1] Many areas of natural vegetation, particularly in low-lying and coastal regions, have thus subsequently been largely or completely destroyed, mostly as a result of the trend towards widespread land development for industrial or housing purposes.

According to a recent report by Yamazaki,[2] which lists rare and unusual plant species in Japan, the following plants (all of which grew in the vicinity of marshes) have been completely extirpated: *Lysimachia leucantha*, *Gratiola japonica*, *Utricularia nipponica*, *Chrysanthemum lineare*, etc. Moreover, *Primula tosaensis* var. *rhodotricha*, which grows in limestone areas, is facing extinction through the massive quarrying of limestone for cement. Some peculiar continental plants growing in limited areas of northern Kyushu are rapidly diminishing under the influence of land development programmes; the species concerned include *Echinops setifer*, *Campanula glomerata* var. *dahurica*, *Polemonium kiushianum*, etc. The combined effects of land development and spraying of agricultural chemicals on paddy fields, etc. have also had a strong deleterious effect on the following other hygrophytic plants of low-lying marshlands: *Adenophora palustris*, *Galium tokyoense*, *Lysimachia candida*, etc.

**Department of Biology, Faculty of Science, Chiba University, 1-33 Yayoi-cho, Chiba-shi 280, Japan*

The Ecological Background to Conservation

The main island of Japan probably separated completely from the Asian mainland towards the end of the Pleistocene, and the Holocene has witnessed the establishment by volcanic activity of many new islands and islets in the region. As a result, there are many endemic species in Japan. Present habitats where these are particularly abundant include the Yaeyama Islands (southern Ryukyus, particularly Iriomote Is.), the Bonin Islands, Yaku Is., the central mountains of Hokkaido, the Izu Peninsula, the Izu Islands, Mt. Yatsugatake, southern Shikoku, Mt. Hayachine, Tsushima Is., Rishiri Is., etc.[3] It can thus be said that many of the most interesting localities, floristically speaking, are small isolated islands or high mountains. However, the rare species present in such places are now in danger of extirpation, partly as a result of increased tourism (e.g. in the Ryukyus and Bonin Islands), and also from the felling of natural mountain forests for the purpose of artificial reforestation and timber production, etc. Clearly, conservation measures must be implemented that are aimed not simply at the preservation of individual plants but rather towards preservation of entire plant communities, indeed entire natural ecosystems.

Typically, the climate of small islands is oceanic. However, the majority of Japan's smaller islands lie in close proximity to a larger land mass, such as one of the four principal islands. Geologically speaking, most are in fact continental, not oceanic. Climatic differences between the large and small islands of Japan thus tend to be minimal in most cases (at least at any particular latitude), although there is of course a strong contrast with the neighbouring Asian continent. These factors give rise to several special problems and circumstances, both from the viewpoint of plant ecology and of conservation.

Among the various plant species common to both northern and southern islands of Japan, two types of distribution are recognizable. In one case, the same species occurs in inland areas in the north and on the coast in the south (e.g. *Sonchus arvensis* var. *uliginosus*, *Zoysia japonica*, etc.).[4] In the other case, the same species occurs on the coast in the north and inland in the south (e.g *Cynodon dactylon*, *Pittosporum Tobira*, *Quercus dentata*, *Q. crispula* and *Pinus densiflora*). Of these, *Cynodon dactylon* for example also occurs on the Asian mainland, dominating over large areas of inland grassland in subtropical and warm-temperate regions of Nepal.[5] Such a species can thus be regarded as having spread out widely over many separated islands distributed along the periphery of its major (continental) area of occurrence. In the case of *Zoysia japonica* (the former distribution type), the normal occurrence is as a pasture

grass of the *Fagus crenata* zone of northern Japan, but there is also a southward extension into coastal areas of the warm-temperate zone of Kyushu.

Another important ecological situation from the viewpoint of plant conservation is seen in a number of small islands such as Awashima Is., Rishiri Is., etc. which have a hilly or mountainous axis. Here, the eastern and western halves display different climates, which in general features resemble those of the Pacific Ocean and Sea of Japan sides of the Japanese mainland, respectively. There is thus a distinct floral and vegetational difference between the eastern and western sides of the islands, equivalent to that between the east and west coasts of the Japanese mainland. In the case of Awashima Is., the eastern half is characterized by bamboo breaks, but these are entirely absent from the western half of the island.

In the warm-temperate zone of Honshu, a clear spatial ordination (from the coast inland) can be distinguished in the climax vegetation, i.e. from *Machilus Thunbergii* forest to *Castanopsis cuspidata* var. *Sieboldii* forest to *Cyclobalanopsis* spp. forest. A typical example of such a forest distribution is seen in Chiba Pref.[6)] However, it is found in the case of certain islets that only the coastal *Machilus Thunbergii* climax forest occurs (e.g. in the Danjo Islands, Nagasaki Pref., Kyushu[7)]). Similarly, there are exceptions to the normal spatial ordination (from the coast inland) of *Miscanthus condensatus* to *M. sinensis* grassland. These deviations from the typical spatial ordination are characteristic of small oceanic islets of area less than 2×2 km^2.

The invasion of exotic plants into an island environment isolated by the sea from the source area, and the ecesis of these plants on the island, constitute some of the most interesting problems of island ecology and conservation. They also demonstrate the influence of human activity on the success or otherwise of new island plants. For example, in the case of the Seto Inland Sea in Japan (which contains more than 500 islets), the uninhabited islands generally have no exotic plants, whereas inhabited islets have widespread exotic weeds such as *Erigeron canadensis*, *E. linifolius* and *Euphorbia supina*, together with other less frequent exotic plants such as *Trifolium repens*, *Oxalis Martiana*, *Aster subulatus*, *Bidens frondosa*, *Physalis angulata*, etc. Moreover, even on the Japanese mainland, the process of naturalization of exotic plants is of similar importance to endemism.[8)]

The effects of human activity are also clear in the case of the Bonin Islands, that is the so-called Volcano Islands (Kita-Iojima, Iojima and Minami-Iojima, part of the Fuji volcanic chain) and the Ogasawara Islands (Chichijima, Hahajima and Mukojima). The flora of the former is in general akin to that of the Izu Islands, Ryukyus and Taiwan (also partly to that of Micronesia, especially the Marianas). The flora of the latter is in part akin to that of the Chinese mainland and more distant localities of the Pacific, as

shown by the presence of *Santalum boninense* (Santalaceae), *Clinostigma Savoryana* (Palmae), *Meterosideros boninensis* (Myrtaceae), *Osteomeles boninensis* (Rosaceae) and *Orobanche boninsimae* (Orobanchaceae). In particular, *Dendrocacalia crepidifolia* (Compositae) is endemic. The flora of the Bonin Islands before World War II was known to include higher plants from 90 families, constituting about 250 genera and 400 species in total, 20% of which were endemic.[9] At that time, only Chichijima Is., Hahajima Is. and Iojima Is. were inhabited, due to the scarcity of drinking water on the other islands. In 1967 (when the islands were returned to Japan) the sites of villages on Chichijima Is. existing prior to the war were covered by jungle containing *Ficus retusa*, *F. elastica*, *Chrysalidocarpus lutescens* (*Areca lutescens*), *Agave americana*, *Leucaena leucocephala*, *Pinus luchuensis*, *Casuarina equisetifolia*, etc. On the other hand, the former vegetation of Mukojima Is. has been seriously depleted through overgrazing by escaped goats.

According to statistics compiled in 1957, the total number of Japanese islands at that time was 3639; however, the Bonin Islands and Ryukyus should now be added to this figure. Excluding all islands in the Seto Inland Sea (525) and other inland seas and bays, the total number of islands was 1025 (of 3639), of which 159 were inhabited and 866 uninhabited. Of the 525 islands in the Seto Inland Sea, 173 were inhabited and 352 uninhabited.

The so-called Remote Island Promotion Law is applied to 845 smaller islands (including many in the Seto Inland Sea), of which a total of 494 are uninhabited. However, this law is aimed primarily at stabilization of the living conditions and improving the welfare of the people, and at developing the islands for the benefit of the national economy. Thus, it is not a law of protection or conservation, and if applied incorrectly can easily become a destructive law insofar as the plant life of the islands is concerned.

Specific Measures for the Conservation of Japan's Flora and Vegetation

As mentioned above, certain plants, etc. were originally designated officially as "national monuments". However, this category now also includes a number of primeval forests, special plant communities, etc. Typical examples are as follows:

(1) Primeval forests on Mt. Fuji, Kasugayama Hill, Shiroyama Hill, etc.

(2) Alpine plant zones on Mt. Apoi, Mt. Hayachine, Mt. Iwate, Mt. Akita-Koma, Mt. Shirouma, etc.

(3) Halophytic plant communities on the banks of the Akkeshi lagoon, etc.
(4) Aquatic and hygrophytic plant communities on the Kirigamine and Ozegahara Moors, etc.
(5) Subtropical plant communities of *Livistona subglobosa* on Aoshima Is., etc.

Apart from these officially designated localities, fragments of certain natural forests (still similar to the regional climax in many cases) have been preserved in the precincts of Shinto shrines and Buddhist temples.

Throughout the country, there are also now many larger areas designated as National Parks (*Kokuritsu-Koen* in Japanese) or Quasi-National Parks (*Kokutei-Koen*) (Table 1). These are illustrated in Fig. 1.

Within the National Parks, certain areas are kept as strict reserves (the Special Protection Areas); these total 222,880 ha, or 11.3% of the total area of the National Parks. In the case of the Quasi-National Parks, the Special Protection Areas constitute 28,247 ha, or only 2.87% of the total. In the case of forested regions, there is a total of 24,022 ha of Protected Forest; however, the area in each locality is rather small.

Following World War II, the Forestry Agency of Japan promoted a policy of felling natural forests (particularly *Fagus crenata* forests in the cool-temperate zone) under a programme for the expansion of man-made forests. This scheme thus led to the destruction of many original, natural biomes in different parts of Japan. Against this destructive policy, the Ecological Society of Japan requested the Government to designate ten areas of primeval

TABLE 1. Number and area of National Parks, Quasi-National Parks and Prefectural Nature Parks in Japan (1972) and their total coverage of Japan

	Number	Area (ha)	% of total area of Japan
National Parks†1	23	19,741,998	5.3
Quasi-National Parks†2	46	10,009,889	2.7
Prefectural Nature Parks	286	19,906,180	5.5
TOTAL	355	49,658,067	13.5

†1 Present total (1973): 26. The parks of Iriomote (Ryukyus) and Ogasawara (Bonin) Islands) are not included in the data. Also, the present Ashizuri-Uwakai National Park (no. 26 in Fig. 1) is omitted; it is, however, included under the Quasi-National Parks as the Ashizuri Quasi-National Park.

†2 Present total (1973): 48. The new parks of Kitakyushu (no. 37 in Fig. 1), Oga (no. 47) and Echigosanzan-Tadami (no. 48) are not included; however, as mentioned above, the data do include the Ashizuri Park.

QUASI-NATIONAL PARKS

1. Rishiri-Rebun
2. Abashiri
3. Niseko-Shakotan-Otaru-kaigan
4. Onuma
5. Shimokita-hanto
6. Kurikoma
7. Chokai
8. Zao
9. Sado-Yahiko
10. Suigo-Tsukuba
11. Minamiboso
12. Tanzawa-Oyama
13. Noto-hanto
14. Echizen-Kaga-kaigan
15. Myogi-Arafune-Saku-kogen
16. Yatsugatake-Chushinkogen
17. Tenryu-Okumikawa
18. Ibi-Sekigahara-Yoro
19. Hida-Kisogawa
20. Mikawa-wan
21. Aichi-kogen
22. Wakasa-wan
23. Suzuka
24. Muroo-Akame-Aoyama
25. Biwa-ko
26. Yamato-Aogaki
27. Kongo-Ikoma
28. Takano-Ryujin
29. Hyonosen-Ushiroyama-Nagiyama
30. Hiba-Dogo-Taishaku
31. Nishichugoku-sanchi
32. Kitanagato-kaigan
33. Akiyoshidai
34. Tsurugiyama
35. Muroto-Anan-kaigan
36. Ishizuchi
37. Kitakyushu
38. Genkai
39. Iki-Tsushima
40. Yaba-Hita-Hikosan
41. Sobo-Katamuki
42. Nichinan-kaigan
43. Meijinomori-Takao
44. Meijinomori-Minoo
45. Okinawa-kaigan
46. Okinawa-Senseki
47. Oga
48. Echigosanzan-Tadami

NATIONAL PARKS

① Shiretoko
② Akan
③ Daisetsuzan
④ Shikotsu-Toya
⑤ Towada-Hachimantai
⑥ Rikuchu-kaigan
⑦ Bandai-Asahi
⑧ Nikko
⑨ Joshin'etsu-kogen
⑩ Chichibu-Tama
⑪ Fuji-Hakone-Izu
⑫ Chubu-sangaku
⑬ Minami-Alps
⑭ Hakusan
⑮ Ise-Shima
⑯ Yoshino-Kumano
⑰ San'in-kaigan
⑱ Daisen-Oki
⑲ Seto-naikai
⑳ Aso
㉑ Unzen-Amakusa
㉒ Saikai
㉓ Kirishima-Yaku
㉔ Iriomote
㉕ Ogasawara (*not shown*)
㉖ Ashizuri-Uwakai

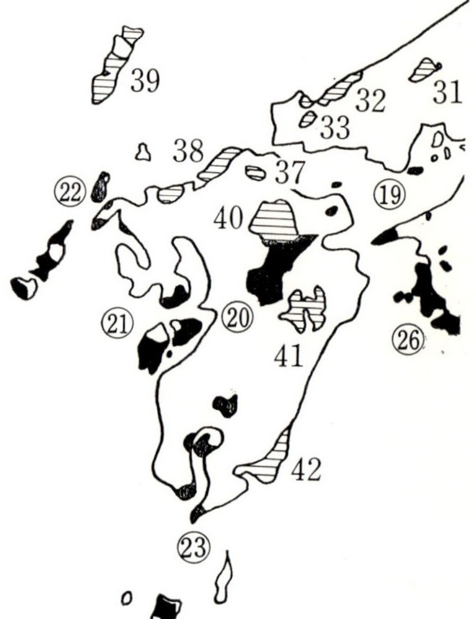

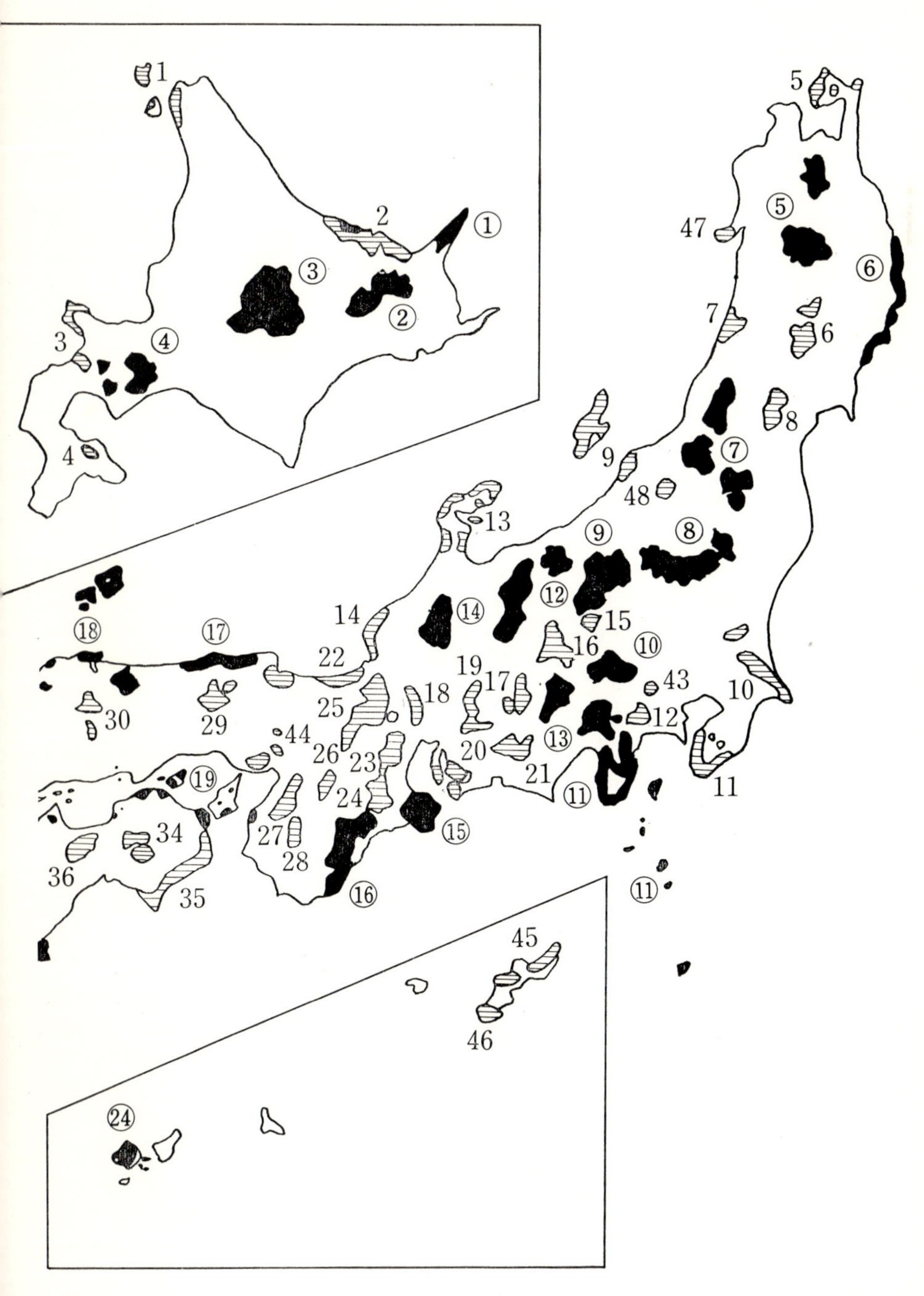

1
2
①
③
②
3
④
4
5
⑤
47
⑥
7
6
8
⑦
9
48
13
⑨
⑧
14
⑭
⑫
15
⑱
⑰
16
⑩
22
19
17
43
18
10
25
12
30
29
44
⑬
26
23
20
21
⑪
11
⑲
24
34
27
⑮
36
28
35
⑯
⑪
45
46
㉔

forest, etc. as National Reserves. These were selected from the most representative vegetational types, and are (from north to south) as follows:[10]

(1) Primeval forests in the upper reaches of the River Tokachi (1000 ha): i.e. subarctic needle-leaved evergreen forests, composed mainly of *Picea jezoensis* and *Abies sachalinensis*.

(2) Mt. Iide (2700 ha): Sea of Japan coastal-type *Fagus* forests and subalpine broad-leaved scrub.

(3) Mt. Shirakura (1200 ha): temperate needle-leaved evergreen forests of the East Japan type, composed mainly of *Tsuga Sieboldii*, plus various kinds of temperate broad-leaved forests.

(4) Mt. Naeba (1200 ha): subalpine needle-leaved evergreen forests, composed mainly of *Abies Mariesii* and *Tsuga diversifolia*, plus Sea of Japan coastal-type *Fagus* forests.

(5) Kamikochi (3000 ha): various kinds of temperate forests and subalpine needle-leaved evergreen forests representative of central Japan.

(6) Mt. Omine and Mt. Odaigahara (3500 ha): Pacific coastal-type *Fagus* forests, plus subalpine needle-leaved evergreen forests, composed mainly of *Picea jezoensis* var. *hondoensis* and *Abies Veitchii*, situated at their southernmost limit of occurrence.

(7) Mt. Ishizuchi (7000 ha): temperate needle-leaved evergreen forests composed of *Abies firma*, *Tsuga Sieboldii*, *Sciadopitys verticillata*, etc., plus Pacific coastal-type *Fagus* forests and subalpine *Abies shikokiana* forests.

(8) Mt. Sobo and Mt. Katamuki (2100 ha): temperate needle-leaved evergreen forests of the West Japan type, composed mainly of *Tsuga Sieboldii*, plus some Pacific coastal-type *Fagus* forests and warm-temperate broad-leaved evergreen forests.

(9) Hilly regions in the southern Osumi Peninsula (1650 ha): warm-temperate broad-leaved evergreen forests, composed of *Machilus Thunbergii*, *Distylium racemosum*, *Cyclobalanopsis* spp., etc.

(10) Yaku Is. (3000 ha): natural *Cryptomeria japonica* forests.

In addition to this primeval forest reserve programme, other steps are also being taken under the auspices of the Ecological Society of Japan for the establishment of grassland reserves, volcanic reserves, wetland reserves, etc. throughout Japan.

So far as conservation-oriented publications are concerned, the Agency for Cultural Affairs has issued vegetation maps, and maps of important, specific plants and animals for all prefectures. The work began in 1969 as an "Urgent Survey of National Monuments", and through it several new and important species and vegetational types in need of preservation have been recognized.

In addition, the Conservation of Terrestrial Communities (CT) Section of the Japanese National Committee for the International Biological Programme (JIBP) has published the following general reports: *Nature Reserves in Japan and Literature on Nature Conservation* (1966), and *Nature Reserves, Designated and Proposed* (1970).

It is thus hoped and expected that measures for the conservation of Japan's natural flora and vegetation will be reinforced along the lines described above. Indeed, in 1972, a Natural Environment Conservation Act was passed under which it is proposed to improve and develop a nation-wide network of scientific nature reserves. The Environment Agency implemented a survey for making a nation-wide vegetation map (1:200,000) on the association basis and a distribution map of floristically important plants, together with various faunistic surveys in 1973, as a scientific basis for deciding environmental policy.

References

1. M. Honda, *Plants Designated as National Monuments* (Japanese), pp. 438, Sanseido, Tokyo, 1957.
2. T. Yamazaki, List of rare and peculiarly distributed plants in Japan, Types and conservation of terrestrial plant communities in Japan, p. 86–92, 1970; p. 74–83, 1971; p. 82–89, 1972.
3. G. Masamune, *New Considerations in Plant Geography* (Japanese), pp. 169, Hokuryukan, Tokyo, 1956.
4. M. Numata, Ecological background and conservation of Japanese islands, *Micronesica*, **5**, 295–302 (1969).
5. M. Numata, Grassland vegetation in eastern Nepal, *Ecol. Study and Mount. of Mt. Numbur in East Nepal* (ed. M. Numata) (Japanese; English summ.), p. 74–94, Chiba University, 1965.
6. M. Numata and S. Asano, Vegetational data for the Boso Peninsula, I. Climax forests in the southern part, *Bull. Marine Lab., Chiba Univ.* (Japanese; English summ.), **7**, 78–92 (1965).
7. S. Toyama *et al.*, Scientific research on the Danjo Islands, Nagasaki Pref., Kyushu, Japan, *Rept. Cult. Prop., Nagasaki Pref.* (Japanese; English summ.), no. 6, 1968.
8. M. Numata and K. Ono, Studies on the ecology of naturalized plants in Japan, I, *Bull. Soc. Pl. Ecol.* (Japanese; English summ.), **2** (3), 117–22 (1952).
9. T. Tsuyama, Higher plants of the Bonin Islands, *Proc. Symp. Nat. Conserv. Bonin Is.* (Japanese), p. 7–10, 1968.
10. Nature Conservation Committee of the Japan Science Council, *Primeval Forest Reserves Scheme of the Ecological Society of Japan* (Japanese), pp. 42, Tokyo, 1962.

Index of Species

Abies, 63, 95, 100, 107, 164, 191, 194, 195
firma, 21, 25, 30, 56, 67, 95, 100, 175, 176, 181, 186, 234, 249, 257, 276
homolepis, 21, 25, 30, 94, 174, 176, 181, 185, 190, 250
Mariesii, 21, 25, 27, 71, 74, 90, 92, 175, 176, 187, 189, 190, 195, 198, 199, 201, 228, 230, 245, 247, 251, 276
Mayriana, 71
sachalinensis, 25, 74, 90, 96, 130, 140, 163, 188, 191, 276
shikokiana, 76, 190, 276
sibirica, 191
Veitchii, 21, 25, 71, 76, 90–2, 95, 175, 176, 189, 190, 276
Acacia, 98
Acanthephippium, 81
Acanthopanax sciadophylloides, 96, 256
Acer, 48, 54, 96, 194
carpinifolium, 70
crataegifolium, 65, 70
diabolicum, 57
japonicum, 90, 96, 179
micranthum, 96, 181
Miyabei, 131, 143, 229
mono, 70, 96, 97, 107, 163, 168, 169, 179, 182, 191, 229, 240–2, 247, 250
palmatum, 57, 65, 250
rubrum, 57, 78
rufinerve, 256
Sieboldianum, 181
Tschonoskii, 178, 181, 193, 195, 196, 199–201, 244, 261
Achlys, 69
Achudemia japonica, 66
Aconitum, 70
Acronychia, 79
Acrostichum, 48, 79
Actinidia, 69
Actinodaphne, 54
lancifolia, 64
longifolia, 64
Adenocaulon, 67
himalaicum, 66
Adenophora, 69
palustris, 269
triphylla, 224
Aesculus, 54
turbinata, 96, 184, 229
Agave americans, 272
Aglaomorpha, 81
Agrimonia Eupatria, 130
pilosa, 136
Agropyron tsukushiense, 170
Agrostis alba, 129
clavata, 139, 141
palustris, 130, 145, 146
Ailanthus, 54
Ainsliaea, 69
apiculata, 66
dissecta, 78
Ajuga elatior, 83
incisa, 74
Akebia, 69
quinata, 65
Alangium, 48, 54
Albizzia Julibrissin, 64, 65
Alchornea liukiuensis, 83
Aldrovanda vesiculosa, 215
Alectorurus, 67
Alisma canaliculatum, 222
Allium Schoenoprasum, 169, 206
Victorialis, 229
Alnus, 48, 54, 71, 183, 192
firma, 247, 256, 258, 262
hirsuta, 192, 240, 241
japonica, 56, 57, 107, 164, 167, 183, 184, 191, 218, 220, 225, 226
Matsumurae, 193, 195, 196
Maximowiczii, 96, 107, 178, 193–6, 199–201, 241, 244, 247
Sieboldiana, 170, 251–5
tinctoria, 130
Alopecurus aequalis, 222
Alpinia japonica, 66
Alsophila, 88
Ambrosia artemisiaefolia, 129, 131, 145
Ampelopsis brevipedunculata, 65
Anapausia, 81
Anaphalis alpicola, 201
margaritacea, 143, 242, 245
Ancistrocarya, 69
Andreaea nivalis, 201
Andromeda polifolia, 226, 231
Anemone flaccida, 218
Anemonopsis, 69
Anemiopsis, 70
Angelica anomala, 168, 192
decursiva, 66
longeradiata, 141, 205
pubescens, 66, 241
Anthelia juratzkana, 201
Anthodendron, 61
Aphananthe, 67
aspera, 65, 106
Apios, 69
Aplectrum, 69
Apodicarpum, 67
Arachnoides aristata, 65
pseudoaristata, 65
Aralia, 54
cordata, 242

elata, 130, 240
Arcteria nana, 199, 203, 251
Arctoa fulvella, 201
Arctous alpinus, 93
Ardisia Sieboldi, 65
Areca lutescens, 272
Arenaria Katoana, 206
Arenga Eugleri, 83
Argostemma, 81
Arisaema, 66, 70
limbatum, 74
monophyllum, 74
Aristida, 82
Aristolochia Kaempferi, 65
Arnica unalaschensis, 201
Artemisia Fukudo, 153
japonica, 136
margaritacea, 130
montana, 130, 143, 168, 169, 239, 242
Stelleriana, 160
vulgaris, 130, 131, 141, 146, 147, 224
Aruncus dioicus, 206
Arundinaria, 65, 140
Arundinella, 138
hirta, 130, 139, 147
Asarum, 83
Asiasarum, 69
Asperula odorata, 240, 241
Aspidistra, 67
Aster, 59
Asa-Grayi, 83
Glehni, 239
Miyagii, 170
scaber, 136
subulatus, 271
Astilbe, 69
odontophylla, 136
Astragalus, 204
sinensis, 222
Athruphyllum neriifolia, 65
Athyrium, 66
deltoidofrons, 220
multifidum, 218
niponicum, 66
Atriplex Gmelini, 160
subcordata, 160
Atylosia, 81
Aucuba, 26, 66, 67
japonica, 26, 61, 64, 180
Aulacolepis, 67
Avicennia, 80, 81
officinalis, 157
Azolla imbricata, 215, 223

Bambusoides, 57
Barringtonia, 82
asiatica, 80
racemosa, 80, 157
Beckmannia erucaeformis, 222
Benzoin erythrocarpum, 100
umbellatum, 96
Berberis, 54, 71
Thunbergii, 139
Berchemia racemosa, 57
Betula, 54, 164, 195
apoiensis, 206
corylifolia, 189, 249
Ermanii, 25, 90, 92, 107, 163, 175, 189–91, 193, 194, 229, 231, 239, 241, 244, 247
grossa, 96, 181, 249
Maximowicziana, 90, 191, 240, 242, 247
platyphylla, 25, 130, 131, 142, 231, 239–42, 251, 252
Bidens frondosa, 271
Bistorta hayachinensis, 206
Bladhia crenata, 65
crispula, 65
japonica, 64
pusila, 65
Sieboldi, 65, 82
Blechnum niponicum, 96
Blyxa caulescens, 222
ceratosperma, 222
Bobua, 65
Bocconia, 70
Boehmeria biloba, 170
Sieboldiana, 66
Boenninghausenia, 67
japonica, 66
Boninia, 79
Boniniella, 79
Ikenoi, 83
Boninofatsia, 70, 79
Botryostege, 70, 72
Boykinia, 69
lycoctonifolia, 201
Brachycyrtis, 69
macrantha, 77
Brasenia Schreberi, 215
Brassica, 170
Bredia yaeyamaensis, 83
Briedelia, 81
Bromus unioloides, 131
Broussonetia, 67
Bruguiera, 80, 82, 108
conjugata, 89, 102, 108, 157
Brylkinia, 72
Buckleya, 68
Bupleurum nipponicum, 206
triradiatum, 204
Buxus, 54, 67
japonica, 57

Cacalia, 70
adenostyloides, 190
delphiniifolia, 66
hastata, 143, 168, 184, 192, 229, 241
idzuensis, 76
Calamagrostis, 126, 133, 201
autumnalis, 205, 256
Epigeios, 147, 155
Fauriei, 201

hakonensis, 147, 242
Langsdorffii, 129, 130, 168, 195, 201, 218, 225
longiseta, 201
Matsumurana, 201
neglecta, 155, 167
sachalinensis, 195
Calanthe, 67
discolor, 64
furcata, 77
gracilis, 77
Calla palustris, 167, 220
Callianthemum, 72, 76
hondoense, 76
insigne, 204
Miyabeanum, 206
Callicarpa, 78
japonica, 65
mollis, 65, 255
Calystegia, 162
Soldanelloides, 160
Caltha palustris, 230
Camaedaphne calyculata, 226
Camellia, 26, 54, 61, 67, 88
japonica, 25, 26, 61, 64, 100, 166, 256, 258
lutchuensis, 82
rusticana, 96, 180
Sasanqua, 57, 61
Campanula glomerata, 13, 269
Cardamine nipponica, 201
Cardiandra, 68
Cardiocrinum, 66, 67
Carex, 66, 155, 161, 167, 191
angustisquamata, 260, 261
Augustinowiczii, 183, 224
blepharicarpa, 202, 203, 256
brunnea, 66
cinerascens, 222
conica, 66, 96
displata, 220
flexuosa, 252
foliosissima, 183
forficula, 222
hakkodensis, 201
heterolepis, 222
Kobomugi, 160–2, 167
lanceolata, 136, 145
leucochloa, 147
limosa, 230
macrocephala, 74, 161, 167
Maximowiczii, 136
Middendorffii, 226, 231
Miyabei, 224
Morrowii, 66
Nakiri, 66
nervata, 130, 139, 141
nubigera, 130, 139
oahuensis, 169
Okuboi, 253
omiana, 203, 227, 231
oxyandra, 242, 252
pisiformis, 205
podogyna, 75
pumila, 155
pseudocuraica, 221
pyrenaica, 201
ramenskii, 153, 155
rhynchophysa, 183, 184
rhynchospora, 230
sachalinensis, 191
subspathacea, 153–5
tenuiformis, 204
Thunbergii, 221
vesicaria, 221
Carpinus, 48, 54, 70, 124
cordata, 25, 191
japonica, 25, 57, 181
laxiflora, 25
Tschonoskii, 25, 64, 65, 181
Carya, 48, 54–7
Cassia mimosoides, 146
Cassytha filiformis, 162, 168
Castanea, 54
crenata, 25, 65, 67, 106, 226, 241
Castanopsis, 54, 61, 63, 67, 76, 77, 98, 124
cuspidata, 25, 59, 64, 82, 100, 111, 131, 140, 253–6, 259, 271
Casuarina equisetifolia, 102–4, 272
Catalpa, 54, 55
Caulophyllum, 69
Cayratia japonica, 65
Cedrela, 54, 55
Celtis, 54
sinensis, 25, 106
Centella asiatica, 66
Centipeda minima, 223
Cephalanthera falcata, 66
Cephalotaxus, 26, 54, 67
drupacea, 94
Harringtonia, 26, 64, 183, 184
Cerastium caespitosum, 130
glomeratum, 139
schizopetalum, 206
Cerastrus orbiculatus, 130
Ceratophyllum demersum, 213–5
Cerbera, 82
Cercidiphyllum, 48, 54, 68
japonicum, 70, 96, 182
Cercis, 54, 55
Cetraria, 203
crispa, 200
Chamaecyparis, 27–9, 68, 185, 187, 188
obtusa, 25, 89, 92, 95, 100, 101, 109, 113, 174, 176, 185–8, 249
pisifera, 25, 61, 174, 176, 185, 187
Chara Braunii, 214
globularis, 215, 216
Chelonopsis, 69
Chenopodium album, 131
virgatum, 170
Chikusichloa brachyanthera, 83

Chioanthus, 54
retusa, 78
Chionographis, 67
japonica, 77
Koidzumiana, 77
Chloranthus, 69
Fortunei, 13, 77
Choerospondias, 56
Chosenia, 68, 192
bracteosa, 98, 192
Chrysalidocarpus lutescens, 272
Chrysanthemum, 169
lineare, 269
nipponicum, 169
pacificum, 169
sibiricum, 13
yezoense, 74, 169
Chrysosplenium, 66, 70, 71
Cibotium, 81
Cicuta virosa, 167
Cinnamomum, 54
Camphora, 64
japonicum, 65, 100, 258
pedunculosa, 65
Cirsium, 70
apoiense, 206
homolepis, 230
hondoensis, 229
japonicum, 136
Sieboldi, 222
Tanakae, 139
tenuifolia, 221
Cladonia, 244
alpestris, 200
rangiferina, 200
sylvatica, 251
Cladopus japonicus, 217
Cladrastis, 54
Clathropteris meniscoides, 46
Clematis Maximowicziana, 66
Cleome, 81
Clethra, 54, 71
barbinervis, 65, 131, 250, 262
Cleyera, 67
ochnacea, 64
Clinopodium sachalinense, 229
Clinostigma Savoryana, 79, 272
Cochlea oblongifolia, 168
Comanthosphace sublanceolata, 147
Comarum palustre, 226, 228, 230, 231
Comptonia, 54, 55
Coniopteris hymenophylloides, 46
Convallaria Keiskei, 163
Coriaria, 44, 67, 71
japonica, 247
Cornus, 54
brachypoda, 65
canadensis, 93, 190, 191, 200
controversa, 57, 65, 240, 241, 252–4
Kousa, 96
Corydalis decumbens, 221
Corylopsis, 48, 54, 68
coreana, 77
Corylus, 48, 54
Sieboldiana, 131, 244
Corymborkis, 81
Cremastra, 67
appendiculata, 64, 66
Crepidiastrum ameristophyllum, 78
linguafolium, 78
Croomia, 69
Crossostephium chinense, 83, 170
Croton, 82
Cryptomeria, 14, 26, 27, 29, 30, 54, 56, 68, 98, 185, 186
japonica, 14, 25, 26, 89, 95, 96, 100, 101, 103, 104, 109, 113, 174, 185, 186, 276
Ctenis japonica, 43
latiloba, 43
Cunninghamia, 54–6
Cyathea, 88, 102
boninsimensis, 102
podophylla, 82
Cycas, 63, 102
revoluta, 25, 83, 102
Cyclobalanopsis, 54, 55, 57, 59, 67, 70, 140, 271, 276
acuta, 64, 131, 186
gilva, 25, 64, 131
glauca, 61, 64, 166, 258
Miyagii, 83
myrsinaefolia, 25, 61, 64
salicina, 64, 186
sessilifolia, 64
stenophylla, 25
Cyclosorus acuminatus, 66, 258
Cymbidium Georingii, 64, 66
lancifolium, 66
Cymbopogon Georingii, 141
Cymodocea, 80, 82
Cynocrambe, 71
Cynodon dactylon, 129, 139, 140, 270
Cynoxylon japonica, 57
Cyperus difformis, 222
polystachyos, 223
Cyrtandra, 79
Cyrtococcum, 81
Cyrtomium falcatum, 169, 170

Dactylis glomerata, 146, 170
Dactyloctenium, 82
Dactylostalix, 72
Dalbergia, 79
Damnacanthus, 83
indicus, 65
Daphne odora, 61
Daphniphyllum, 26
macropodum, 26, 64, 75, 180, 255
Teijsmanni, 65, 170
Debregesia edulis, 65
Deinanthe, 69
bifida, 77

Dendrocacalia, 79
crepidifolia, 272
Derris, 82
Deschampsia caespitosa, 203
flexuosa, 247, 252, 260
Desmodium, 138
racemosum, 136, 258
Deutzia crenata, 131, 250
polita, 250
scabra, 65, 255
Sieboldiana, 65, 258
Dianthus superbus, 168, 206
Diapensia lapponica, 199, 203
Diarrhena, 69
Dicalyx, 65
glauca, 65
lancifolia, 57, 65
lucida, 65
myrtacea, 57, 65
prunifolia, 65
Dicentra peregrina, 205, 246, 247
Dichondra repens, 138
Dicranella secunda, 201
Dicranopteris linearis, 65
Dicranum, 161
falcatum, 201
fuscescens, 200
japonicum, 251
majus, 200
Dictyophyllum Nilssoni, 43
Dieranella tosaensis, 257
Digitaria adscendens, 129, 130, 141, 145, 146, 170
violascens, 130
Diospyros, 54
maritima, 83
Diphylleia, 69
Grayi, 75, 183, 190
Diplanthera, 80, 82
Diplazium, 66
Dipteris, 81
Disanthus, 48, 54, 68
cercidifolia, 77
eocercidifolia, 48
Disporum, 67
sessile, 240, 241
Distylium, 78, 83
racemosum, 64, 82, 100, 186, 276
Distylopsis, 54, 57
Dodonea, 54, 55, 81
Dopatrium junceum, 222
Draba borealis, 168
Drepanocladus, 216
fluitans, 216, 217
Drosera anglica, 226, 229, 231
peltata, 221
rotundifolia, 139, 221, 222, 225, 227, 231, 261
spathulata, 221
Dryopteris amurensis, 191
austriaca, 191
crassirhizoma, 183, 191, 218, 220, 229, 241
erythrosora, 64, 66, 258
lacera, 66
monticola, 183, 229
polylepis, 183
Thelypteris, 145
tokyoensis, 218
varia, 66
Duchesnia indica, 66
Dunbaria, 67
Dystaenia, 69

Ecdysanthera, 81
Echinochloa crus-galli, 222
Echinops setifer, 269
Ehretia, 54
Eichhornia crassipes, 223
Elaeagnus glabra, 65
macrophylla, 65, 170
Yoshinoi, 77
Elaeocarpus japonicus, 65, 82, 102
sylvestris, 65, 83
Eleocharis japonica, 225, 227
pellucida, 223
Elliotia, 70
Elodea canadensis, 214
Elymus, 161
arenarius, 153
mollis, 153, 160–2, 167
Empetrum nigrum, 169, 199, 203, 226, 234, 245–7, 251, 252, 261, 262
Engelhardtia, 54, 55
Enhalus, 80, 81
Enkianthus, 68
campanulatus, 262
cernuus, 262
perulatus, 62
Eoeuryale, 56
Eotrapa, 56
Ephippianthus, 72
Schmidtii, 190
Epigaea, 68
Epilobium angustifolium, 130, 242
Epimedium, 70, 71
grandiflorum, 75
sempervirens, 75
Epipremnum, 81
Equisetum fluviale, 226, 229, 230
ramosissimum, 216
Eranthis, 71
Erigeron, 129, 130, 137, 145–7
annuus, 130, 131, 145, 166
canadensis, 129–31, 145, 146, 271
linifolius, 271
sumatrensis, 131, 146
Thunbergii, 206
Eriocaulon nudicuspe, 78, 221
Sieboldtianum, 221
Eriophorum gracile, 224
vaginatum, 167, 226, 230, 231, 234, 261
Erythrina, 82

Eucommia, 54, 55, 57
Eukianthus, 67
Eulalia speciosa, 78
Euonymus, 54, 78, 96
japonica, 61, 65, 166, 170
Sieboldiana, 241
Eupatorium japonicum, 136
sachalinense, 143
Euphorbia adenochloa, 221
Sieboldiana, 230
supina, 271
togakuensis, 229
Euptelea, 68
polyandra, 70, 96
Eurya, 78
emarginata, 65, 100, 170
japonica, 64, 166, 255, 258
Euryale, 67
ferox, 214
Euscaphis japonica, 57, 65
Evolvulus, 81

Fagaria ailanthoides, 57, 65
Fagus, 12, 29, 30, 54, 63, 68, 76, 88, 98, 107, 124, 138, 177, 185–7, 276
crenata, 21, 25, 56, 70, 90, 93, 94, 106, 126, 131, 132, 140, 165, 174, 176, 178–81, 185–7, 193, 220, 228, 241, 245, 247, 248, 250, 251, 260–2, 270
ferruginea, 56
Hayatae, 56
japonica, 25, 93, 106, 107, 181, 182, 250
palaeocrenata, 54
Sieboldi, 70
Farfugium japonicum, 170, 258
Fatsia, 70, 78, 79
japonica, 61, 65, 166
Fatoua villosa, 66
Fauria, 72
crista-galli, 202, 203, 234
Festuca, 126, 133
ovina, 129, 138
rubra, 170
Ficus crenata, 258
elastica, 272
retusa, 88, 102
Wightiana, 25, 83, 102
Filipendula kamtschatica, 168, 184
Fimbristylis, 156
complanata, 221
dichotoma, 221, 262
diphylloides, 223
sericea, 167
tristachya, 139
Fiwa japonica, 65
Flagellaria, 81
Fokienia, 54, 55
Forrestia, 82
Forsythia, 54, 71, 72
japonica, 77
suspensa, 62
viridissima, 62
Fortunearia, 57
Fragaria, 71
Fraxinus, 54
insularis, 83
japonica, 57
lanuginosa, 96
mandshurica, 183–5, 191, 218–20, 225, 226, 228, 229
Sieboldiana, 244
Spaethiana, 182
Frecynetia, 81

Gahnia, 82
Galium japonicum, 242
tokyoense, 269
Garcinia, 81
subelliptica, 102
Gardenia, 67, 78
Gardneria multiflora, 77
Garnotia, 81
Gastrochilus japonicus, 77
Gastrodia, 67
Gaultheria adenothrix, 247
Miqueliana, 200, 261
Gentiana algida, 204
axillariflora, 227
shikokiana, 77
Thunbergii, 227, 231
triflora, 139
Geonomites, 48
Geotaenium, 83
gelasinum, 83
Geranium nepalense, 66
Thunbergii, 130, 139
Geum japonicum, 66
pentapetalum, 201
Ginkgo, 56
biloba, 61
Ginkgoites, 43
digitata, 46
Glaucidium, 69
diphyllum, 75
Glaux maritima, 153–5
Gleditschia, 54
Glehnia littoralis, 160, 161
Gleichenia dichotoma, 131
glauca, 65
japonica, 65
Glyceria alnasteretum, 195
Glyptostrobus, 48, 54–6
europaeus, 48
Glysine Soja, 146
Gratiola japonica, 269
Guettarda, 81
Gymnadenia conopsea, 16
Gymnaster, 59
Gymnotheca, 70

Habenaria linearifolia, 16
radiata, 221

Hakonechloa, 69
macra, 147
Haloragis micrantha, 141, 145
Hamamelis, 54, 68
bitchuensis, 77
japonica, 179, 181
obtusata, 96
Hausmannia, 43
nariwaensis, 43
ussuriensis, 43
Hedera rhombea, 64
Hedyotis biflora, 170
prostrata, 223
Hedysarum, 204
Helacleum dulce, 192
Helicteres, 82
Helminthostachys, 81
Heloniopsis, 67
Helwingia, 66, 67
japonica, 65
Hemerocallis esculenta, 231
Middendorffii, 163, 197, 221, 225
Thunbergii, 222
yezoensis, 163
Hemiboea, 81
Hemigraphis, 81
Hemigramma, 81
Hemitrapa, 54–6
Heritiera, 82
Hernandia, 81
peltata, 167
Heterosmilax, 67
Heterotropa, 15, 36, 66, 67, 69, 77, 83
albivenia, 74
asaroides, 77
aspera, 77
Fudsinoi, 83
Minamitaniana, 77
nipponica, 34–6, 74
sakawana, 77
Hibiscus Hamabo, 162
Hieracium japonicum, 247
Hipparis vulgaris, 229
Honkenya peploides, 153, 160
Hosiea, 67
Hosta, 69, 70
longipes, 74
longissima, 221
rectifolia, 221, 225
undulata, 136
Houttuynia, 67, 70
cordata, 66
Hovenia, 67
Hugeria japonica, 75
Humulus japonicus, 66
Hydrangea, 54, 67, 71
involucrata, 74
macrophylla, 65, 75, 183, 184, 218, 255
paniculata, 130, 136, 220, 225, 226, 231, 240–2, 244, 247, 248, 250, 252, 256, 258, 262
shikokiana, 77
Hydrilla verticillata, 214, 217, 223
Hydrobryum, 67
Hydrocharis asiatica, 215
Hydrocotyle maritima, 131
ramiflora, 130, 145
sibthorpioides, 66
Hylocomium splendens, 200
Hylomecon, 69
Hypericum erectum, 136, 143, 146
samaniense, 206
Hypnum, 166
plicatulum, 200
Hypochoeris crepidioides, 206

Ichinanthus, 81
Idesia polycarpa, 65
Ilex, 26, 54, 78
cornuta, 56
crenata, 26, 61, 75, 184, 226, 249, 260
dimorphophylla, 83
integra, 25, 26, 61, 64
leucoclada, 26, 180
macrophylla, 64
macropoda, 96
Nemotoi, 226
pedunculosa, 249
rugosa, 200
Sugeroki, 75, 181, 187, 188, 190, 200, 201, 260, 261
Illicium, 67
Illigera, 79
Impatiens noli-tangere, 218
Textori, 218
Imperata, 138, 166
cylindrica, 129, 131, 147, 166, 256
Intsia, 79
Inula britannica, 222
glandulosa, 229
salicina, 136
Ipomoea pes-caprae, 162, 168
Iris ensata, 136, 221, 222
japonica, 66
laevigata, 221, 230
Isachne globosa, 221
Ischaemum, 156, 162
anthephoroides, 161
anthroides, 147
aristatum, 221
Isodon, 70
Isopyrum, 69
Itea, 68
Ixeris, 161
dentata, 136
japonica, 239
repens, 160, 161, 167

Japonolirion, 72
Juglans, 54, 71
cinerea, 56
mandshurica, 56

Juncus beringensis, 203
effusus, 223
gracillimus, 154, 155
yokoscensis, 221
Jungermannia infusca, 263
Juniperus, 165
chinensis, 168, 169, 205
communis, 200
conferta, 165, 166
Jussiaea prostrata, 223

Kadsura, 67
japonica, 64
Kalopanax, 54
innovans, 96, 97
pictus, 163, 179, 240, 241
Kalimeris, 59
Kandelia, 80, 108
Candel, 89, 102, 108, 157, 158
Kengia, 71
Kerria, 68
Keteleeria, 54–7
Kinugasa, 72
japonica, 75
Kirengeshoma, 69
palmata, 77
Kleinhovia, 81
Koelreuteria, 54
Kummerowia striata, 139

Lactuca dentata, 139, 145
Lagotis glauca, 206
Laportea macrostachya, 183, 229
Larix, 71, 194
Gmelini, 56, 58, 71
leptolepis, 25, 56, 71, 74, 88–90, 93, 106, 109, 120, 175, 190, 192, 228, 233, 247, 249, 251–3
Lastrea quelpaertensis, 201, 202
Thelypteris, 136, 183, 221, 226
Lathyrus, 162
maritimus, 161, 168
Lecanorchis, 67
Ledum palustre, 200, 226, 247, 260–2
Lemmaphyllum microphyllum, 66, 255
Lemna minor, 215, 223
paucicostata, 223
trisulca, 223
Lentinus edodes, 124
Leontopodium, 204
Fauriei, 204, 206
hayachinense, 204, 206
shinanense, 204
Lepidagathis, 81
Lepistemon, 81
Leptidictyum riparium, 218, 220
Lespedeza, 68, 129, 144
bicolor, 65, 130, 131, 135, 136, 147
Buergeri, 65
cuneata, 145
cyrtobotrya, 131, 141
homoloba, 65
pilosa, 136
Leucaena leucocephala, 101, 272
Leucothoe, 68
Grayana, 262
Ligularia Fischeri, 230
Ligustrum japonicum, 61, 65, 258
pauciflcum, 255
Lilium cordatum, 240, 241
longiflorum, 83
maculatum, 163, 169
Limnophila sessiliflora, 222
Limonium tetragonum, 153
Wrightii, 83
Linaria, 162
japonica, 161
Lindera, 54, 68
citriodora, 57
membranacea, 179, 244
sericea, 250
umbellata, 57, 181, 250
Lindernia verbenaefolia, 223
Liparis, 67
formosana, 77
Liquidambar, 48, 54–7
Liriodendron, 54, 55
Liriope, 64, 67
Lissopepo, 57
Listera nipponica, 190
Lithocarpus edulis, 64
glabra, 64
Litsea aciculata, 100
japonica, 65
Livistona, 67, 78
subglobosa, 65, 102, 273
Lobelia boninensis, 79
chinensis, 223
sessilifolia, 222, 230, 234
Loiseleuria procumbens, 199, 203, 234, 245, 247, 251, 252
Lomariopsis, 81
Loropetalum, 67
Lotus corniculatus, 147, 168, 239
Ludwigia ovalis, 223
Luisia teres, 77
Lumnitzera, 80, 81, 108
racemosa, 157
Luzula campestris, 130
capitata, 242
Lycium griseolum, 83
Lycopodium, 192, 264
cernuum, 262, 263
inundatum, 226, 227, 231
serratum, 190
sitchense, 201
Lycopus lucidus, 218
Maackianus, 145, 221
parviflorus, 218
Lycoris sanguinea, 66
Lyonia, 67
Neziki, 65, 131, 249

Lysichiton, 72
camtschatense, 72, 183, 184, 218, 220, 230
Lysimachia candida, 269
clethroides, 136, 141
Fortunei, 223
leucantha, 269
mauritiana, 169
thyrsiflora, 167
vulgaris, 218, 221
Lythrum salicina, 221

Maackia, 54
amurensis, 244, 247
Macaranga, 82
Machilus, 54, 61, 63, 76–8, 83, 166
japonica, 64, 258
Nathorsti, 48
Thunbergii, 25, 61, 64, 82, 100, 102, 131, 166, 253–6, 258, 259, 271, 276
Macleaya, 69, 70
Macroclinidium, 67
robustum, 66
Macrodiervilla, 68
Macropodium, 72
Maesa, 54
japonica, 65
Magnolia, 54, 67, 68, 70–2, 78
denudata, 62
Kobus, 62, 226
obovata, 96, 179, 182, 191, 241
parviflora, 256
salicifolia, 57, 75, 96, 179
stellata, 78
Mahonia, 54
Majanthemum bifolium, 244
dilatatum, 233
Mallotopus, 75
japonicus, 75
Mallotus, 54
japonicus, 64, 65
Malus baccata, 162
Sieboldii, 139
Toringo, 231
Mangifera, 48
Marattiopteris asiatica, 43
Marsupella sphacelata, 201
Matteuccia Struthiopteris, 231, 233
Maytenus, 82
Meehania, 69
urticifolia, 185
Melanolepis, 81
Melia Azedarach, 57, 83
Melicope, 79
Meliodendron, 54, 55, 57
Meliosma Oldhami, 13
rigida, 57
Menyanthes, 56
trifoliata, 226, 228, 230, 231
Menziesia, 68
pentandra, 191
Mertensia asiatica, 160
Messerschmidia argentea, 167
Metanarthecium luteo-viride, 139
Metasequoia, 48, 54–7
Meterosideros boninensis, 79, 272
Michelia, 67
compressa, 64
Microgonium, 79
Microstegium vimineum, 131, 141
Microtis, 67
Milletia japonica, 66
Mimulus sessilifolius, 202
Minuartia arctica, 204
hondoensis, 204
Miscanthus, 102, 103, 126, 129, 132, 134–6, 138, 139, 143, 145
condensatus, 169, 271
floridulus, 129, 134
Matsumurae, 256
oligostachys, 134
sacchariflorus, 135, 217
sinensis, 126, 129–37, 139–43, 145, 146, 166, 169, 174, 221, 239, 240, 242, 244, 247, 248, 252, 256–8, 262, 265, 271
tinctorius, 135
Mitchella repens, 66
undulata, 96
Mitella, 69, 70
stylosa, 77
Mitrastemon, 67
Yamamotoi, 77
Moerckia nivalis, 201
Molinia coerulea, 224, 225
Moliniopsis, 227
japonica, 202, 224–7, 230–2, 260, 261
Monochoria vaginalis, 222
Monotropastrum, 67
Morus bombycis, 169, 240
tiliaefolia, 77
Muhlenbergia, 69
Murraya, 82
Musa, 72
Musophyllum, 48
Myrica Gale, 225, 226, 230, 231
rubra, 64
Myriophyllum spicatum, 213–6
Myrsine Seguinii, 65, 166

Najus marina, 213, 214
Nandina, 67
Nanocynide, 67
Narthecium, 72
asiaticum, 202, 231, 234
Nathorstia alata, 46
Oishii, 46
Nelumbo nucifera, 214, 223
Neofinetia, 67
falcata, 77
Neolitsea, 78
sericea, 64, 166
Sieboldii, 64
Nephrolepis auriculata, 255, 257–9

Nervilia Aragoana, 81
Nilssonia, 43
orientalis, 43
Nipa, 81
Nitella flexilis, 214–6
Nothapodytes, 79
Nothofagus, 44
Nuphar akashiensis, 56
japonicum, 213, 214, 220
pumilum, 167, 216, 226, 229, 230
Nymphaea japonica, 230
tetragona, 213, 214, 216, 226, 234
Nymphoides indica, 214, 223
peltata, 213, 223
Nyssa, 54–7

Oberonia japonica, 77
Ochrosia, 79
Odontoschisma denudatum, 263
Oenothera biennis, 239, 240
Lamarckiana, 242
parviflora, 131
Oishia, 43
Onoclea sensibilis, 221
Ophiopogon, 64, 67
Jaburan, 83
Oplopanax, 72
japonicus, 72, 190
Oreomyrrhis, 44
Oreopanax, 70
Orobanche, 79
boninsimae, 272
Orostachys, 69
Oryzopsis, 81
Osmanthus, 54, 78
aurantiaca, 61
Osmorrhiza, 71
Osmunda cinnamomea, 147, 183, 225–7, 230–3
japonica, 218
Osteomeles, 78
anthyllidifolia, 83
boninensis, 272
subrotunda, 170
Ostericum Sieboldii, 66
Ostrya, 54
Oxalis Acetosella, 93, 190
amamiana, 83
Martiana, 271
Oxycoccus, 227
quadripetalus, 167, 225, 227, 230, 231, 234, 261
Oxytropis, 204
japonica, 204
Kudoana, 204
megalantha, 204
rishiriensis, 204
shokanbetsuensis, 204

Pachysandra terminalis, 241
Paederia chinensis, 66
Palaeodavidia, 57
Paliurus, 54
nipponicus, 56
Palura chinensis, 65
Pandanus, 63, 88, 102
boninensis, 102
tectorius, 157, 167
Parabenzoin, 54, 67, 68
eotrilobum, 48
trilobum, 77, 181
Paraixeria Yoshinoi, 77
Parnassia alpicola, 201, 202
Parrotia, 54, 55
Parthenocissus tricuspidata, 65
Pasania edulis, 131, 258
glabra, 57
Paspalum dilatatum, 129
Thunbergii, 146
Patrinia scabiosaefolia, 145
sibirica, 204
triloba, 205
villosa, 139, 240
Pelea, 79
Pellionia radicans, 66
Peltoboykinia, 69
Watanabei, 77
Penthorum, 69
Peracarpa, 67
Persicaria Blumei, 131, 141, 145, 146
nodosum, 130, 146
perfoliatum, 130
Sieboldi, 223
vulgaris, 129, 130, 145, 147
Pertya, 69
glabrescens, 65
scandens, 65
verticillata, 75
Petasites japonicus, 136, 143, 168, 192, 239–42
Peucedanum japonicum, 169, 170
multivittatum, 195, 202
Phaenosperma, 67
Phleum pratense, 130, 143
Pholiota nameko, 124
Photinia glabra, 61, 64
Phragmites, 167, 227
communis, 155, 156, 167, 183, 213, 214, 216–20, 224, 226, 227, 230, 242
Karka, 156
Phryma, 69
Phyllodoce aleutica, 201, 202
nipponica, 251
Phyllostachys, 121
bambusoides, 121, 122
nigra, 121
pubescens, 121, 122
Physalis angulata, 271
Picea, 54, 63, 71, 88, 93, 107, 163, 164, 191, 194, 195
bicolor, 189
Glehnii, 74, 90, 163, 165, 188, 189, 192, 225

jezoensis, 25, 56, 71, 90, 130, 140, 163, 176, 189, 191, 228, 230, 249, 276
Koribai, 56
Koyamai, 90, 189
latibracteata, 56
Maximowiczii, 56, 90, 189
polita, 100, 181, 249, 250
Shirasawae, 189
sitchensis, 191
Picris japonica, 239
Pieris, 67
japonica, 57, 61, 131, 187, 188, 249, 250, 263
Pilea petiolaris, 66
Pileostegia, 82
Pinus, 54, 102, 165, 166
densiflora, 25, 27, 30, 61, 88, 89, 99, 101, 105, 106, 109, 116, 117, 122, 131, 141, 165, 166, 168, 169, 181, 226, 244, 247–9, 251, 252, 262, 270
Fujiii, 56
koraiensis, 56
lutchuensis, 83, 102–4, 272
parviflora, 187, 262
pentaphylla, 92, 181, 182, 187–9, 233, 245–7, 249
pumila, 28, 63, 72, 98, 175, 194–201, 234, 261
Thunbergii, 25, 61, 88, 89, 100, 104, 105, 109, 122, 131, 158, 159, 162, 165, 166, 168, 255, 258, 259
trifolia, 55–8
Piper Kadzura, 65
Pipterus, 81
Pithecelobium, 81
Pittosporum, 54, 78
Tobira, 65, 100, 105, 166, 170, 270
Plagiogyria Matsumureana, 96
Plagiothecium roeseanum, 201
Plantago asiaticus, 131, 146
camtschatica, 139
hakusanensis, 202, 203
Platanus, 48, 54, 55
Chaneyi, 48
Platanthera hologlottis, 222
ophrydioides, 190
sachalinensis, 221
Platycarya, 54
Platycrater, 67, 68
arguta, 77
Platypholis, 79
Pleioblastus, 65, 95, 126, 129, 132, 133, 138–40, 142, 143
Chino, 65, 131, 140–2, 146
communis, 65
distichus, 129, 131–3, 137, 140–2, 144
linearis, 83
Simonii, 65
variegatus, 140
Yoshidake, 140
Pleopeltis Thunbergiana, 255
Pleurotus ostreatus, 124
Pleurozium Schreberi, 200
Poa, 126
annua, 129, 131, 146, 170, 239, 242
pratensis, 129, 132, 133, 138, 240
Podocarpus macrophylla, 25, 64, 100, 102–4
Nagi, 25, 64, 100
Pogonatum, 166
alpinum, 201
Pogonia, 67
japonica, 225, 227
Polemonium kiushianum, 269
Pollia, 67
Polygala japonica, 141
Polygonatum sachalinense, 136
Polygonum, 59
Polystichum, 64, 66
retroso-paleaceum, 183
tripteron, 183
Polystichopsis Standishii, 183
Polytrichum, 166
commune, 241, 242
Ponerorchis, 69
Pongamia, 82
Populus, 93, 192, 194
Maximowiczii, 93, 192, 193, 239–42
Sieboldi, 244, 247
tremula, 163
Portulaca oleracea, 170
Potamogeton alpinus, 220
biwaensis, 214
capiscus, 223
compressus, 215
crispus, 215, 217
distinctus, 223
heterophylla, 215
Maackianus, 213–5
malaianus, 214, 215, 217
oxyphyllus, 213, 217, 220, 226
pectinatus, 215, 226
perfoliatus, 213–6
Potentilla egedei, 153, 154
Freyniana, 136, 139, 145
Primula cuneifolia, 202
hidakana, 206
macrocarpa, 206
modesta, 168
nipponica, 202
Sieboldi, 16, 220–2
tosaensis, 269
yuparensis, 206
Pristia, 81
Protosequoia, 56, 57
Prunella vulgaris, 145
Prunus, 54
apetala, 70
Grayana, 231
incisa, 70, 250
Jamasakura, 62, 65, 70
Lannesiana, 62, 253–5
macrophylla, 64

nipponica, 231, 244
pendula, 62
Sargentii, 70, 240, 242, 244
spinulosa, 64
Ssiori, 229, 241
verecunda, 70, 226, 240, 241, 252
yedoensis, 62
Zippeliana, 64
Pseudolarix, 54–6
Pseudosasa Owatarii, 14
Pseudostellaria, 71
Pseudotsuga, 54, 68
japonica, 76, 186
Psilotum nudum, 255
Pteridium, 132, 144, 145
aquilinum, 130, 131, 136, 137, 141–5, 231
Pteridophyllum, 72
racemosum, 190
Pteris cretica, 66
multifida, 66
Pterocarpus, 79
Pterocarya, 54, 71
paliurus, 56
rhoifolia, 182, 184, 220, 229
Pterygopleurum, 67
Ptilium crista-castrensis, 200
Puccinellia, 154
kurilensis, 153–5
nipponica, 169
Pueraria lobata, 66
Pyrola japonica, 96, 240
renifolia, 190
Pyrrosia lingua, 66, 255

Quercus, 54, 59, 88, 98, 107, 164, 165, 194
acuta, 25, 100
acutissima, 25, 89, 106, 107, 111, 124, 174
aliena, 75
crispula, 25, 93, 97, 105–7, 124, 131, 142, 229, 270
dentata, 25, 97, 105, 107, 131, 163, 165, 166, 168, 169, 191, 270
gilva, 100
glauca, 100
ilex, 100
mongolica, 70, 130, 142, 163, 165, 166, 168, 169, 174, 175, 178, 179, 181, 191, 193, 195, 196, 201, 240, 241, 244, 247, 249, 250, 252, 256, 260
myrsinaefolia, 100
phillyraeoides, 65, 100–2, 105, 170
salicina, 100, 102
serrata, 25, 64, 65, 67, 89, 106, 111, 124, 131, 166, 168, 174, 226, 252
sessilifolia, 100
variabilis, 25, 64, 65, 106

Ranzania, 69, 72, 75
japonica, 75
Ranunculus, 70, 71
acris, 202
japonicus, 130
sceleratus, 222
ternatus, 221
Rapanea neriifolia, 100, 105
Raphiolepis, 78
umbellata, 65
Rehderodendron, 57
Reynoutria, 59, 69, 256
hachidyoensis, 253, 254
japonica, 136, 252, 256, 257, 259, 262
sachalinensis, 168, 239, 240, 242, 244
Rhacomitrium, 166, 255
anomondontoides, 201
canescens, 242, 244
heterostichum, 201
hypnoides, 251
Rhamnus crenata, 183, 226
japonica, 183
Rhaphidophora, 81
Rhaphiolepis umbellata, 170
Rhizophora, 80, 81, 108
mucronata, 89, 102, 108, 157
Rhododendron, 54, 61, 181
Albrechti, 179, 187
boninense, 78
brachycarpum, 200
dilatatum, 65, 131
Fauriae, 191, 244, 251, 252, 260–2
hortense, 61
indicum, 61
japonicum, 131, 139, 231, 244, 252
Kaempferi, 131, 258
kiusianum, 256, 262
macrosepala, 61
Makinoi, 78
Metternichii, 187, 205
mucronatum, 61
nipponicum, 75
nudipes, 75
pulchrum, 61
reticulatum, 65
scabrum, 61
Tanakae, 76
Wadanum, 75, 249
yedoense, 77
Rhodomyrtus, 81
Rhodotypos, 68
scandens, 77
Rhus, 54
javanica, 65, 131, 146
orientalis, 130
silvestris, 65
succedanea, 65
trichocarpa, 96, 231, 242, 249, 260
Rhynchospora, 227
alba, 226, 227, 231–4
chinensis, 222, 223
Fujiiana, 221
Umemurae, 221
Yasudana, 234
Rhyssopterys, 81

Ribes, 71
Robinia, 54, 55
Rodgersia, 69
  podophylla, 183
Rosa, 162
  multiflora, 65, 131, 139, 240
  rugosa, 130, 162, 164
  Wichuraiana, 162, 169
Rubia Akanae, 66
Rubus amamianus, 83
  corchorifolius, 65
  crataegifolius, 146
  ikenoensis, 190
  palmatus, 65
  parvifolius, 145
  pedatus, 200
  phoenicolasius, 130
  Sieboldi, 65
  trifidus, 65
Rudbeckia laciniata, 130
Rumex Acetosa, 221
  Acetosella, 130, 242
  obtusifolius, 130, 145, 146
Rumohra, 65
  mutica, 188

Sabalites, 48
Sabia japonica, 57
Sagittaria pygmaea, 222
Salicornia europea, 153, 155
Salix, 93, 192
  Bakko, 131, 143, 163, 239–42
  gracilistyla, 107
  integra, 139, 239, 242
  japonica, 74
  jessoensis, 192
  Nakamurana, 202
  petsusu, 192
  Reinii, 200, 201, 244, 247
  sachalinensis, 93, 107, 192, 239, 241, 242
  Saidaeana, 141
  serissaefolia, 248
  Shiraiana, 74
  Sieboldiana, 256
  subfragilis, 220
  subopposita, 141
  vulpina, 131
  yezoalpina, 202
Salsola Komarovi, 160
Salvia pygmaea, 83
Salvinia natans, 222, 223
Sambucus Sieboldiana, 130
Sanguisorba albiflora, 218
  obtusa, 206
  officinalis, 139, 141, 230, 231, 233
  tenuifolia, 221, 225
Sanicula chinensis, 240
Santalum boninense, 79, 272
Sapindus, 54
Sapium, 54
  sebiferum, 56
Sarcandra, 67
Sarcanthus scolopendrifolius, 77
Saricornia, 20
Sasa, 26, 89, 93, 95, 98, 126, 129, 132, 137, 140, 142–4, 169, 181, 190–2, 194, 195
  hirtella, 190
  ishizuchiensis, 190
  kurilensis, 26, 75, 89, 95, 130, 178–81, 190, 191, 193, 195–7, 200, 234, 247
  nipponica, 26, 89, 95, 130, 133, 142, 181
  owatarii, 205
  ozeana, 231, 233
  palmata, 26, 225
  paniculata, 89, 95, 191
  senanensis, 142, 169
  Veitchii, 130, 133, 142, 143
Sasaella, 140
  ramosa, 95, 130, 141
Sasamorpha, 26
  purpurascens, 26, 75, 89, 180, 181
Sassafras, 54, 55
Satakentia, 81, 83
  liukiuensis, 83
Saurauia, 81
Saururus, 69
Saussurea, 70
  Riederi, 206
Saxifraga Merkii, 201
  Nishidae, 204
  sendaica, 77
Scaevola taccada, 167
Scheuchzeria palustris, 227, 229, 231
Schima, 78, 82
Schisandra, 54
Schizaea, 81
Schizocodon, 69
  rotundifolia, 83
Schizophragma, 54, 69
Sciadopitys, 28, 56, 68, 95, 185
  verticillata, 76, 95, 100, 181, 185, 187, 276
Scirpus caespitosus, 203
  lacustris, 220, 226
  Maximowiczii, 203
  tabernaemontani, 216
  triqueter, 213
Scolopia, 81
Sedum, 66
  Ishidae, 168
  verticillatum, 169
Selaginella selaginoides, 202
Senecio palmatus, 218
Sequoia, 54–6
Sesleria tesselata, 222
Setaria viridis, 131, 145, 146
Shibataea, 67
Shiia, 88, 100
  cuspidata, 100
  Sieboldii, 100, 102
Shisandra, 69
Shortia, 69
  ilicifolia, 96

soldanelloides, 205, 247, 251
uniflora, 96
Siegesbeckia pubescens, 146
Sieversia pentapetala, 231, 234
Sinomenium, 69
acutum, 57
Skimmia, 67
japonica, 75, 180, 249
Smilacina yezoensis, 195
Smilax China, 131
Solidago Virga-aurea, 225, 242, 244
Sonchus arvensis, 270
Sonneratia, 79, 80
alba, 157
Sophora, 54
Sorbus, 54, 93
alnifolia, 96, 240, 244
commixta, 130, 181, 193, 195, 199, 200, 231, 241, 244, 260–2
Matsumurana, 195, 199, 200
sambucifolia, 107, 199, 200
Spathoglottis, 81
Spergularia marina, 153, 155
rubra, 153
Sphagnum, 72, 167, 201, 227
amblyphyllum, 230, 231
compactum, 203, 226, 231, 232
cuspidatum, 227
fimbriatum, 167
fuscum, 226, 231
Girgensohnii, 167
magellanicum, 226
palustre, 226
papillosum, 226, 227, 231
pulchrum, 230, 231
recurvum, 261
squarrosum, 167, 225
tenellum, 234
Spinifex littoreus, 162, 168
Spiraea japonica, 139
Spiranthes sinensis, 139, 227
Spirodela polyrhiza, 215, 222
Spodiopogon sibiricus, 130
Spuriopimpinella, 69
Stachyurus, 67, 78
praecox, 65, 255
Statice arbuscula, 170
Stauntonia, 67
hexaphylla, 65
Staurochilus, 81
Stellaria longifolia, 184
ruscifolia, 168
Stephanandra, 68
incisa, 131
Tanakae, 76
Stereocaulon exutum, 251
vesvianum, 242, 244
Stereosandra, 81
Stewartia, 54, 68
monadelpha, 181
pseudo-Camellia, 181, 256
Streptolirion volubile, 77
Streptopus streptoides, 190
Styrax, 54
japonica, 64, 65
Suaeda asparagoides, 153
japonica, 153
maritima, 153
Suzukia luchuensis, 83
Swertia bimaculata, 66
Tashiroi, 83
Symplocos, 54, 65
chinensis, 96, 131
coreana, 181
prunifolia, 100
Syringa, 71
reticulata, 184
Syzygium, 58
buxifolium, 58

Taiwania, 54, 55
Tanakaea radicans, 77
Tapeinidium, 81
Taraxacum officinale, 143, 239
Tashiroea, 83
Taxodium, 48, 54, 55
Taxus, 26, 194
cuspidata, 26, 75, 130, 194, 196, 200
Terminalia cattapa, 167
Ternstroemia, 54
gymnanthera, 61, 64, 170
Tetraplasia, 83
Tetrastigma, 57
Thalassia, 80, 82
Thalictrum, 71
simplex, 221
Thunbergii, 130, 147
Thelygonum, 71
Thelypteris bukoensis, 229
decursivepinnata, 66
palustris, 224
Thermopsis lupinoides, 163
Thespesia, 81
Thuarea involuta, 162, 168
Thuja, 54, 68, 185
Standishii, 92, 181, 182, 185–90, 192, 233
Thujopsis, 28, 30, 54, 67, 112, 165, 185–7
dolabrata, 25, 75, 95, 165, 185–7, 220, 260
Tiarella, 69
Tilia, 54, 165
japonica, 96, 107, 165, 169, 191, 229
Maximowicziana, 97
Tilingia holopetala, 195, 202
Tachiroei, 204
Tipularia, 69
Titanotrichum, 81
Tofieldia coccinea, 206
Okuboi, 202
Toisusu, 68
Urbaniana, 93, 192, 193
Torreya, 26, 67, 68
nucifera, 25, 26, 75, 100, 180

Tovara, 59, 69
Trachelospermum, 78
asiaticum, 64
Trachycarpus, 67
Trapa japonica, 216, 226
natans, 214, 223
Trapella sinensis, 214
Trautvetteria, 69
japonica, 184, 190, 202
Triadenum japonicum, 221, 222
Tricyrtis, 69
Trifolium lupinaster, 168
repens, 130, 139, 145, 146, 239, 240, 271
Triglochin, 154, 155
maritimum, 153, 154
Trillium kamtschaticum, 185, 218
Smallii, 241
Tschonoskii, 218
Tripetaleia, 54, 67, 70
bracteata, 200, 201
paniculata, 244
Tripterygium Regelii, 75
Tristellateia, 81
Tritomodon, 68
subsessilis, 74
Trochodendron, 54, 67
aralioides, 186
Trollius Riederianus, 202
Tsuga, 28, 54, 68, 100, 177, 195
diversifolia, 25, 90, 91, 175, 176, 181, 182, 186–90, 192, 193, 195, 199, 233, 251, 276
oblonga, 56
Sieboldii, 25, 67, 100, 175, 176, 181, 185–7, 249, 276
Tsusiophyllum, 76
Tutcheria, 81
Typha angustifolia, 214
congesta, 217
latifolia, 214, 217
orientalis, 217

Ulmus, 48, 54, 194
Davidiana, 97, 107, 182, 184, 218, 229, 241
laciniata, 97, 107, 191
Uncaria rhynchophylla, 65
Urtica platyphylla, 168, 192
Thunbergiana, 66
Utricularia bifida, 221, 222, 227
exoleta, 215
intermedia, 227
japonica, 215
nipponica, 269
racemosa, 222, 227
yakushimensis, 221, 222

Vaccinium, 78, 88, 93, 181
amamianum, 83
axillare, 93, 191, 261, 262
bracteatum, 258
ovalifolium, 200, 201
quadrifolium, 226
Smallii, 130, 260
uliginosum, 203, 252
Vitis-Idaea, 93, 200, 226, 234, 261
Vallisneria asiatica, 213–5
denseserrulata, 214
Vancouveria, 70
Vanda, 81
Veratrum japonicum, 230
nigrum, 230
stamineum, 184
Viburnum, 54, 78, 96
Awabuki, 61, 65
Awabukioides, 48
Carlesii, 77
dilatatum, 64, 65, 131
erosum, 65, 131
furcatum, 179, 247
japonicum, 57, 65
Sargentii, 226, 241
suspensum, 83
Viola, 66, 70
amamiana, 83
Bisseti, 75
crassa, 205
grypoceras, 66
hidakana, 206
Iwagawai, 83
Keiskei, 66
mandshurica, 136, 139
obtusa, 130
pumilio, 141
Raddeneana, 16
Tashiroi, 83
vaginata, 75
verecunda, 143
yubariana, 206
Vitex rotundifolia, 162, 166, 168

Wedelia prostrata, 162, 167
Weigela, 68
coraeensis, 62, 75, 253, 254
floribunda, 65
hortensis, 131, 244, 247, 248
Wenlandia, 82
Wikstroemia Gampi, 131
Wisteria, 54, 67, 69
brachybotrys, 66
floribunda, 62, 66
Woodwardia orientalis, 66

Zabelia integrifolia, 77
Zelkova, 48, 54, 70, 71
serrata, 21, 25, 62, 65, 70, 89, 106, 107, 256
Zingiber Mioga, 66
Zizania latifolia, 213–7
Zizyphus, 54, 55
Zoysia, 126, 129, 132–4, 136, 138, 139, 141, 143, 145, 161

japonica, 128–33, 136–40, 142, 144, 145, 174, 256, 270
macrostachya, 161, 167
sinica, 153, 162, 167
tenuifolia, 170

Subject Index

Alpine desert, 196, 203–7
Alpine zone, 63, 72, 175, 196–207
Altitudinal vegetation zones, 174–8
Aquatic vegetation, 211–7
Arctic zone, 63
Arcto-Tertiary flora, 48–9
Artificial forests, 88–9, 113–22, 162
Artificial regeneration, 93, 113–22
Atetsu region, 77

Bamboo, 28, 121–2, 271
Beech forests, 93–5, 179–81
Biomass of forests, 97
Black soil, 27
Blanket bog, 233–4
Bogs, 192, 202, 216, 223, 230–4
Bonin region, 78–9
Brown forest soil, 27, 29–30
Bryophytes, 201
Butsuzo line, 40–1, 46–7

Camellia class, 63, 88
Chishima Current, 6–7
Clear-cutting, 93
Cliff vegetation, 152, 168–70
Climatic zones, 20, 63, 129, 138
Coalfields, 47–9
Coastal vegetation, 102, 104–5, 152–70
Cold index, 21, 23–6, 97, 176
Cool-temperate zone, 63, 67–71, 93–8, 130–1, 153, 181, 185
Conservation, 269–77
Coppices 107, 109, 129
Coriaria type, 44, 67
Crassinodi line, 180
Creeping trees/shrubs, 26, 96, 107

Degree of succession, 133–4, 137
Destruction of ecosystems, 152, 158, 272–3
Dominance ratio, 134
Dry-land vegetation, 88, 105–6
Dune vegetation, 150, 158–68, 192
Dwarf-bamboo, 26, 89, 93, 95, 133, 140–4, 181, 190, 193, 197

Edaphic-biotic factors, 132–3, 138
Eel-grass communities, 158
Emersed plants, 213, 217, 222
Exotic plants, 271

Fagus-Quercus class, 63, 88

Fens, 167, 223, 230
Fixed dune communities, 162–7
Floating leaf communities, 213
Floristic regions, 72–83
Foreshore communities, 159, 167
Forest limit, 177, 196, 199
zones, 21, 26, 63, 87–104, 126–7, 153
Forestry reserves, 109–12, 116–20, 276
Fossa Magna, 4, 10–2, 40, 47
Fossa Magna region, 75
Fossil floras, 43–58
Fumarole vegetation, 262–5

Gallery forest, 229
Geohistory, 34–58
Geological structure of Hokkaido, 18–9
of Japan, 3–5, 10–20, 35–42
of Ryukyus, 14–5
Grassland types, 126–9, 134–47
successions, 129–34
Grazing, 126, 132–3, 138–9, 144–5, 154–5
Ground cover ratio, 134

Halophytes, 152, 160
Heath, 196–9, 203
Hilly zone, 63, 174
Himalayo-Sino-Japonica type, 66–71, 82
Honshu geosyncline, 36
Hygrophilous types, 201–3

Irrigation ponds, 223

Japan Current, 4–7
Japanese Alps, 12, 21, 90, 176–7, 189

Kano line, 81
Kanto Loam, 16
Kanto region, 74
Kuromatsunai line, 18–9, 96, 178

Lake vegetation, 213–5
Larch forests, 106, 109, 112, 120–1, 190
Lava flows, 244–59, 264
Lowland zone, 63

Macroclimate, 128–9, 152, 270
Magnolia type, 67–71
Makinoesia flora, 48
Mangrove forest, 87, 102, 107–8, 157–8
Man-made forests, 88–9, 109–24, 273
Maritime vegetation, 151–70

Marsh vegetation, 220–2
Meadow vegetation, 126, 132
Median Line, 12–3, 40–1
Metasequoia fossil group, 55
Mino-Mikawa region, 78
Mobile dune communities, 161–2, 167
Montane zone, 63, 174, 178–88
Moors, 167, 224–32
Mountain vegetation, 173–207
Mountaintop effect, 14, 19
Mud flows, 211, 247–8, 252

National monuments, 269, 272
National Parks, 159, 169, 273–5
Natural regeneration, 89, 109, 113, 122–4
Net production of forests, 98
Nitrophilous vegetation, 170

Orogenic cycles, 38–41, 46–8, 52–3
Orthoseral grassland, 129–31, 139

Pacific Ocean side (*Omote-Nihon*), 8, 21, 114, 176, 180, 271
Paddy fields, 99, 183, 222–3, 269
Palaeoequatorial distribution, 44–6, 49–52, 67, 70–1
Pan-mixed forest zone, 191
Pasture vegetation, 126, 132–3
Pinus trifolia fossil flora, 55
Pioneer stages, 132, 145, 240–59
Plagioseral grassland, 129–31
Plate tectonics, 3–5, 36–42
Podsol, 27–9, 93, 187
Pond vegetation, 214–7
Precipitation, 20-3
Protected Forests/Areas, 273
Psammophytes, 152
Pseudo-subalpine zone, 27
Pumice flows/fields, 241–4, 252

Quasi-alpine zone, 178, 194–5, 197
Quasi-National Parks, 273–5

Raised bog, 224, 231
Reclamation, 153, 156, 212
Red soil, 27–8, 101
Reforestation, 93
River vegetation, 213, 217, 229
Ryukyu region, 79–83
  structure of, 14–5
  vegetation of, 101–4

Salt marshes, 152, 153–8
Sandy beaches, 152, 158–68
Scree, 152, 168–70
Scrub, 162, 199–201, 254
Sea of Japan region, 74–5
Sea of Japan side (*Ura-Nihon*), 8, 21, 113–4, 174, 176, 179–81, 193–6, 271
Selective cutting, 117
Serpentinitic areas, 28, 189, 205
Shifting cultivation, 99, 111, 118
Snow-bed grassland, 194, 201–3
Snowland vegetation, 21, 26–7, 92, 96, 115, 180, 193, 196–8, 263–4
Sohayaki region, 76–7
Soil types, 27–30
Solfatara vegetation, 200, 205, 259–62, 265
Standing volume, 110
Subalpine zone, 63, 90–3, 175, 188–96
Subarctic zone, 63, 71–2, 90–3, 130, 153
Submerged communities, 213–7
Subtropical zone, 63, 101–4, 152
Successions of vegetation, 129–34, 139, 143, 145–7, 165, 217, 238–59, 264–5
Swamp forest, 93, 102, 107, 192, 217–20
Swamps, 213–20, 230

Tidal drift, 161
Tidal floods, 161–2
Timber line, 91–2, 107
Trampling, 138–9, 145
Transitional communities, 87, 97, 106–7, 134, 226
Transmigration routes, 2–3
Tsushima Current, 6–7
Tussock, 154, 218, 224

Vaccinium-Picea class, 63, 88, 93
Valley forests, 182
Vegetation zones, 20, 26, 63, 88–9, 126–7, 174–8
Vegetative propagation, 114, 123–4
Volcanic soil, 16, 27, 133
  vegetation, 27, 75, 174, 192, 205, 212, 235, 237–65
Volcanoes of Japan, 8–10, 27, 174, 238

Warm-temperate zone, 63–7, 98–101, 131, 152, 166
Warmth index, 21, 23–6, 90, 97, 175–6
Windward grassland, 195, 203

Xerophytic forest, 181

Yezo-Mutsu region, 74

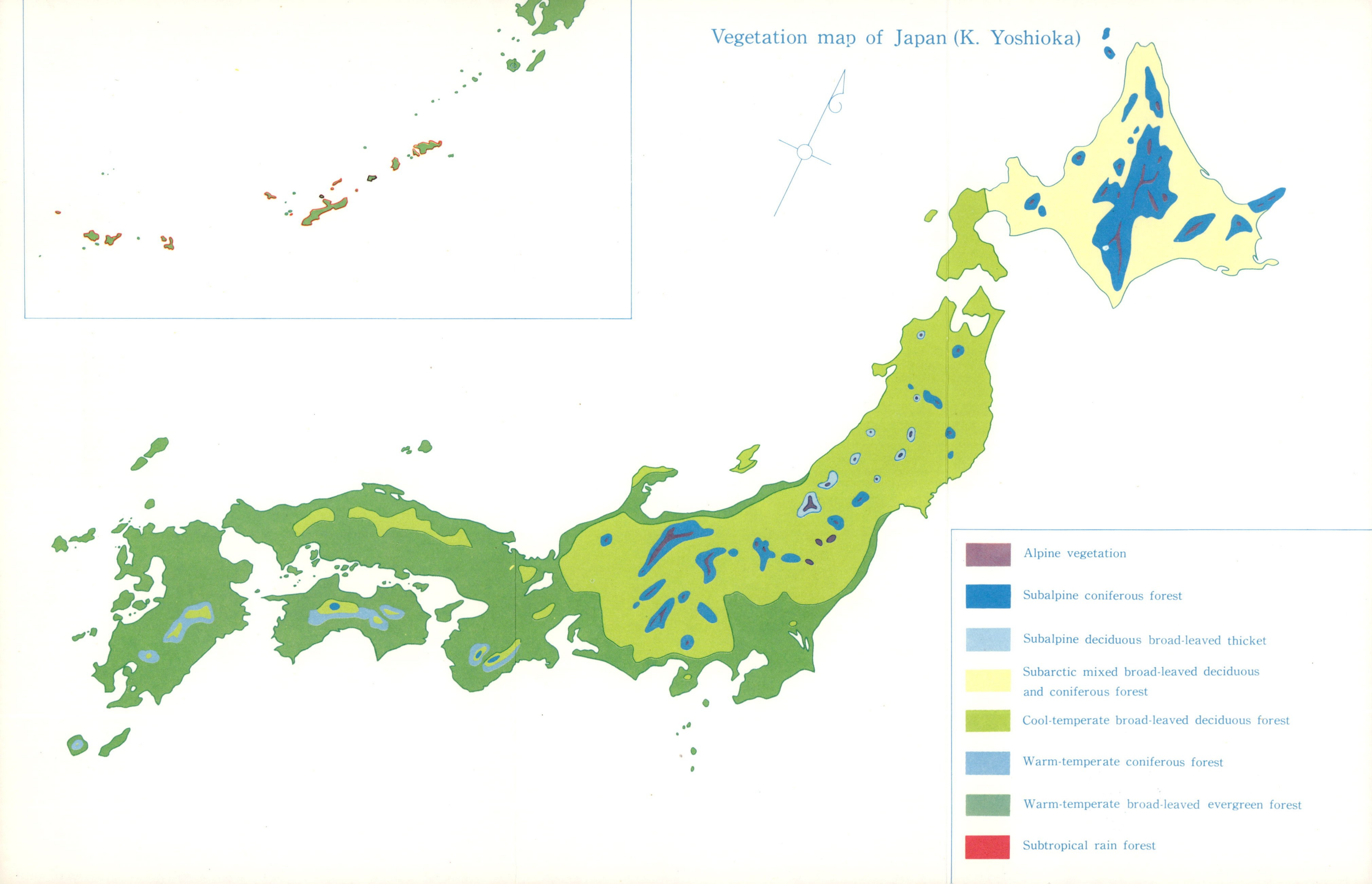

Vegetation map of Japan (K. Yoshioka)
Alpine vegetation
Subalpine coniferous forest
Subalpine deciduous broad-leaved thicket
Subarctic mixed broad-leaved deciduous and coniferous forest
Cool-temperate broad-leaved deciduous forest
Warm-temperate coniferous forest
Warm-temperate broad-leaved evergreen forest
Subtropical rain forest

Map of Geographical Localities in Japan

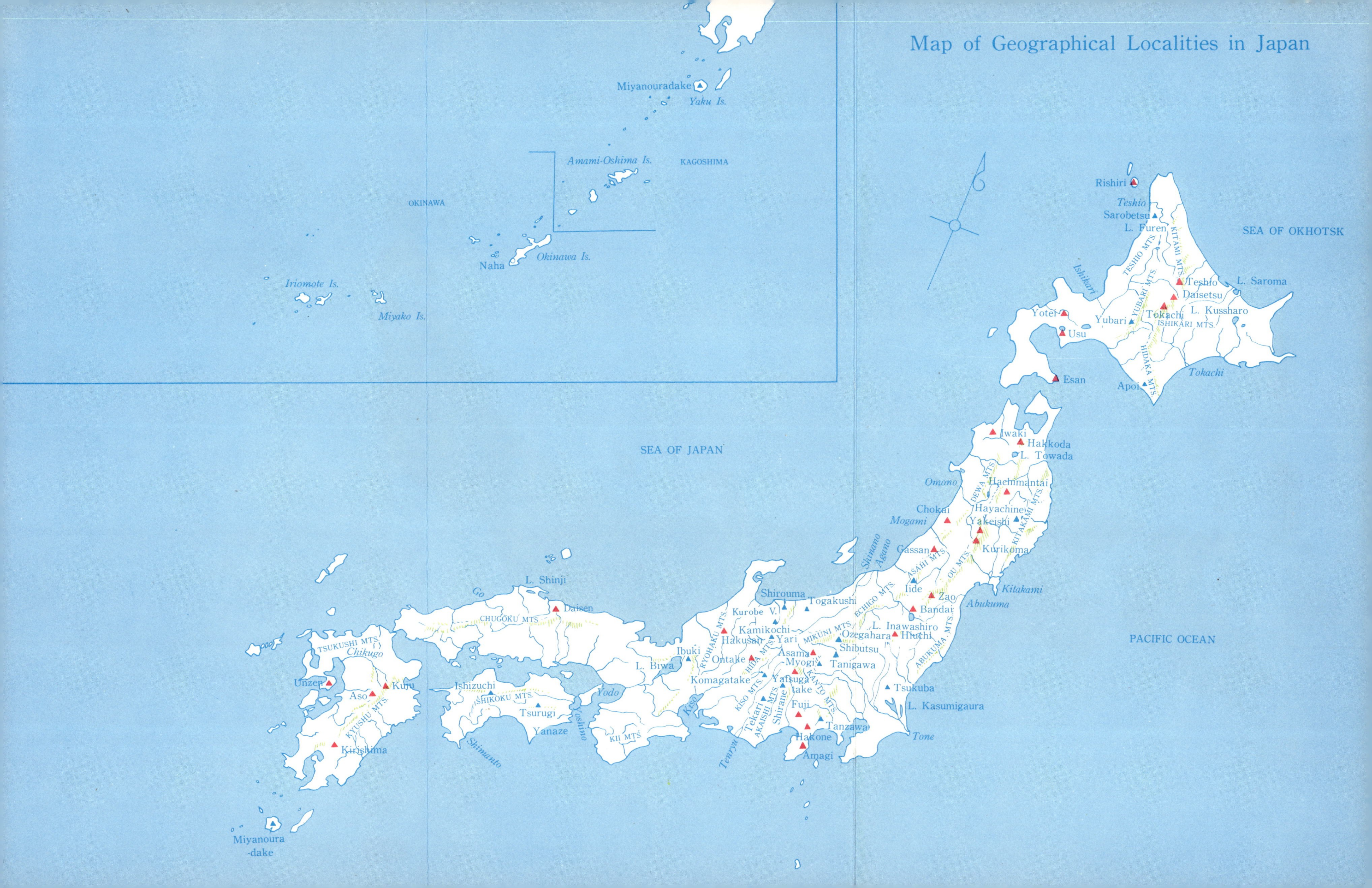

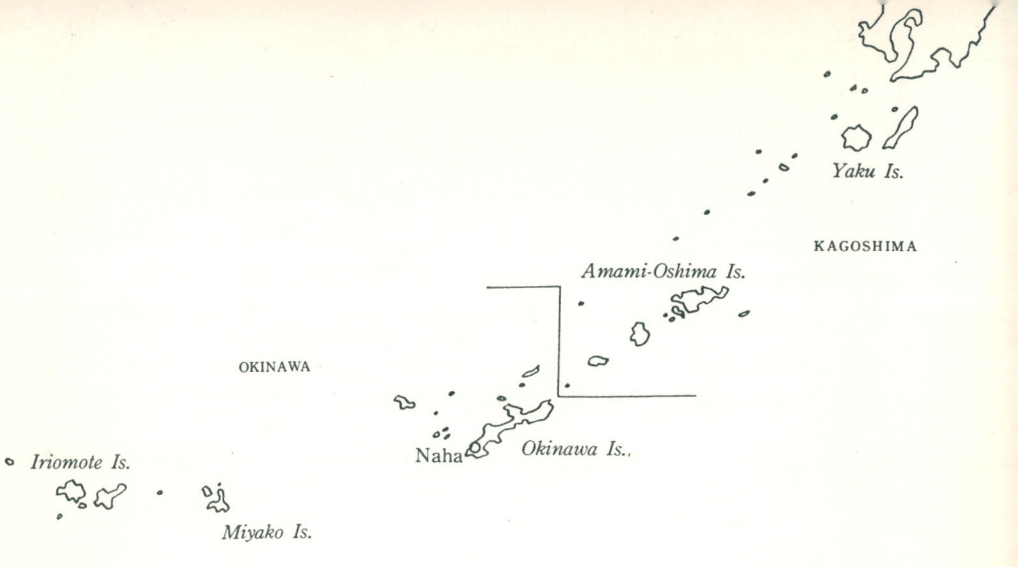

Yaku Is.
KAGOSHIMA
Amami-Oshima Is.
OKINAWA
Okinawa Is.
Naha
Iriomote Is.
Miyako Is.

SEA OF

Tsushima Is.
Oki Is.
Noto Pe
Iki Is.
YAMAGUCHI
Matsue
SHIMANE
TOTTORI
Tottori
Kanazawa
Yamaguchi
Hiroshima
HIROSHIMA
OKAYAMA
Okayama
HYOGO
KYOTO
ISHIKAWA
TO
Toy
Fukui
FUKUI
SAGA
Saga
Fukuoka
FUKUOKA
NAGASAKI
Nagasaki
OITA
Oita
SETO INLAND SEA
Kobe
Kyoto
SHIGA
Otsu
Gifu
GIFU
Kumamoto
KUMAMOTO
Matsuyama
EHIME
Kochi
KOCHI
KAGAWA
TOKUSHIMA
Osaka
OSAKA
Nara
NARA
Tsu
MIE
Nagoya
AICHI
Shizu
SHIZUOK
KAGOSHIMA
Kagoshima
MIYAZAKI
Miyazaki
Wakayama
WAKAYAMA
Tokushima
Takamatsu
Sakurajima Is.
Osumi Pen.
Kii Pen.
Izu-Osh
Yaku Is.
Tanegashima Is.